P9-AEW-473

GLACIAL GEOMORPHOLOGY

Glacially striated and grooved sandstone surface of Upper Ordovician age in the central Sahara (eastern Hoggar, Algeria). The fine rippling, sometimes at right-angles, sometimes oblique to the grooving, is believed due to soft sediment deformation in the sands of the continental shelf over which the ice was advancing from south to north. The deep transverse fractures are due to fractures related to subsequent minor vertical adjustments of the earth's crust in this region. Photo by Rhodes Fairbridge. Also see his article, this volume.

GLACIAL GEOMORPHOLOGY

Donald R. Coates,
Editor

A proceedings volume of the Fifth Annual Geomorphology Symposia Series, held at Binghamton New York
September 26-28, 1974

London
GEORGE ALLEN & UNWIN
Boston Sydney

George Allen & Unwin (Publishers) Ltd,
40 Museum Street, London WC1A 1LU, UK

George Allen & Unwin (Publishers) Ltd,
Park Lane, Hemel Hempstead, Herts HP2 4TE, UK

Allen & Unwin Inc.,
9 Winchester Terrace, Winchester, Mass 01890, USA

George Allen & Unwin Australia Pty Ltd,
8 Napier Street, North Sydney, NSW 2060, Australia

First published in 1974
Second impression 1982

Cover photographs of Wind River Mountains, Wyoming.
Courtesy Austin Post, University of Washington.

British Library Cataloguing in Publication Data

Glacial geomorphology. – (The 'Binghamton' symposia in geomorphology: international series; no. 5)
1. Glaciers – Congresses
I. Coates, Donald R. II. Series
551.3'1 QE576
ISBN 0-04-551045-8

Printed in Great Britain by Short Run Press Ltd, Exeter

CONTENTS

This book is dedicated to two "so-called" retirees—George W. White and Robert F. Legget—who continue to be as busy as ever and continue their impact on the profession. Their warmth as colleagues will always be remembered and as senior statesmen for "Friends of the Pleistocene" their contributions have been legion.

INTRODUCTION

This proceedings volume is the fifth in our continuing publication series that result from the annual geomorphology symposiums conducted in the Department of Geological Sciences, State University of New York at Binghamton. The First proceedings *Environmental Geomorphology* spoke to an emerging field that is becoming ever more popular and necessary in today's complex world. The Second proceedings, *Quantitative Geomorphology,* again cross-cut many of the geomorphic subdisciplines and united them with one of the most important methodologies of the science. The Third and Fourth proceedings, *Coastal Geomorphology* and *Fluvial Geomorphology,* zeroed in on analysis of the special processes that comprise the fundamental building blocks of geomorphic research. The present volume continues this trend in showing how the dynamic processes associated with glaciation transform the landscape.

There are many different avenues for expression of scientific ideas, but the knowledge and publication explosion creates hardships for those who attempt to keep in tune with their specialties. It is not our purpose to add an unnecessary burden to this verbage increase. Instead we feel there comes a time when reassessment of the vital fabric of geomorphology is necessary and where geomorphologists can gather as a group to share their newest ideas. The more than 300 participants who have been attending these yearly symposia attest that this type of event helps fill a communications gap. Thus one of the advantages of these particular meetings is the opportunity for specialists to interact with some depth with colleagues over a two-day period in an atmosphere conducive for discussion, deliberation, debate, and dialogue.

One fault with many meetings, including some symposia, is that they do not generate a summing-up in terms of a publication ., ... or if a book is finally produced it often takes so long to be printed that some of the impact of the meeting may have been lost. To correct such shortcomings we have endeavored to publish the proceedings volumes as rapidly as possible. This year a new experiment is being attempted, and it will be most interesting to watch the results. The proceedings volume has been published in time for distribution at the symposium. This has been done with the principal aim of increasing the level of participation along with the chance to develop meaningful and lively repartee, during and after the meeting. The themes of several chapters particularly lend themselves to close scrutiny and interplay among colleagues because of the controversial or new nature of the ideas being presented. There was even a logistical reason for accelerating production of this volume. In these days of rapidly rising costs a 20 percent savings on publication expenditures was made possible by obtaining early bids and purchase of materials ahead of time. Happily these economies have been passed on to the consumer . . . you.

The composition of this book, the authors, and the topics reflect some of the prejudices of the editor. The study of glaciers and the effects of glaciation is such a broad area that justice cannot be done in a single symposium, or a single proceedings volume. Therefore no attempt has been made to be inclusive, comprehensive, or to cover all aspects of what might be called the science of "glacial geology". It should be clear that this book cannot be all things to all people . . . there are other glacial books that cover much more ground. Furthermore it was felt that emphasis would be misdirected if some of the "geologic" types of topics were to be discussed in any depth. Thus there are at least three subjects that do not play a major role in this volume: (1) the petrology of glacial deposits, (2) the stratigraphy of materials, and (3) glaciology. Students interested in the first subject would do well to read such books as Goldthwait's edited volume on *Till.* Those who wish further information on correlation and chronology of glacial materials should start their study with Flint's *Glacial and Quaternary Geology*. The Cold Regions Research and Engineering Laboratory of the U.S. Army is the best source of information on the physics and mechanics of ice. Although the book does not cover such aspects of glaciology as the analysis of mass budgets and ice accumulation, the chapters by Boulton and by Clayton and Moran do discuss the processes of ice motion as it specifically relates to landform changes.

Having announced what this volume *is not,* it is time to tell what *it is.* The principal focus of *glacial geomorphology* is to treat terrain changes that have occurred because of glaciation. Therefore discussion is primarily centered on the development of landforms and those processes that have been influential in their sculpture. Glacial erosion and deposition processes are analyzed along with the topography they create or change. Several chapters refer to the idea that there is a continuous hierarchy of features that develop from the erosion- deposition regime of glacial ice . . . . a type of hybridized landscape. The majority of chapters are concerned with areas that were covered by glaciers, but some chapters (Moss) deal with landforms (terraces) that occur beyond the ice margin and others (Black) investigate terrain (patterned ground) that forms after the glacier has disappeared from the land. With one exception the chapters are concerned with Quaternary time and use data from this most recent ice age, Fairbridge, however, dramatically shows the importance of Ordovician glaciation in an area that is today's best-known desert.

The significance of glaciation may not always be obvious or understood by those who are not conversant with the "Friends of the Pleistocene". It may prove helpful to provide several points of interest why today it is worthwhile to have some knowledge of glaciers and of the Ice Ages. So although the following remarks may be redundant to my peers, they are mentioned as a rationale for showing the importance of glaciation.

1. *Historical importance.* The concept of glaciation and the data it helped generate provided some of the instrumental ideas that aided in dispelling the dogma of the Noachian Deluge and the Doctrine of Catastrophism that was often linked with it. For example erratics could now be explained by a new process that did not have to resort to world-wide floods that were supposed to have drifted alien rocks on top of a new locale. It is paradoxical, however, that glaciation . . . which is itself an extraordinary event (should it be called "catastrophic"?) . . . . should be the means to give uniformitarianism a helpful boost. In contrast to the other geomorphic agents that operate on the earth's surface . . . . rivers, oceans, groundwater, winds, and gravity . . . . glaciers have not acted continuously throughout geologic time. Instead there have been only 5 or 6 ice ages throughout the 4 billion years of recorded earth history andeachlastedonly a few million years. Thus ice-age duration is a mere geologic drop in the bucket and can be considered almost a unique event.

Another historical aspect of glaciation that is expecially timely today concerns the important role it played in Alfred Wegener's documentation of continental drift. His careful linkage of former ice ages in different continents now far-removed from arctic regions provided some of the most important clues for continental mobility. The regeneration of the concept for fragmentation of continents has now found nearly universal acceptance under such rubrics as "polar wandering", "seafloor spreading", and "global tectonics". Thus it is important not to forget that the unusual geography of pre-Quaternary ice ages, and this would now include the Ordovician glaciation, played a significant role in the belief of wandering continents until the new types of data (such as palemagnetism) emerged to provide overwhelming support.

2. *Economic importance.* Glaciers have left a heritage or made possible the development of important resources. Numerous groundwater supplies for cities in and near glaciated terrain are developed in glaciofluvial sands and gravels that provide the best aquifers in the metropolitan areas. Wells drilled into such sediments, as in the Binghamton region, yield thousands of gallons a minute at shallow depths and with little drawdown. A typical case history for the importance of such water resources involves the selection of Binghamton as the site for the Ansco plant because such glacial aquifers provide a ready source for large quantities of cool, clean, and clear water which are vital requisites for the photographic industry. Glacial sands and gravel also provide important rock products for mineral and rock industries. They make possible the relatively inexpensive construction of roads and buildings. Although of little overall value, glaciers were responsible for transportation of diamonds from Canada into the midcontinent states. The attempts to trace such diamond occurrences back to a 'mother lode' such as the finds in Wisconsin, as related by William Hobbs, makes very interesting reading.

Even gold occurs in many glaciofluvial deposits in such states as Michigan and Indiana. If the price of gold continues to escalate, who knows but that one day the search may be on for recovery of such deposits!

3. *Importance to earth science and related fields.* Glaciation and its effects have profoundly altered not only the landscape over which ice lay, but have produced notable impact on contiguous lands as well as throughout the world. Thus a study of glaciers and ice ages transcends many fields and includes such disciplines as geology, climatology, pedology, anthropology, and archeology. During their maximum development in the Quaternary, glaciers covered 30 percent of the earth's land surface, and even today 10 percent of the lands are still entombed under ice. Glaciers create massive terrain adjustments where they reside, as emphasized in this volume, and also induce manifold changes sometimes referred to as "indirect effects", outside the ice boundaries. These include such phenomena as modification of erosion rates, changes in sedimentation rates, altering the flora and fauna (such as foraminifera), deposition of loess, and changes in sea level. Glacial climate changes are likewise of great magnitude. The climate zonal belts of today's world are unusual because during most of earth history the climate has been more equitable and uniform. Antarctic glaciation is so powerful that its climatic effects are felt throughout the world. Some observers believe that it is currently self-sustaining; that is, instead of the present-day climate determining the ice regime, the reverse is true .... Antarctica creates its own weather patterns. Other climate effects include the pluvial climates that brought greater moisture retention on many lands far-removed from the glaciers. This permitted the formation of pluvial lakes, shallower groundwater levels, and increased streamflow. Such climates were apparently related to the rise of early man in Africa. The type of climate also influenced the distribution of Neanderthal Man, and determined the settlement patterns and habits of Cro-Magnon Man. Man's dispersal to North America was dependent on sea level changes that made possible his migration by means of the Bering Strait . Indeed when the total picture is analyzed we are still in the Ice Ages today!

4. *Aesthetic importance.* Glaciation enhances many commonplace landscapes and provides dramatic effects generally not possible or so spectacularly sculptured as with other geomorphic processes. This scenic aspect has been capitalized and commercialized for economic purposes in many areas, and in others such localities have been enshrined into magnificent parks and recreation sites. Think of what our country would be like without Glacier National Park, Yosemite National Park, Niagara Falls, the Great Lakes, the Finger Lakes, etc. The earth's surface features would be considerably duller if alpine glaciers had not carved cirques, materhorns, hanging valleys, and fiords. Such major vistas have given these regions magnificent topographic grandeur. Locally the Appalachian Plateau would be only routine lands if it did not contain such sites as Watkins Glen and the "Grand Canyon of Pennsylvania".

Although much editorial license has been exercised in this volume it must be pointed out that the selection of topics that are presented and authors that were chosen were not hastily or randomly done. There are many important factors that must be blended in attempting to produce a well-balanced treatment of glacial geomorphology. The goal was to obtain the widest possible coverage of ideas and scientists as possible. For speaker selection it was decided to obtain broad geographic representation, both in terms of parent institution and in subject area. Thus there is appropriate representation of the East, the Midwest, and the West, as well as Canada, England, and Africa. Likewise there is also coverage of two basic types of glaciers . . . . alpine and ice sheets. Glaciers as a process are discussed in several chapters, wherein their erosion, method of material entrainment and transportation, and deposition are analyzed. Glacial landforms are also discussed by several authors. The impact of glaciers over lands they traverse is especially detailed but there is also discussion of features created beyond the ice margin or during the frozen ground regime. Quaternary glaciation is the theme of most chapters but Ordovician glaciation forms the topic for one chapter. The speakers are some of the outstanding names in the science and represent small and large institutions and government agencies. The style of presentation is as variable as the number of authors. Some chapters will undoubtedly be the definitive works on the topic in which new ground has been paved. Other chapter provide very useful reviews and summations of controversial or of far-scattered literature. Several topics have been brought into different focus and a new framework instituted. Three chapters show the importance of glacial data and how it can be transcribed into the practical affairs of the modern world.

This volume is not aimed at any single or particular audience, but it should be of interest to professionals and new students alike. Most chapters are geared to the type of information that is very necessary in course work, and all chapters provide fundamental ideas that can easily become part of special reading assignments in glacial and in geomorphology courses. The organization plan of this proceedings was to divide the subject matter into four principal headings, and these four parts emphasize this approach. Although certain chapters lend themselves to placement within one of the parts, owing to the integrative nature of the discipline, the ideas in many often transcend into other topical areas.

PART 1 Glacial Models. The chapters in this section show the scientific aspect for the development of conceptual models and illustrate the application of models in the formation of ideas that have general relevance for many different regions.

PART 2 Glacial Landforms. The emphasis in this part is on the shape of the terrain that has been transformed by the glacial processes. It forms a close knit companion with the dynamic aspects of glaciation which comprise an integral portion of Part 1.

PART 3 Other Terrain Considerations. This part contains those chapters which suggest some of the many different types of features that constitute the complex array of glacially-related phenomena.

PART 4 Practical Applications. Chapters in this part indicate some of the typical ways in which knowledge of glaciation is useful and necessary when man operates in a glaciated region.

It would be possible to give many acknowledgements, but the greatest debt is to the authors. Their willingness to undertake the assignment of this early publication is greatly appreciated, and the outstanding character of their work will assure the success of this venture.

Donald R. Coates

AUTHORS AND AFFILIATIONS

Black, Robert F. Department of Geology. University of Connecticut, Storrs, Conn. 06268

Boulton, Geoffrey S. School of Environmental Sciences. University of East Anglia, Norwich, England

Clayton, Lee. Department of Geology. University of North Dakota, Grand Forks, North Dakota 58201

Coates, Donald R. Department of Geological Science, State University of New York at Binghamton, Binghamton, New York 13901

Fairbridge, Rhodes W. Department of Geology, Columbia University, New York, New York 10027

Goldthwait, Richard P. Department of Geology, Ohio State University, Columbus, Ohio 43210

Heroy, William B. Adjunct Professor, Southern Methodist University, Dallas, Texas 75222 (deceased)

King, Cuchlaine, A.M. Department of Geography, University of Nottingham, Nottingham, England

Koteff, Carl, 80 Broad Street, U.S. Geological Survey, Boston, Massachusetts 02110

LaFleur, Robert G. Department of Geology, Rensselaer Polytechnic Institute, Troy, New York 12181

Legget, Robert F. 531 Echo Drive, Ottawa, Ontario, Canada (National Research Council of Canada, Ret.)

Mears, Brainerd, Department of Geology, University of Wyoming, Laramie, Wyoming 82070

Moran, Stephen R. Department of Geology, University of North Dakota, Grand Forks, North Dakota 58201

Moss, John H. Department of Geology, Franklin and Marshall College, Lancaster, Pennsylvania 17604

Muller, Ernest H. Department of Geology, Syracuse University, Syracuse, New York 13210

White, George W. Department of Geology, University of Illinois, Urbana, Illinois 61801

Topographic effects of landforms created by ice loading and glacial rebound of the Bearpaw Shale, Saskatchewan, Canada. (See Legget, Figure 2 and pages 362-3 for credits and further information).

PART 1

GLACIAL MODELS

Models have always formed an integral part of the natural sciences . . . . indeed they are the basic building blocks of science. Models are important because they provide a universality to the principle upon which they are based. Their purpose is the exposition of ideas that have wide application, and by their fabric they are ideal for testing and can even provide an element of prediction. The scale of models range from those that deal with the universe to those of the DNA molecule, the atom, and even sub-atomic particles. Because of the complexity of science, there must be a large variety of models that include conceptual models, physical-working models (such as modelling the dynamics of rivers), geometric models (such as crystals) etc. The ascendancy of computers during the past 25 years has revolutionized modelling. The earliest and biggest impact has been in science and engineering, but now computerized models are widely used in all spectra of society including academia, industry, government, and many aspects of the decision-making process. It is now commonplace to develop systems approaches in all avenues of life and digital and analog models are household names. There is a danger in so much automation, however, because the opinion is held by some that if it hasn't been computerized then "it isn't science". Another error can occur if computerized results are taken in blind faith, without a constant surveillance of the basic assumptions that were fed into the mathematical game. It would have been possible to use a different range of models than was selected for this part, but I believe the ones that are included are the most fundamental and provide the clearest insight into the discipline of glacial geomorphology.

Chapter 1 by Mears contains an historical approach to the development of a conceptual model for glaciation and its chronology in the Rocky Mountains. He also provides a scientific philosophical tone that gives a framework for the book and would be useful for many other endeavors.

Chapter 2 by Boulton illustrates the interplay of field work and experimentation in the development of a model for glacial erosion. He shows the manner of the abrasion and plucking processes that can be integrated into erosion at various scales. The mechanics and mathematics of the system he describes will have extraordinary relevance to a wide set of conditions and environments.

Chapter 3 by Clayton and Moran unites the best current thinking of such workers as Boulton and Nye into a glacial model of erosion, transportation, and deposition that explains landforms in the midcontinent region of North America. This chapter provides a fitting expansion of the detailed model explained in Chapter 2.

Chapter 4 by Koteff synthesizes three decades of research by U.S. Geological Survey personnel in New England. The idea of depositional and landform sequences during valley deglaciation has already been applied with great success in other northeastern states such as New York and Pennsylvania.

CHAPTER 1

THE EVOLUTION OF THE ROCKY MOUNTAIN GLACIAL MODEL

Brainerd Mears, Jr.

INTRODUCTION

"Buffalo", Bull Lake, Pinedale; Temple Lake and Gannett Peak--Wyoming's glacial geomorphology provides a scheme, derived largely from and near the Wind River Mountains, that is now used far beyond the provincial boundaries of the state. Although basically sound, existing controversy suggests a current need for evaluation of this chronological model, whose history gives a nice example of the nature of scientific work.

Geomorphic Setting

The structural-geomorphic framework (Fig. 1) controlling Wyoming Quaternary glaciation involves broad anticlinal mountains separated by wide intervening basins--except in the western-most part of the state where a typical orogenic belt (the Wyomides),a block fault mountain (the Tetons), and volcanic plateaus (Yellowstone and Absaroka) exist. The "typical" [1] Wyoming mountains (the Wind Rivers, Owl Creek-Bighorns, Sierra Madre, Medicine Bow, Laramie Range, and Black Hills) are broad, flat-topped, eroded anticlines flanked by flatirons and hogbacks (Fig.2) of Paleozoic and lower Mesozoic sedimentary rocks (where not overlapped by Tertiary sediments or hidden beneath thrust or high-angle reverse faults). These ranges have plateau-like, high-level erosion surfaces, beveling Precambrian crystalline rocks, and surmounted by higher peaks.

The crystalline rocks, in some places, contain possible evidence of a very ancient glaciation. Eliot Blackwelder (1926) proposed Precambrian glaciation from meta-conglomerates and laminated phyllites in the Headquarters Formation of the Medicine Bow Mountains. Although many recent workers in areas outside Wyoming have embraced the now fashionable view that such diamictites and laminites are largely turbidites or other non-glacial deposits –Sylvester (1973) made a good case, from drop-stone markings and other evidence, that the Headquarters Formation contains glaciomarine deposits.

Tertiary History

The following synopsis of Wyoming Tertiary geomorphic history largely reflects my long exposure to S.H. Knight's ideas from his 40 years work (1953). The Laramide orogeny (late Cretaceous through late Eocene) established the existing

1. "Typical" here means the unusual Cenozoic mountains that rose from the Rocky Mountain geosyncline on the North American craton in Colorado and Wyoming.

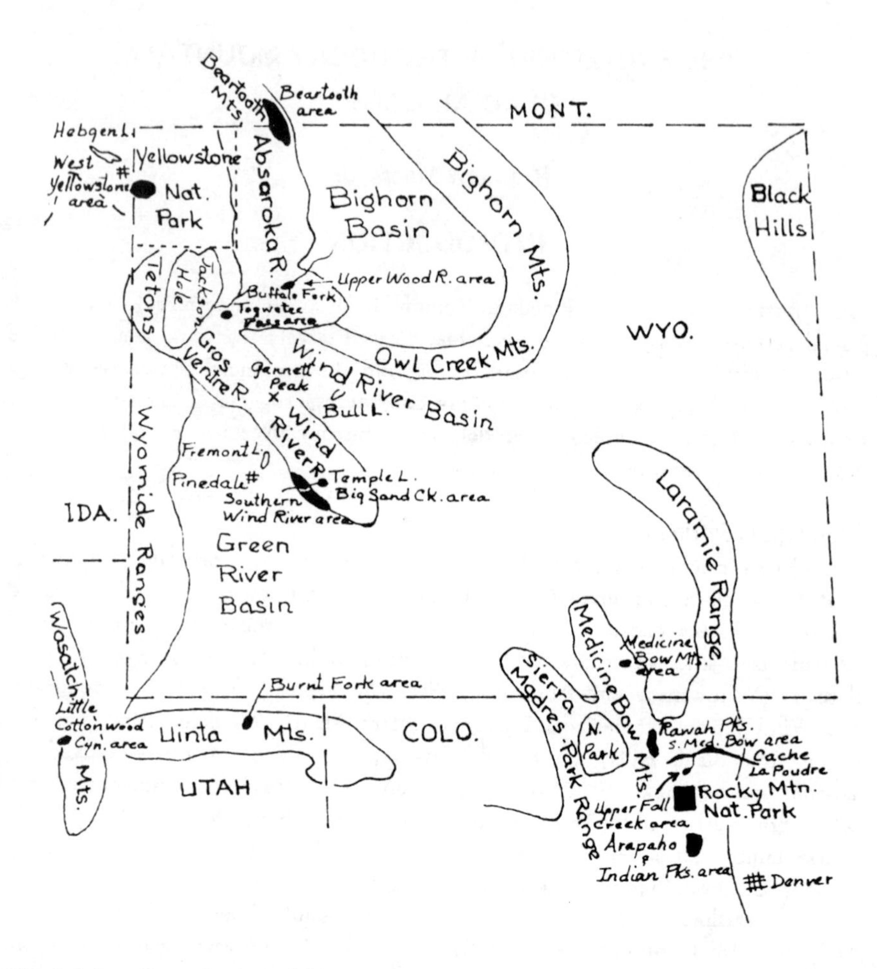

Figure 1. Locations of principal Wyoming mountains and some localities discussed in the text.

Wyoming structural-geomorphic pattern of mountains and basins (except for the Tetons and Yellowstone-Absaroka plateaus which are later developments). Laramide pulsations created early Cenozoic ranges with local relief comparable to that of today; however, the region as a whole was closer to sea level. A pronounced topography developed whose relict valleys still contain Oligocene and younger sediments. Whether all the existing high-level erosion surfaces are also Eocene relicts as Knight, recently supported by Scott (1973) from work in Colorado, believes is still controversial. Love (1960) and I think that some extensive surfaces may be as young as Pliocene.

Post-Laramide (Oligocene through Pliocene) Tertiary development of Wyoming involved dominant basin filling by fluviatile sediments greatly supplemented by eolian, volcanic, ash falls. Degradation of the mountains accompanying the basin-filling resulted in a subdued topography, by late Pliocene time, with scattered peaks standing above a graded surface extending from the planated mountain cores across the Tertiary-filled basins (Fig.2). During Tertiary time following the Laramide orogeny, epeirogenic uplift broadly elevated the region; and the climate--from vertebrate and floral paleontologic evidence--"deteriorated" from subtropical to a more rigorous sort such as we enjoy today. Thus regional uplift with climatic deterioration culminated in Quaternary glaciation.

Hares (1948) suggested Eocene glaciation in the Medicine Bow Mountains. However, my examination of his "Eocene" glacial locality disclosed only Pleistocene deposits; moreover, the generally accepted nature of Wyoming's early Tertiary climate makes any such glaciation in the Rockies suspect.

Quaternary Basin Excavation

Either climatic change or epeirogenic uplift, or a combination of both, caused rejuvenation of the region's streams which initiated the dominant excavation of the Tertiary-filled basins that has characterized the region's development since mid-Pliocene or the beginning of Quaternary time. Progressive basin excavation has left a series of benches and terraces, the interpretation of which has reflected two schools of thought. One, fathered by Mackin (1937), favors continuous basin degradation involving graded streams and capture --a view fitting Hack's dynamic equilibrium concept (1960) for valley surfaces in the Appalachians. The other school, stemming from Blackwelder (1915), relates the benches and terraces to climatic cycles and glaciation which might modify possible downcutting reflecting epeirogenic uplift. Scott (1965) compiled available information stressing Quaternary stratigraphy in non-glaciated regions of the Rockies. He strongly supports the climatic interpretation, but following Blackwelder considers the basin surface-deposits as mainly interglacial; whereas others, such as Moss (1951, and this volume) consider the deposits as glacially-related. A primary climatic control of cyclic downcutting with associated captures and drainage diversions seems a reasonable hypothesis.

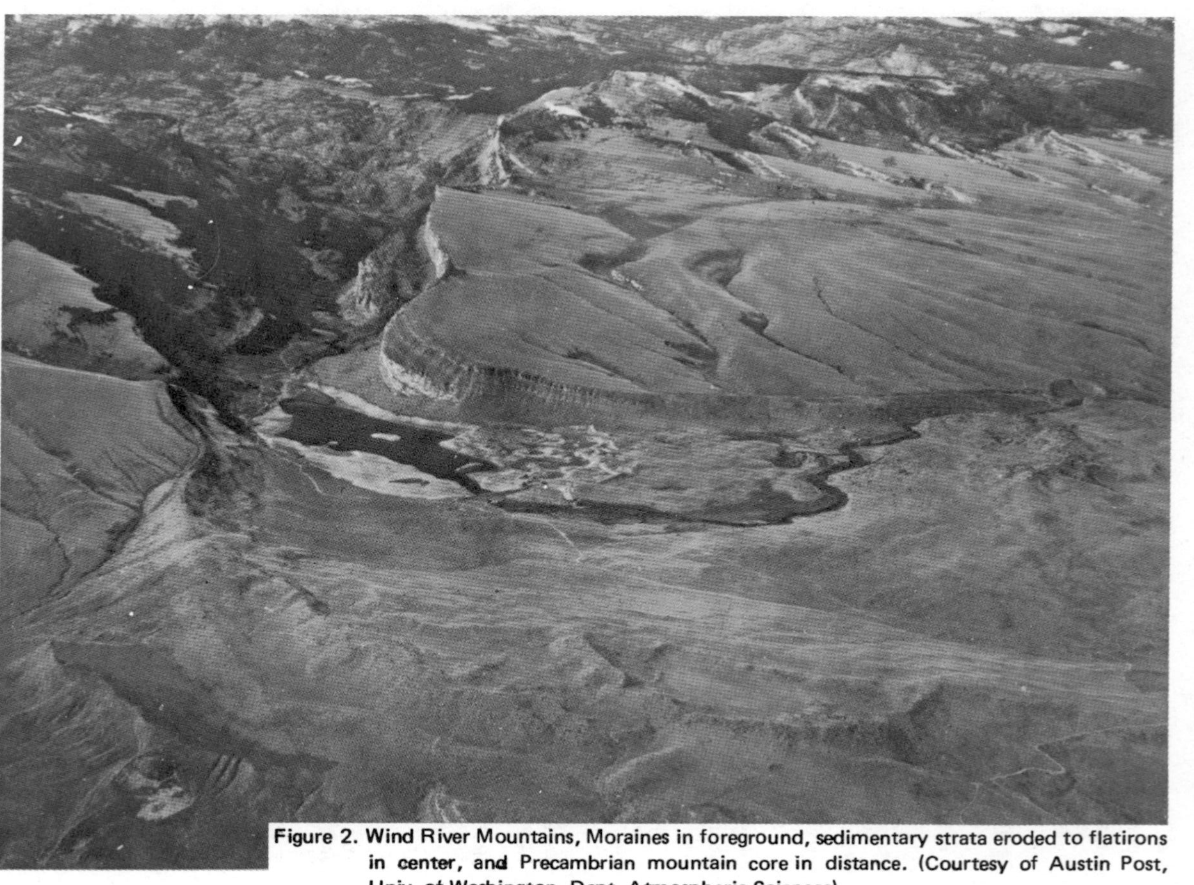

Figure 2. Wind River Mountains, Moraines in foreground, sedimentary strata eroded to flatirons in center, and Precambrian mountain core in distance. (Courtesy of Austin Post, Univ. of Washington, Dept. Atmospheric Sciences).

THE WYOMING GLACIAL MODEL

Quaternary glaciation left many scenic (Mears, 1972), but little studied erosional landforms (Fig.3) in all high Wyoming mountains, with only the relatively low Owl Creek Mountains, Black Hills, and Laramie Range unaffected. However, most published studies deal mainly with depositional landforms and their associated stratigraphy in developing a Quaternary chronology. Among early work, that of Eliot Blackwelder (1915) is outstanding and established the basis for most later thinking.

The Prototype

Blackwelder suggested three major glacial expansions based on depositional landforms and tills in and around the Wind River Mountains. The earliest ice advance, which he considered to now lack relict morainic topography, he called the "Buffalo" from tills along Buffalo Fork of the Snake River on the west side of Togwotee Pass into Jackson Hole. Scattered patches of till suggested to him that this advance was "pre-canyon", and not related to existing canyons from which later glaciers issued, and probably involved piedmont glaciers more extensive than later ice advances. He considered the Buffalo as pre-Wisconsin (Kansan or Illinoian), and noted the possibility of multiple tills within this early glaciation.

Largely on geomorphic evidence, Blackwelder proposed two later glacial advances from distinct moraines looping around the mouths of major Wind River Mountain canyons (Fig.4). The earliest of these he named "Bull Lake", from a type locality at that moraine-dammed lake on the northeast side of the Wind Rivers. The moraines have a somewhat subdued, but clear, knob-and-basin topography, and lie on terraces from 80 to 200 ft. above the present level of the Wind River. He tentatively suggested they were early Wisconsin in age.

Closer to canyon mouths, directly behind (in the type localities, and in places overlapping) the Bull Lake moraines are younger moraines whose more pronounced

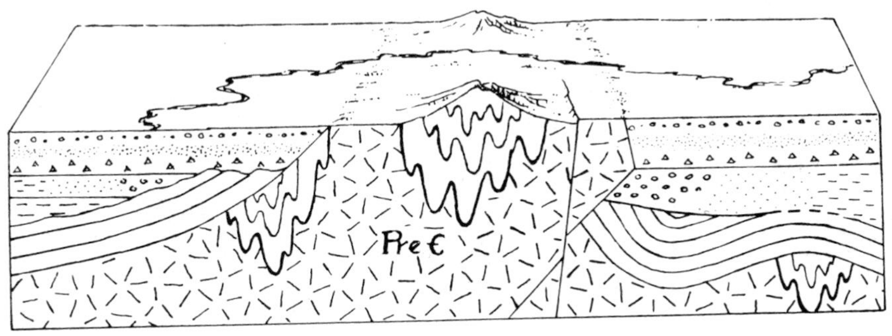

Figure 3. Wyoming range and adjacent basins with surface of low relief in Pliocene time.

Figure 4. Dinwoody Canyon and Gannett Peak with glacier and small Neoglacial moraine (upper center). Photo courtesy of the Wyoming Travel Commission.

topography seems little reduced by erosion. Blackwelder named them "Pinedale" from a type locality, near that town, where representative moraines lie around the south end of Fremont Lake. The more youthful form, less mature soils, and extension down into valleys trenching Bull Lake moraines and terraces suggested to him a late Wisconsin age.

Pre-Bull Lake ("Buffalo") Glaciations

Additions to his basic scheme were probably anticipated by Blackwelder. Holmes and Moss (1955) made the significant discovery, in the southwestern Wind River Mountains, that Buffalo-type tills (some of which exhibit a very subdued morainic topography) are associated with terraces not far above stream level along the margin of the adjacent Green River Basin. Thus it was demonstrated that the original "Buffalo" represented at least two glaciations (as Blackwelder had suspected), and they were separated by an interval of erosion which produced deep mountain canyons.

Gerald Richmond has been the leader in developing and refining Blackwelder's model (numerous papers summarized 1965; with additions in Birkeland, Crandell, and Richmond, 1971). Richmond (1962) recognized two buried soils, which he interpreted as representing major non-glacial intervals, within "Buffalo" tills along the north side of Bull Lake beneath Washakie Point. Based on these and his similar findings in the La Sal Mountains of Utah (1962) and in Saint Mary Ridge (1957) near Glacier Park in Montana – he proposed three distinct pre-Wisconsin glaciations with type localities at Washakie Point, Cedar Ridge, and Sacagawea Ridge on the northeast side of the Wind River Mountains (Fig.5).

Of the three separate tills, the Sacagawea represents a post-canyon phase of "Buffalo" glaciation, which advanced farther from source areas than any other Quaternary glaciation. Gravels and fine alluvium representing either Cedar Ridge or ensuing interglacial deposits contain one of the several Pearlette-like volcanic ashes which indicate a possible equivalence with Kansan or Yarmouth events in the continental interior. Tentatively, Richmond (1964) correlated the Washakie Point, Cedar Ridge, and Sacagawea Ridge glaciations with the Nebraskan, Kansan, and Illinoian respectively.

Bull Lake Glaciation

Subsequent work has also refined Blackwelder's Wisconsin model. Geomorphically, two sets of Bull Lake moraines and two related outwash terraces have long been suggested – stemming from Fryxell's (1930) Jackson Hole studies. Richmond (1948) reported two outwash-capped terraces in the northern Wind River Mountains that were related to Bull Lake moraines. In the southern Wind River Mountains, Holmes and Moss (1955) also concluded that the Bull Lake involved two distinct episodes of glaciation. Their evidence included the nearly equal development of the two sets of moraines, and the broad extent of associated outwash-capped terraces at two different levels. Richmond (1960) discovered buried soil remnants separating two Bull Lake tills in the Wind River Range area of Wyoming, the La Sal Mountains of Utah and the Glacier Park vicinity in Montana;

thus he suggested a warmer soil-forming episode, and the probability of marked glacial retreat, between early and late Bull Lake till deposition. In some localities, Richmond (1964) recognizes a third Bull Lake moraine interpreted as representing a second episode during the late stade of Bull Lake glaciation.

Pinedale Glaciation

The Pinedale was also a complex glaciation (Figs. 4-6) In the southern Wind River Mountains, Holmes and Moss (1955) distinguished an outer massive moraine complex associated with a broad Pinedale outwash terrace, and an inner group of recessional moraines having a narrow composite terrace. Richmond's (1948) recognition of three Pinedale moraines with related outwash terraces led to his designation of three Pinedale stades, or glacial advances. However, the magnitude of glacial recessions between the Pinedale stades has never been clearly established. Geomorphic relations of the moraines and successive terraces produced by downcutting are evident. However, soils developed on the moraines are relatively similar, and the buried soils reported are weak and limited to a few outwash deposits.

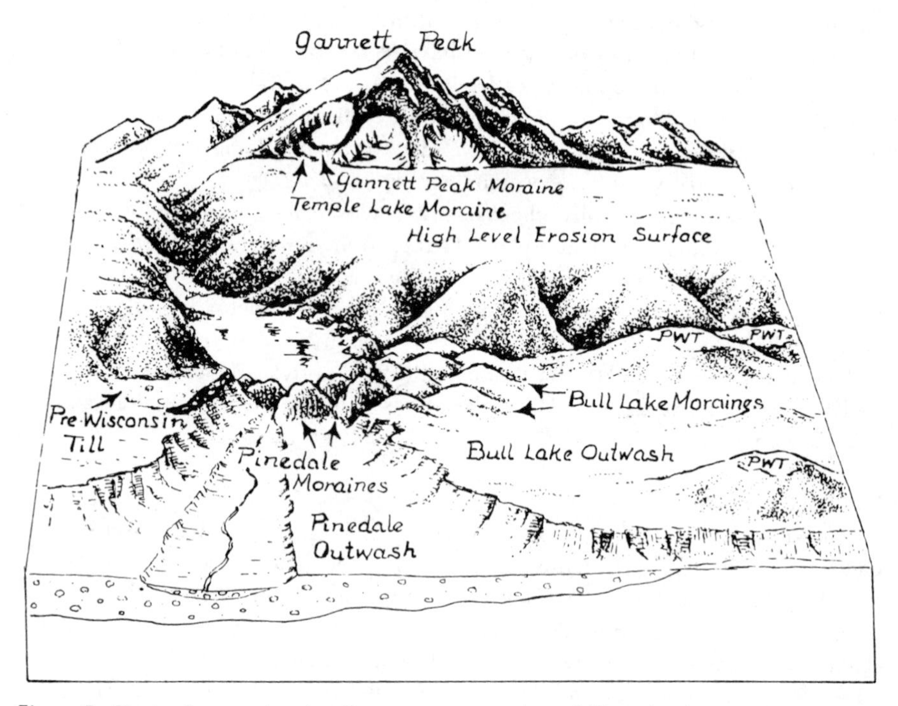

Figure 5. Block diagram showing Quaternary excavation of Wyoming basins with moraines reported by Blackwelder in the foreground, and high country cirque moraines studied by later workers.

Bull Lake and Pinedale Correlations

Since Blackwelder's time, the early Wisconsin age of Bull Lake moraines has been generally accepted (until very recently), although Alden (1953) and Atwood (1937) considered such moraines as pre-Wisconsin. Richmond (1965) correlated the Bull Lake stades with the early Wisconsin of the Midcontinent, and the Pinedale with later ("classical") Wisconsin. He also correlated the Bull Lake with the Riss, and the Pinedale with the Wurm of the classical European glacial sequence in the Alps (1970). Thus from extensive field work and the logical assumption of world-wide climatic changes – Richmond proposed a coherent Rocky Mountain pattern, and broad regional relationships.

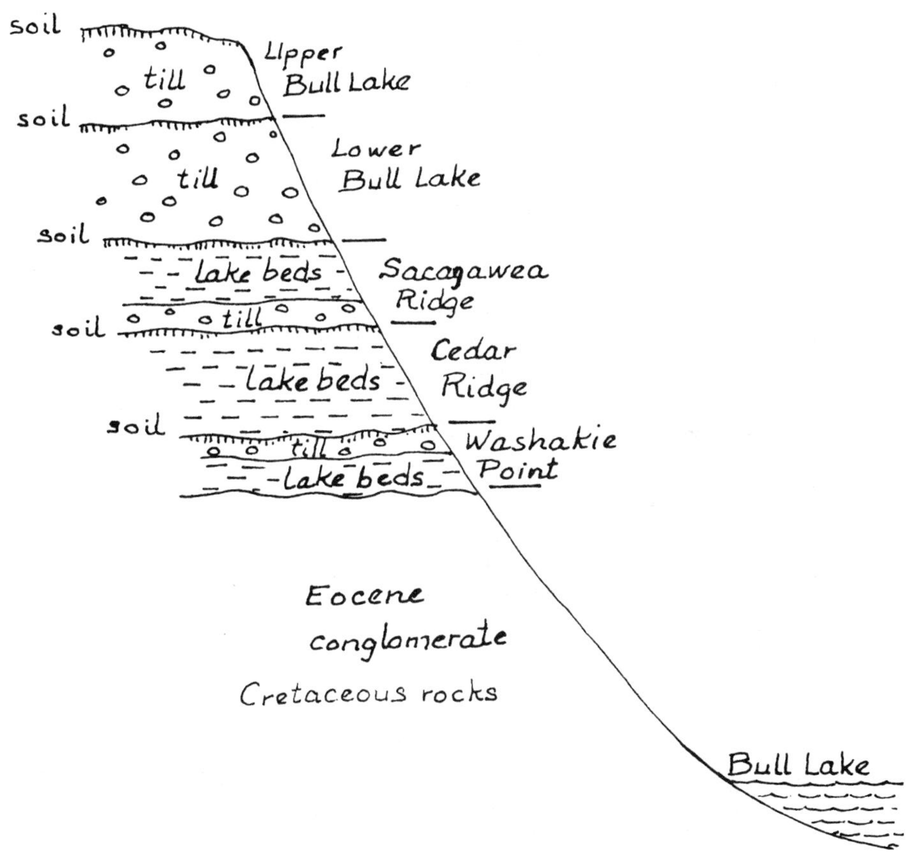

Figure 6. Schematic representation, from data of Richmond, of Bull Lake and older glacial deposits below Washakie Point on the north shore of Bull Lake. Washakie Point cliff to Bull Lake surface is 922 ft.

Late Minor Glaciations

A new aspect, not developed in Blackwelder's productive work, is the recognition of late Quaternary glaciations in or near cirque heads of glaciated valleys. Initially, two separate moraines were recognized in the "high country". Hack (1943), from a brief reconnaissance, reported a distinct moraine, looping around Temple and Miller Lakes at the head of the Big Sandy Creek drainage in the southern Wind River Mountains, which he named the Temple Lake moraine and attributed to a distinct glacial advance (Fig. 9).

Moss (1951), in a later study of the Temple Lake area, distinguished a morainic mass not far from a relict glacier. This moraine reflects another very recent, possibly compound, ice advance occurring in the cirque-head area. Based on Matthes' reports from the Sierra Nevada (1942), historic records from the Alps, and reports of Wyoming pioneers – Moss considered these very fresh unstable deposits as products of Matthes' (1939) "Little Ice Age", of the last few centuries, for which Moss suggested the term "Neoglaciation".

The Temple Lake moraine was interpreted by Moss as representing a post-Pinedale but pre–Altithermal glacial advance. In considering multiple hypotheses, he rejected a "Little Ice Age" designation because of the soil development and limonitic staining of the moraine (the smooth grassy condition of the lateral moraine there also supports this conclusion). The idea of an earlier "Little Ice Age" advance, then current in European discussions, he tentatively abandoned. Moss settled on a late Pleistocene ice pulsation preceding the "Climatic Optimum" which is now recognized as the Altithermal in the Rocky Mountain region – a time of marked stress for prehistoric man who retreated from basin areas northward and into the mountains.

Richmond (1960 b, 1965) gave the name "Gannett Peak" to fresh cirque moraines – "upslope from those of the Temple Lake Stade" – beneath Gannett Peak, Wyoming's highest pinnacle. He introduced the term into Rocky Mountain geologic literature as the type for late Neoglacial deposits. Richmond also designated the Temple Lake moraine as the type early Neoglacial in the Rockies--from evidence in Rocky Mountain National Park of Colorado and in the La Sal Mountains of Utah.

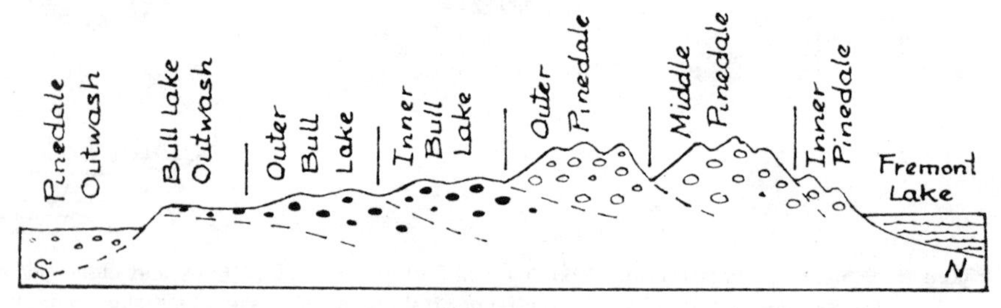

Figure 7. Schematic profile of multiple Bull Lake and Pinedale moraines at the south end of Fremont Lake, near Pinedale, Wyoming (after Richmond). Section length is 3.5 mi. Pinedale outwash surface (left) is 7,180 ft, and highest moraine is 7600 ft.

Figure 8. Pinedale moraines at Boulder Lake, Wind River Mountains. (Courtesy of Austin Post, Univ. of Washington, Dept. Atmospheric Sciences).

Subsequent workers in the high Rockies have described abundant early Neoglacial moraines as "Temple Lake". They are rough, bouldery, steep-sided ridges with weak soil development but relatively stable slopes, ranging from 20–50 ft in height. Commonly the moraines are double, indicating two minor glacial pulsations which Richmond (1972, designated as Temple Lake "a" and Temple Lake "b". He assigned dates of 3,100–2,800 and of 1,800–1,000 years-ago to these respective glacial pulsations by correlation with two radiocarbon dated alluviums east of Rocky Mountain Park, and one date in alluvium of the Gold Basin Formation of the La Sal Mountains, and one from Navajo County, Arizona.

THE BULL LAKE PROBLEM

The current Wyoming model has been a useful conceptual scheme. However new findings and re-interpretations suggest a lively future for glacial and related Quaternary geomorphic studies in Wyoming and the Rocky Mountains.

Absolute Datings of Bull Lake and Pinedale Glaciations

Prior to the U.S. Geological Survey's recent Yellowstone Park project, absolute dates for Bull Lake and Pinedale events were few and indirect, with none from tills.

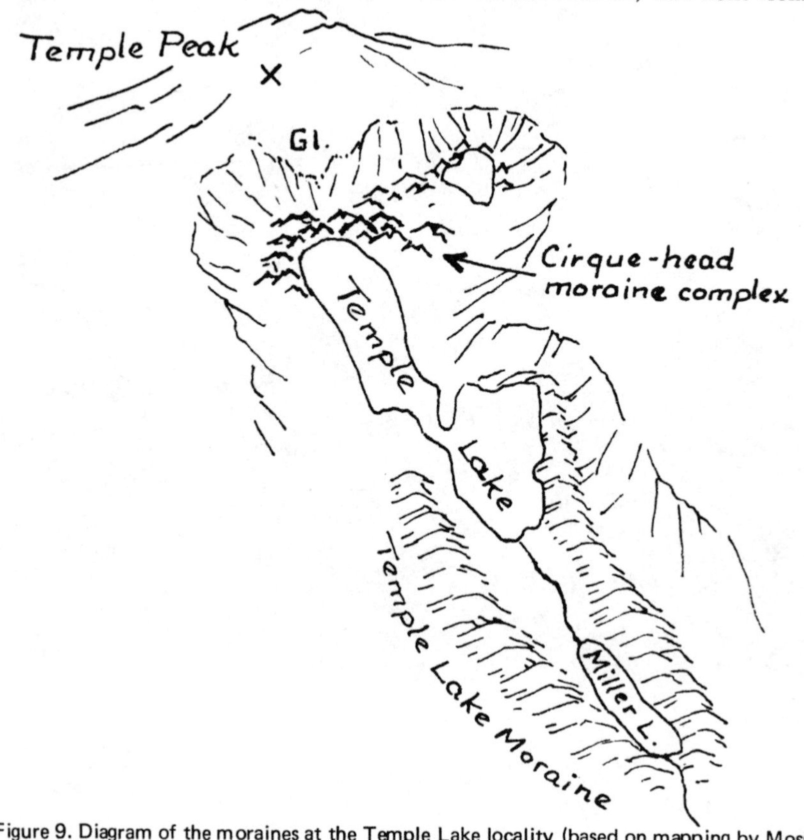

Figure 9. Diagram of the moraines at the Temple Lake locality (based on mapping by Moss). Distance from Head of Temple Lake to lower end of Miller Lake ≈ 1.5 mi.

In a summary (1965), before the project, Richmond tentatively bracketed Bull Lake glaciations between 70,000 and 30,000 years B.P. He cited a radiocarbon date of greater than 42,000 years (Trimble and Carr, 1961) for peat in lake deposits, near American Falls in Idaho, which were correlated with late Bull Lake events. The dated lake deposits were overlain by gravels interpreted as Lake Bonneville flood deposits of late Bull Lake time. Terrace deposits lying on the bevelled surface of the possible flood gravels contained shells having a radiocarbon date of about 29,000 years, which was thought to provide a minimum terminal date for the Bull Lake.

Pinedale glaciation (Richmond, 1965) was approximated as from 25,000 to 6,500 years ago. This glaciation was dated from interfingering till and lake deposits at the mouth of Little Cottonwood Canyon in Utah where Lake Bonneville radiocarbon dates ranging from 25,400 to 11,300 yr. B.P. were correlated with the early and middle stades of Pinedale glaciation. The end of Pinedale glaciation, 7,280 years ago at lower elevations and 6,190 years ago at higher elevations, was based on radiocarbon dates from several Rocky Mountain sources.

The first fairly "solid" absolute dates for late Pleistocene events in Wyoming came from the U.S.G.S. Yellowstone project. Richmond thereafter (1972) considered early Bull Lake glaciation as spanning 125,000 to 115,000 yrs. B.P. with the late Bull Lake from 100,000 to 85,000 yrs. B.P. He cited three K-Ar dated volcanic flows as evidence. The oldest (120,000yr.) in places has sharply concave, steep margins that are interpreted as ice-contact faces of eruptions against Bull Lake ice. The second volcanic flow (105,000 yr.) invaded lake deposits indicative of an interglacial episode interpreted as separating Bull Lake glaciations. The youngest volcanic flow (70,000 yr) is overlain by Pinedale deposits but not Bull Lake, despite the presence of the latter materials on a higher adjacent mountain face; thus indicating a minimum termination for Bull Lake glaciation. Pinedale glaciation, Richmond (1972) considered as younger than 29,000 yr based on a radiocarbon date from non-glacial lake beds beneath Pinedale till. In 1973 (INQUA presentation at New Zealand), he revised the figure to 40,000 yr.

A Possible New Interpretation

Despite the new Yellowstone information, correlations of moraines and tills with the type localities in the Wind River Mountains, over 100 mi away, remain indirect. Moreover, unlike the glaciers moving down valleys in the stable crystalline-cored Wind River Mountains – Yellowstone ice advances spread in part as sheets over a structurally and volcanically active plateau. Thus topographic outlets controlling the dispersion of Bull Lake glaciers on the west side of Yellowstone Park were so filled by later volcanic eruptions that subsequent Pinedale ice was diverted north and south in somewhat different patterns (Pierce, pers. comm., 1974). Despite some lack of agreement in the details of mapping, all Yellowstone workers accept the Bull Lake and Pinedale broad designations at critical localities (on traditional geomorphic and stratigraphic grounds).

Pierce, Obradovich, and Friedman (1972 AMQUA presentation; 1974 written comm.,ms. in preparation) challenge the current Bull Lake and Pinedale correlations based on their work in the U.S.G.S. Yellowstone project. They suggest that Bull Lake moraines near West Yellowstone, Montana are of late Illinoian age; and that Pinedale glaciation spans all or much of the Midcontinent's Wisconsin (Table 1). Their case is based on K-Ar and obsidian hydration dating, comparisons with the marine paleoclimatic record, and re-evaluation of previous corrlations of Rocky Mountain and Midcontinent glaciations.

Pierce et al. (1974) start their case with the geologic sequence involving glacial deposits and the West Yellowstone rhyolite flow. All agree with Richmond and Hamilton (1960) that: first the Bull Lake tills were deposited; thereafter the West Yellowstone flow was erupted; and later Pinedale materials were deposited. Pierce et al. estimate the flow as about 107,000 years old by averaging four separate K-Ar analyses of 117,000, 105,000, 114,000, and 93,000 yr. They suggest that a late Bull Lake age is equally if not more likely than an early Bull Lake age for the pre-flow glacial deposits in the critical area; therefore, no Bull Lake materials are younger than the 107,000 yr rhyolite flow. The ice which contacted the eastern margin of the flow, causing lobate reentrants, they interpret as a restricted mass far back from Bull Lake terminal moraines at Hebgen Lake to the west of the Park – either recessional Bull Lake ice, or the glacier of a later limited advance.

They consider that Pinedale glaciation began more than 45,000 yr ago. The original radiocarbon date for the pre-Pinedale lake, Pierce notes, has been revised to "greater than" 29,000 yr. And, another date from comparable lake materials is greater than 38,000 yr, which Pierce believes represents a minimal estimate for the time of the advance of Pinedale glaciers.

Obsidian Hydration Data

The obsidian hydration technique, first used to date artifacts (Friedman and Smith, 1960), is a focal point in the Yellowstone debate. Since this rather new technique is limited to obsidian-bearing regions, and involves some dispute over sampling and laboratory procedures – the method merits discussion. For the Yellowstone project, a time curve for local hydration rates along minute cracks in obsidian was first established using the K-Ar dated, 107,000 yr old West Yellowstone Flow and the K-Ar dated, 176,000 yr old Obsidian Cliff Flow as points of reference. The technique was further adapted to studies of glacial deposits by restricting laboratory measurements to hydration rinds along traces of conical fractures attributed to pressure between glacially transported fragments (and to fractures in bedrock samples attributed to shearing caused by thick over-riding ice). Cracks resulting from non-glacial causes (such as: contraction during the cooling of flows, frost action, and earthquakes) were avoided. From many hydration measurements made upon glacially abraded fragments – Pierce et al. propose dates of around 145,000 yr for West Yellowstone Bull Lake moraines; and of 40,000 to 30,000 yr for early Pinedale moraines, and 20,000 to 12,000 yr B.P. for middle Pinedale ("recessional") deposits.

	Blackwelder (1915)	WIND RIVER RANGE Moss; Holmes & Moss (1951,1955)	Richmond (1948 to 1971)	MED. BOW MTS. Kiver ('68)	COLO. FRONT R. Benedict (1973)	JACKSON HOLE AREA Love (1974)	WEST YELLOWSTONE Pierce, Obradovich, Friedman (1974)
NEOGLACIAL		"Little Ice Age"	Gannett Peak Temple Lake "a" "b"	G.P. T.L. late T.L. early	Arapaho Peak Audubon Triple Lakes		
WISCONSIN	Pinedale	Temple Lake recessional P. main	late stade P. middle stade early stade	"Pinedale 4"	Satanta Peak	Pinedale	P. late recessional main
						&	
	Bull Lake	B.L. II B.L. I	B.L. late stade 2nd B.L. late stade 1st early stade			Bull Lake	Bull Lake
ILLINOIAN	Buffalo	Buffalo	Sacagawea Ridge			Buffalo	
KANSAN			Cedar Ridge				
NEBRASKAN			Washakie Point			"Ghost Glaciation"	

Table 1.
Selected terminology for Wyoming glaciations and various interpretations

Correlation with the Marine Record

Pierce, Obradovich, and Friedman (1974 written comm.) also compared their Yellowstone data with marine paleoclimatic records. Using Broecker and Van Donk's time scale (1970), they plotted oxygen-isotope temperature curves from four Atlantic and Caribbean deep sea cores (Emiliani, 1955, 1964, 1966, 1972), foraminiferal data from a core in the western Gulf of Mexico (Kennett and Huddlestun, 1972), and sea level stands based on dating of corals from Barbados by the uranium-series disequilibrium method (Mesolella et al., 1969; James et al., 1971). Pierce et al. (1974) claim good agreement between the Yellowstone and marine records for major climatic fluctuations. They consider the cold marine episode prior to the last interglaciation of the marine record as late Illinoian and equivalent in age to their 145,000 yr Bull Lake determination from obsidian hydration. High sea levels at 125,000, 105,000, and 82,000 yr B.P. are viewed as Sangamon, and correspond to the interglacial interval between the last two major Yellowstone glaciations. The last major cold, marine episode, they class as Wisconsin and relate to their Yellowstone dates for Pinedale glaciation.

Geomorphic Relations

In considering topography and soils, Pierce notes that Leverett's (1899) description of the Midcontinent Buffalo Hart moraines (considered Illinoian by Frye, Willman, and Black, 1965) mentions subdued knob-and-basin topography, having a contrast with Wisconsin moraines that resembles the difference between Bull Lake and Pinedale moraines. Johnson (1964) confirmed Leverett's description and notes a Sangamon soil on the Buffalo Hart moraines. Midcontinent workers (R.B. Morrison, comm. to Pierce) report the Sangamon soil as better developed than post-glacial soils or any buried Wisconsin ones (with the possible exception of the problematic Chapin soil). Since Bull Lake soils in the Rocky Mountains are better developed than post-Pinedale ones (Richmond, 1965), Pierce suggests that the Bull Lake soil logically correlates with the Sangamon, rather than an intra-Wisconsin soil-forming interval.

Quite aside form the Yellowstone controversy, the geomorphic contrast between the relatively subdued Bull Lake and youthful Pinedale moraines of typical Wyoming mountains has long merited some consideration. It seems reasonable to assume that the Bull Lake moraines were initially at least as rugged and bouldery as those of the Pinedale – because both were deposited by glaciers travelling similar distance across identical bedrocks in the same confining canyons. Thus the simplest interpretation of the contrasting morainic topographies is that a significant episode of weathering, erosion, and loess deposition intervened between the two glaciations – an interval better accomodated by Pierce et al's correlations and time spans than the long intra-Wisconsin interglacial required by the current model.

Status of Pre-Bull Lake Glaciations

If the Bull Lake should prove to be Illinoian, the present correlations (Richmond, 1965) of earlier Wyoming glaciations with the Midcontinent will need re-evaluation. Outside the type localities, the assignment of pre-Bull Lake deposits to Washakie Point, Cedar Ridge, and Sacagawea Ridge episodes has been difficult in the Rocky Mountain region. The various Pearlette-type ashes are not always found in critical places; the range of radiocarbon dating is inadequate; vertebrate fossils have so far not been helpful in relevant dating: and lithologic contrasts of the deposits have not been definitive. Thus outside the type area, correlation of deposits with the Wind River sections has relied heavily on height above stream levels and geomorphic position (where available).

At the Washakie Point type locality in the cliff along the north shore of the present-day Bull Lake (Fig.5), all the reported pre- Wisconsin glacial deposits are exposed in normal stratigraphic succession. Because the successively deposited Washakie Point, Cedar Ridge, and Sacagawea Ridge materials all contain lake beds interspersed with tills in the cliff face, their deposition could have been in earlier moraine-dammed lakes along the same axis as the existing one. Thus all these deposits, if related to glaciers issuing from the mountain canyon of Bull Lake Creek, might possibly be classed as Blackwelder's "post-canyon" glaciations. Washakie Point deposits exposed in the cliff are at least 2,000 ft below the edge of the high level erosion surface on the adjacent mountains, indicating extensive basin excavation before the earliest reported glacial deposition. The basin deepening might have begun in the Pliocene and possibly include pre-glacial Quaternary excavation.

J.D. Love, however (pers. comm.), suspects a pre-"Buffalo" glacial event – informally called the "Ghost" glaciation – that preceded any extensive excavation of Jackson Hole. He reports at least four localities in the adjacent mountains where erratics and cemented tills at high elevations contain rock types that are now separated from possible bedrock sources by deep valleys. Thus Love, from extensive field work, tentatively proposed three main phases of glaciation: the Pinedale and Bull Lake which left moraines on valley floors; the "Buffalo" whose deposits are at somewhat higher elevations in basins and on valley walls, and on valley floors many miles downstream from Bull Lake and Pinedale former glacial fronts; and the "Ghost glaciation" whose features are the highest and oldest. Relicts of this oldest glaciation (following Love's "law of obliteration") have been largely destroyed by subsequent weathering and erosion as well as by later glaciations. As a working hypothesis, the possibility of a very early pre-Washakie Point (pre-Buffalo), glaciation deserves attention.

NEW INTERPRETATIONS OF LATE GLACIATIONS

Until recently, the "high country" glaciations seemed adequately accommodated in the Gannett Peak and Temple Lake categories. Now at least four minor late episodes seem possible.

Neoglaciation

Kiver (1968) proposed three-fold Neoglaciation of the Rawah Peaks in the southern Medicine Bow Mountains, northernmost Colorado. (See Figs. 10 and 11 for typical neoglacial features). There, outer stable cirque moraines (beyond unstable cirque-head features) commonly bear stabilized patterned ground of an ensuing cold period, along with double moraines having geomorphic and vegetational contrasts. He mapped cirque-head features as Gannett Peak; and, to avoid undue proliferation of local names, the outer stable deposits were mapped as "Temple Lake a and b". However, based on the morainic contrasts, and accepting the published radiocarbon dates for early and late "Temple Lake" alluvium (Richmond, 1962), Kiver suggested that a 1,000 yr interval between "a" and "b" indicated three Neoglacial advances rather than minor pulsations of a single glaciation. It now seems probable that the dates of 3,100-2,800 yr B.P. and 1,800-1,000 yr B.P. attributed to "Temple Lake a and b" moraines, and which were obtained from alluviums some distance from actual tills, do not both apply to the double, stable, early Neoglacial, cirque moraines.

Benedict (1968) also proposed three Neoglacial episodes in the Front Range of Colorado. In addition to the Temple Lake and Gannett Peak advances, he introduced the Arikaree episode that was distinguished and dated (1,900-1,000 yr B.P.) by the application of lichenometry. Since the type locality was protalus, and "Arikaree" was pre-empted by Darton's (1899) Miocene Formation – Mahaney (1972) selected a type till and renamed the advance Audubon, for a prominent peak in the area. Richmond (1971) classed the Audubon as Temple Lake "b", because of comparable radiocarbon and lichometric dates. However, Benedict (1972) pointed out that Audubon and Temple Lake moraines are geomorphically distinct, as Audubon features are in the cirque-head complexes, often over-ridden and masked by later Gannett Peak deposits, and are only distinguishable by lichen contrasts. In Wyoming, the Audubon was first noted by Birkeland and Miller (1973) within the "cirque moraine" complex originally described by Moss at the southeast end of Temple Lake in the Wind River Mountains.

A Pre-Altithermal Late Glaciation

A fourth "high country" glacial pulsation (reflecting Moss' original Temple Lake designation) seems distinctly possible. It would be pre-Altithermal, but clearly younger than moraines of the Pinedale complex in and around the lower ends of

Figure 10. Neoglacial features at the head of Dinwoody canyon, Wind River Mountains.

Figure 11. Late Neoglacial moraines near Mt. Lander, Wind River Mountains. (Courtesy of Austin Post, Univ. of Washington, Dept. Atmospheric Sciences).

glaciated valleys. Kiver (1968, 1972) reported a till sheet with distinct morainal fronts on the high-level Rawah (subsummit) erosion surface, and comparable features confined in the deep adjacent West Branch valley. The morainal fronts are generally 1-2 mi from cirque headwalls, and 3.5-5 mi inside late Pinedale ("Pinedale III") moraines at lower elevations.

Based on maximum frontal slope angles, relations to alluvial terraces, differences in amounts of soil fines, and boulder weathering contrasts – as well as examination and re-evaluation of moraines in the nearby Cache la Poudre drainage (Ray, 1940; Richmond, 1960, 1965) – Kiver suggested a pre-Altithermal advance in latest Pinedale time (perhaps, unfortunately designated as "Pinedale 4"). he refrained, however, from formal designation of another glacial advance pending further stratigraphic studies to determine if it represented more than a time of stability in general Pinedale ice recession. In the nearby Cache la Poudre drainage, Kiver considers Ray's (1940) Long Draw moraine as latest Pinedale. If the interpretation is correct, Rubin and Suess' date (1955) of about 6,170 yr for peat in a bog associated with Long Draw deposits gives a minimum age that is pre-Neoglacial.

Breckenridge (1969) reports a prominent end moraine well up-valley from massive Pinedale moraines but 2.5 mi from cirque heads and outside the Neoglacial complex which he studied at the head of Fall Creek in the Cache la Poudre drainage. Schoenfeldt (1969) in a **reconnaissance** of the upper Burnt Fork drainage (on the north flank of the Uinta Mountains) tentatively designated moraines, just beyond those he considered Neoglacial, as "Pinedale 4". Graf (1970) reports "Pinedale IV" moraines in 15 valleys of the Beartooth Mountains of Montana and Wyoming. His regression equation for the down-valley extent of "Pinedale IV" moraines relative to "Pinedale III" gave a regression line slope that cannot be separated at the .05 confidence interval from a similar computation by Kiver (1972) for the southern Medicine Bow Mountains, 350 mi away. Such calculations are suggestive, but hardly prove a late glacial pulsation just prior to the Altithermal.

Breckenridge (1974a) in a detailed Quaternary study at the head of the Wood River Drainage (southern Absaroka Range, east of Yellowstone National Park) found moraines rimming cirque thresholds several miles up glacial troughs, but just outside features he interprets as Neoglacial. No radiocarbon dates are yet available, but in one cirque (Cascade Creek site) pond deposits of a tarn just outside Neoglacial deposits yielded a probable Early Middle Period point. The artifact (identified by M. Wilson) is a large dart point of the "Oxbow type" which has been dated at about 5,000 yr in several Canadian plains localities. Thus a pre-Neoglacial age for deposits lower in the Cascade Creek cirque seems possible (Breckenridge, 1974b).

The Temple Lake type locality has received considerable attention recently. My casual reconnaissance with Kiver in 1967 left the impression that Moss' pre-Altithermal dating of the right lateral moraine there could be correct. From detailed study, Birkeland and Miller (1973) suggest that the "type early Neoglacial"

Temple Lake moraine is pre-Altithermal; and that Moss' "cirque moraines" contain early Neoglacial, Audubon, and Gannett Peak equivalents. Currey (1974) also believes that the Temple Lake moraine is pre-Altithermal and obtained a radiocarbon date of about 6,500 yr for overlying bog deposits. The date is minimal for the moraine, but clearly older than any obtained from Neoglacial deposits. Currey's suggestion of two substadial Temple Lake advances is, however, questioned by Birkeland and Miller (1974).

Overall, the evidence for a very late but pre-Altithermal glacial event in the high country is becoming more than speculation (Fig. 12).

Benedict (1973) proposes a chronology of cirque glaciation involving four, possibly five, pulsations for the northern Colorado Front Range. (Fig.8) It is based on many careful studies and more than 20 radiocarbon dates from the high country. Using local names--to avoid specific correlation with the Wyoming model--he proposes the following glacial expansions: Satanta Peak (shortly before 9,900 yr B.P.); a possible un-named, 8,000 yr advance; Triple Lakes (5,000-3,000 yr B.P.); Audubon (1850–950 yr B.P.); and Arapaho Peak (300–100 yr B.P.). The Satanta Peak and Triple Lakes advances left similar appearing moraines which Benedict states have previously both been correlated with "Temple Lake", although one advance preceded and the other followed the Altithermal interval.

Satanta Peak glaciers moved farthest from cirque walls to about the position of present-day timberline and left moraines of two closely spaced advances. The somewhat less extensive Triple Lakes glaciers had at least two minor fluctuations recorded in moraines or rock glaciers, with a possible third advance represented by a rock glacier at the type locality. Satanta Peak and Triple Lakes features are difficult to distinguish by relative dating techniques. Lichenometry is not useful beyond about 3,000 yr B.P.; and both complexes are stable, well-vegetated, bouldery moraines and periglacial features, in some cases marked by stabilized patterned ground. Both have loess mantles, and overlapping soil characteristics. Benedict's intervening "possible 8,000 year advance" is suggested by sand

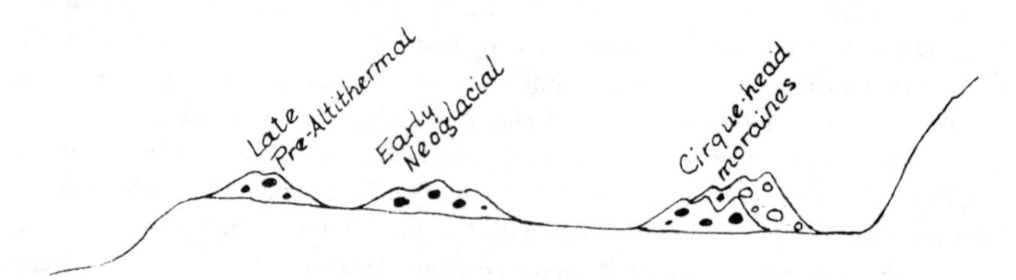

Figure 12. Schematic profile suggesting relations in cirques of one late-pre Altithermal glacial advance, and three Neoglacial pulsations. Cirque head moraines include Audubon-type tills with overlying Gannett Peak type deposits.

interpreted as an outwash or nivation deposit within dated bog sediment; moraines of this age have not been positively identified in the Front Range.

The Audubon and Arapaho Peak advances produced the fresh, unstable, ice-cored moraines and associated features near cirque walls and relict glaciers and snow fields. Benedict suggests that the advances followed a non-glacial interval of soil formation and cavernous weathering which may, in the Front Range, have rivaled the Altithermal. The Arapaho Peak episode seems correlative with the Gannett Peak, based on preliminary observations by Birkeland in the type area of the Wind River Mountains.

SUMMARY

Blackwelder's concept of Buffalo, Bull Lake, and Pinedale glaciations provided direction for later investigations which: added Washakie Point, Cedar Ridge, and Sacagawea Ridge as pre-Bull Lake glaciations; subdivided the Bull Lake and Pinedale into multiple advances; and added Temple Lake and Gannett Peak Neoglacial fluctuations. Ongoing investigations suggest that some Bull Lake moraines are Illinoian instead of Wisconsin in age, and that at least one pre-Altithermal and three Neoglacial advances occurred in and around cirque areas.

CONCLUSIONS

The development of Wyoming's glacial model makes a nice case history in scientific operations. As a prelude – from the many short statements attempting to explain what science is,[2] perhaps the most helpful is Conant's (1948) definition: "an interconnected series of concepts and conceptual schemes that have developed as a result of experimentation and observation and are fruitful of further experimentation and observation." The statement does reflect a chemist's preoccupation with laboratory investigations which are not basic to glacial geomorphology. But the definition fits – if "techniques" is substituted for "experimentation" and observation emphasized, considering the dynamic situation in the Rockies as well as our broad backgrounds in Quaternary schemes.

Concepts

A broad conceptual framework, the mark of a relatively mature science, reduces empirical groping where all observations seem equally important and gives direction to investigations. Our "conceptual schemes" are well-tested and accepted doctrines

2. Some examples are: the concise "controlled imagination", the pungent "doing your damndest to solve your problems", the circular but revealing "what people who call themselves scientists do professionally"; the pompous "using my techniques to solve the problems of the Earth", and the misleading "knowledge, as of general truths or particular facts, obtained and shown to be correct by accurate observation and thinking".

such as "Ice Ages" and multiple glaciations, which are usually taken for granted. Among somewhat less well-established concepts in the framework (Conant calls them grand working hypotheses) would be such an ingrained idea as world-wide synchronous climatic changes through time. It allows correlation of deep sea and terrestrial glacial records (as separate converging lines of evidence) as in the current Yellowstone debate.

Grand "speculative ideas", the least-well established concepts according to Conant, include most explanation for the causes of Ice Ages and global climactic changes; for example, the CO_2 or the aerosol-smog hypotheses. They form the bases for well-intentioned speculation on the environmental impact of air pollution. For instance: Wyoming basins could soon become altithermal deserts from global warming; or the reverse, Wyoming basins will become tundras during a new glaciation. Since "Buffalo", Bull Lake, and Pinedale glaciations–separated by climates fully as warm as the present–long preceded man's tampering through the Industrial Revolution, CO_2 and smog mechanisms need "more work" before inclusion in the established conceptual framework.

Conant's repetition of such words as "concepts" and "schemes" emphasized the tentative nature of scientific generalizations.[3] Thus the term "model"--although somewhat ostentatious and overworked-- is worthwhile in describing theories and hypotheses; as well as geologic maps, profiles, equations, graphs, correlation charts, and the like. All are simplifications (absolutely essential to our thinking) that ignore irrelevant details. But if significant details *were* overlooked, a current model may need revision. This seemingly semantic and "self-evident" discussion may have a down-to-earth bearing in Rocky Mountain glacial geology.

Despite our scientific humility in using such words as model, however; T.S. Kuhn (1962) "outrageously" suggests that scientists usually reject novelty and attempt to force nature into "conceptual boxes". Hack (1960) did reinterpret Appalachian erosion surfaces by purposely ignoring the Davisian cyclic erosion scheme; and recently (1974) Birkeland and Shroba stimulatingly challenge the Quaternary doctrine requiring warm soil-forming intervals – but the approach is not common. Certainly the revision of a current model rightfully puts the burden of proof on the innovator. On the other hand, could existing correlation charts have so strongly influenced thinking in the Rockies that evidence for an Illinoian Bull Lake and at least four glacial fluctuations in the high country was forced into established "conceptual boxes"?

Techniques

The role of techniques (Conant's "tactics" in the grand "strategy" of improving scentific concepts) is demonstrable in the Rocky Mountain-glacial case history. Soil stratigraphy has aided in correlations and demonstration of interglacial episodes of weathering. The introduction of lichenometry led to discovery of a minor cirque

3. Probably stemming from Whitehead's (1925) discussion of "the fallacy of misplaced concreteness" in which he points out that scientific concepts are actually mental abstractions from sense perception of external reality.

glaciation, the Audubon, previously masked in Gannett Peak deposits. The adaptation of the obsidian hydration technique may change the dating of the Bull Lake.

On the other hand, the difficulties with techniques are also demonstrable. Birkeland (1973) discusses some general problems derived from his Neoglacial investigation of rock glaciers in Colorado. Since radiocarbon (whose use requires careful appraisal) is often missing in critical localities, various relative dating techniques are required. He stresses careful consideraiton in their application because: weathering, soil development, and lichen growth may strongly reflect such local factors as environmental differences and rock type; and past "operator variance" has resulted from inexperience, failure to quantify, and selection of differing parameters. Moreover, different relative dating techniques have different effective time spans. Birkeland's recommendation of multiple techniques without undue reliance on one or two, as panaceas, seems valid. Soil stratigraphy has not solved the Bull Lake problem; lichenometry even after establishing a local growth rate is limited to the Neoglacial; obsidian dating requires special bedrock; and so forth. Thus his advice recalls past recommendations of "mental procedures" using multiple working hypotheses.[4]

Multiple Hypotheses

The familiar classics of Gilbert(1886) and Chamberlin (1897) still seem relevant since they stress: careful selective observation, logical reasoning, the tentative nature of conclusions, and avoidance of "undue affection for a beloved intellectual child". Moss (1951) did present multiple working hypotheses in tentatively designating the Temple Lake moraine as pre-Altithermal. And, despite its later selection as the type early Neoglacial, the most recent work there confirms his original conclusion. The Gilbert and Chamberlin papers remain excellent operational guides in the field when confronting an enigmatic landform or deposit (such as a diamicton). However, discussions of alternative interpretations are now rare in published papers (perhaps from editorial emphasis on brevity) – and the method of multiple working hypotheses has been attacked by the philosophically-minded in recent times as a gross over-simplification of scientific endeavor.

Conant considers the method as common sense procedures, using "limited working hypotheses", for an immediate problem whose solution closes the

4. Lord Bertrand Russell (1960) aptly discusses the method:
"The people who first think of these hypotheses can't say, 'this is the truth'—they can only say, 'this may be the truth.' And if you have a good scientific imagination you can think of all sorts of things that might be true, and that's the essence of science. You first think of something that might be true—then you look to see if it is, and generally it isn't".

investigation – little different from trouble-shooting when your car won't start. What old classics on scentific method lack is attention to the sophisticated frame of reference.[5] A broad theoretical background is needed to recognize worthwhile scientific problems, and to aid in their solutions, which in turn may change a conceptual framework and lead to new problems. Even to recognize the question of whether Bull Lake moraines are early Wisconsin or late Illinoian takes a sophisticated geologic background; and the answer could change the Wyoming glacial model, and a raft of related deductions.

Normal Investigations

Despite the Bull Lake controversy however, the Wyoming glacial scheme is not involved in a revolution (as when plate tectonics replaced Dana's doctrine of permanency of continental and oceanic blocks in the mainstream of geologic thinking). Kuhn (1962) clearly distinguishes "scientific revolutions" from "normal science" in which research findings are smoothly fitted into the accepted grand framework of concepts, which he calls a paradigm (roughly, a pattern).

Past and present glacial investigations in the Rockies seem normal operations with three main trends: refining Blackwelder's original model, extending it to new areas, and relating it to the broader Quaternary paradigm. Field work refined the model by providing evidence for multiple Bull Lake, multiple Pinedale, and Neoglacial pulsations. Deduction from the Quaternary paradigm suggested a search for evidence of pre-Wisconsin glaciations within the "Buffalo", and a more aesthetically pleasing correlation with the Mid-continent. The present widespread application of the Wyoming model throughout the Rockies reflects observations far from the type localities that fit the scheme(it has allowed prediction). That the model fits the general paradigm ("corresponds to nature") involves attempted correlations with the Mid-continent and European glacial records, and the deep sea findings. Since a revolution only occurs when normal investigation turns up so many anomalies that they cannot be accomodated by merely modifying an established paradigm – Wyoming developments seeem only a normal healthy evolution. Neo-uniformitarianism, multiple glaciations, global climatic changes, and other basic Quaternary doctrines remain unchallenged.

A Human Endeavor

Our present view of the Rocky Mountain's complicated glacial history represents, of course, a collective enterprise. Yet in post-Blackwelder time, no one has worked harder in the field and office than Gerald Richmond.[6] If the

5. Conant's criticism may be academic for professionals who, from long training, instinctively operate within their scientific frame of reference.

6. A foreign, 1965 INQUA, Rocky Mountain-excursion participant's glowing accolade "Penck of the Rockies" may have been close to the mark.

stimulation of present investigations largely reflects challenges to his interpretations, it was his leadership that focused attention on worthwhile problems. A fondness for "conceptual boxes" is not necessarily undue affection for a "ruling theory" (Chamberlin's "hastily proposed, then stoutly defended explanation")–because the Wyoming glacial model is a sophisticated scheme within the Quaternary paradigm. Wisconsin and Neoglacial thinking may need revision. But if it comes to that–Albert Heim, distinguished structural geologist, changed his prediliction for a prior hypothesis on admitting to assembled Swiss geologists, "there is no double-fold in Canton Glarus". In retrospect, the Wyoming glacial model never was "the Gospel"--rather it represents the collective scientific impressions of a delightful region, sifted from the dedicated work of highly argumentative men.

ACKNOWLEDGEMENTS

For helpful discussions and the weeding of technical misconceptions as well as "muddy" prose, I am indebted to: J.B. Benedict, D.L. Blackstone, R.M. Breckenridge, E.P. Kiver, J.D. Love, J.H. Moss, and K.L. Pierce. The flaws remaining in science and rhetoric are purely my own.

REFERENCES

Alden, W.C. 1953.Physiography and glacial geology of western Montana and adjacent areas: U.S. Geol. Survey Prof. Paper 231, 200 p.

Atwood, W.W, Jr. 1937.Records of Pleistocene glaciers in the Medicine Bow and Park Ranges: J. Geol, v. 45, n. 2, p. 113-140.

Benedict, J.B. 1968.Recent glacial history of an alpine area in the Colorado Front Range, U.S.A., Pt. II Dating the glacial deposits: J. Glaciology, v. 7, n. 49, p. 77-87.

——————— 1973.Chronology of cirque glaciation, Colorado Front Range: Quat. Research, v. 3, n. 4, p. 584-599.

Birkeland, P.W. 1973.Use of relative age-dating methods in a stratigraphic study of rock glacier deposits, Mount Sopris, Colorado: Arct. Alp. Res., v. 5, n. 4, p. 401-416.

Birkeland, P.W., Crandell, D.R. and Richmond, G.M. 1971.Status of Correlation of Quaternary stratigraphic units in the western conterminous United States: Quat. Research, v. 1, n. 2, p. 208-227.

Birkeland, P.W. and Miller, L.D. 1973.Re-interpretation of the type Temple Lake moraine, and other Neoglacial deposits, southern Wind River Mountains, Wyoming: Geol. Soc. Amer. Absts. with Programs, Rocky Mtn. Sect., v. 5, n. 6, p. 465-466.

Birkeland, P.W. and Shroba, R.R. 1974.Quaternary soil-forming intervals: stratigraphical, pedological, and geochemical arguments cast doubt on their validity: Geol. Soc. Amer. Absts. with Programs, Rocky Mtn. Sect., v. 6, n. 5, p. 426.

Blackwelder, E. 1915.Post-Cretaceous history of the mountains of central western Wyoming: J. Geol, v. 23, n. 4, p. 307-340.

——————— 1926.Pre-Cambrian geology of the Medicine Bow Mountains: Geol. Soc. Amer. Bull. v. 37, N. 4, p. 615-618.

Breckenridge, R.M. 1969.Neoglacial geology of upper Fall Creek Basin, Mummy Range, Colorado: M.A. thesis, Univ. of Wyo., Laramie, 59 p.

——————— 1974a.Quaternary and environmental geology of the upper Wood River area, Absaroka Range, Wyoming: Ph.D. Dissertation, Univ. of Wyoming, Laramie, 138 p.

——————— 1974b.The use of archaeology in dating Quaternary deposits in the Upper Wood River area, Absaroka Range, Wyoming: in Archaeology and the Holocene record in Wyoming: Wyo. State Geol. Survey Bull. (in press).

Broecker, W.S. and van Donk, J. 1970.Insolation changes, ice volumes, and the O^{18} record in deep-sea cores: Revs. of Geophysics and Space Physics, v. 8, n. 1, p. 169-198.

Chamberlin , T.C. 1897.The method of multiple working hypotheses: J. Geol, v. 5, n. 8, p. 837-848.

Conant, J.B. 1947.*On understanding science:* Yale Univ. Press (also Mentor Book paperbacks, 1951), 144 p.

Currey, D.R. 1974.Probable pre-Neoglacial age of the type Temple Lake moraine, Wyoming: Arct. Alp. Res., v. 6, n. 3, (in press).

Darton, N.H. 1897-1898.Preliminary report on the geology and water resources west of the 103rd meridian: U.S. Geol. Survey 19th Annual Rept., pt. 4, p. 719-814,

Emiliani, C. 1955.Pleistocene temperatures: J. Geol, v. 63, n. 6, p.538-578.

——————— 1964.Paleotemperature analysis of the Caribbean cores A 254-BR-C and CP-28: Geol. Soc. Amer. Bull., v. 75, n. 2, p. 129-143.

——————— 1966.Paleotemperature analysis of Caribbean cores P 6304-8 and P 6304-9 and a generalized temperature curve for the past 425,000 years: J. Geol, v. 74, n. 2, p. 109-126.

——————— 1972.Quaternary paleotemperatures and the duration of the high- temperature intervals: Science, v. 178, n. 4059, p. 398-401.

Friedman, I. and Smith, R.L. 1960.A new dating method using obsidian, Pt. I the development of the method: Amer. Antiquity, v. 25, n. 4, p. 476-522.

Fryxell, F.M. 1930.Glacial features of Jackson Hole, Wyoming: Augustana Library Pub. 13, 128 p.

Gilbert, G.K. 1886.The inculcation of scientific method: Amer. Jour. Sci., 3rd Series, v. 31, n. 184, p. 284-299.

Graf, W.L. 1971.Quantitative analysis of Pinedale landforms, Beartooth Mountains, Montana and Wyoming: Arct. Alp. Res., v. 3, n. 3, p. 253-261.

Hack, J.T. 1943.Antiquity of the Finley Site: Amer. Antiquity, v. 8, no. 3, p. 235-241.

——————— 1960.Interpretation of erosional topography in humid temperate regions: Amer. Jour. Sci., Bradley vol., v. 258-A, p. 80-96.

Hares, C.J. 1948.Striated boulders on the Medicine Bow Mountains, Wyoming: Geol. Soc. Amer. Bull., v. 59, n. 12, pt. 2, p. 1329.

Holmes, G.W. and Moss, J.H. 1955.Pleistocene geology of the southwestern Wind River Mountains, Wyoming: Geol. Soc. Amer. Bull., v. 66, n. 6, p. 629-654.

James, N.P., Mountjoy, E.W. and Omura, A. 1971.An early Wisconsin reef terrace at Barbados, West Indies, and its climatic implications: Geol. Soc. Amer. Bull., v. 82, n. 7, p. 2011-2017.

Johnson, W.H. 1964.Stratigraphy and petrography of Illinoian and Kansan drift in central Illinois: Illinois State Geol. Sur., Circular 378, 38 p.

Kennett, J.P. and Huddlestun, P. 1972. Late Pleistocene paleoclimatology, foraminiferal biostratigraphy and tephrochronology, western Gulf of Mexico: Quat. Research, v. 2, n. 1, p. 38-69.

Kiver, E.P. 1968. Geomorphology and glacial geology of the southern Medicine Bow Mountains, Colorado and Wyoming: Ph.D. Dissertation, Univ. of Wyoming, Laramie, 129 p.
——— 1969. Neoglaciation of the Rawah Peaks, Colorado.: Geol. Soc. Amer. Absts. with Programs, Rocky Mtn. Sect., pt. 5, p. 41
——— 1972. Two late Pinedale advances in the southern Medicine Bow Mountains, Colorado: Contributions to Geol. of the Univ. of Wyoming, v. 11, n. 1, p. 1-8.
Knight, S.H. 1953. Summary of the Cenozoic history of the Medicine Bow Mountains, Wyoming: Wyo. Geol. Assoc. Guidebook, 8th annual field conf., p. 65-76.
Kuhn, T.S. 1962. The structure of scientific revolutions: Phoenix Books, Univ. of Chicago Press, 172 p.
Leverett, F. 1899. The Illinois glacial lobe: U.S. Geol. Survey Monograph 38, 817 p.
Love, J.D. 1960. Cenozoic sedimentation and crustal movement in Wyoming: Amer. Jour. Sci., Bradley vol., v. 258-A, p. 204-214.
Mackin, J.H. 1937. Erosional history of the Big Horn Basin, Wyoming: Geol. Soc. Amer. Bull., v. 48, n. 6, p. 813-894.
Matthes, F.E. 1939. Report on glaciers, April 1939: Amer. Geophys. Un. Trans., 20th Ann. Mtg., pt. 4, p. 518-523.
Mears, B. Jr. 1972. Wyoming's glaciers past and present: Wyo. Game & Fish Comm., Wyo. Wildlife, Cheyenne, v. 36, n. 4, p. 26-34.
Mesolella, K.J., Matthews, R.K., Broecker, W.S. and Thurber, D.L. 1969. Astronomical theory of climatic change: J. Geol, v. 77, n. 3, p. 250-274.
Miller, L.D. and Birkeland, P.W. 1974. Discussion of probable pre- Neoglacial age of the type Temple Lake moraine, Wyoming: Arct. Alp. Res., v. 6, n. 3, (in press).
Moss, J.H. 1951. Late glacial advances in the southern Wind River Mountains, Wyoming: Amer. Jour. Sci., v. 249, n. 12, p. 865-883.
Pierce, K.L., Obradovich, J.D. and Friedman, I. 1972. Oral presentation: Amer. Assn. Quat. Envir., Miami, Florida.
Pierce, K.L. Obradovich, J.D. and Friedman, I. 1974. Obsidian-Hydration dating of Yellowstone glaciations, Bull Lake end moraines may be pre-Wisconsin, for they predate the last interglaciation of the marine record: publication authorized by the Director of the U.S. Geol. Survey Written Communication.
Richmond, G.M. 1948. Modification of Blackwelder's sequence of Pleistocene glaciation in the Wind River Mountains, Wyoming: Geol. Soc. Amer. Bull. Absts., v. 59, n. 12, pt. 2, p. 1400-1401.
——— 1960a. Glaciation of the east slope of Rocky Mountain National Park, Colorado: Geol. Soc. Amer. Bull., v. 71, n. 9, p. 1371-1381.
——— 1960b. Correlation of alpine and continental glacial deposits of Glacier National Park, Montana: U.S. Geol. Survey Prof. Paper 400-B, p. B-223-224.
——— 1962. Quaternary stratigraphy of the La Sal Mountains, Utah: U.S. Geol. Survey Geol. Survey Prof. Paper 324, 135 p.
——— 1962. Three pre-Bull Lake tills in the Wind River Mountains, Wyoming: U.S. Geol Survey Resch. p. D-132-135.
——— 1964. Three pre-Bull Lake tills in the Wind River Mountains, reinterpreted: U.S. Geol. Survey Prof. Paper 501-D, p. 104-109.
——— 1965. Glaciation of the Rocky Mountains: in Wright and Frey, ed. *The Quaternary of the United States,* Princeton Univ. Press, p. 217-230.
——— 1972. Appraisal of the future climate of the Holocene in the Rocky Mountains: Quat. Research, v. 2, n. 3, p. 315-322.

Richmond, G.M. and Hamilton, W.B. 1960. The late Quaternary age of obsidian-rhyolite flows in the western part of Yellowstone National Park, Wyoming: U.S. Geol. Survey Prof. Paper 400-B, p. B-224-225.

Russell, B. 1960. *Bertrand Russell speaks his mind:* Bard Books, Avon Book div. of Hearst corp., 144 p.

Schoenfeld, M.J. 1969. Quaternary geology fo the Burnt Fork area, Uinta Mountains, Summit County, Utah: M.A. thesis, Univ. of Wyo., Laramie. 75 p.

Scott, G.R. 1965. Nonglacial Quaternary geology of the southern and middle Rocky Mountains: in Wright and Frey, eds. *The Quaternary of the United States,* Princeton Univ. Press, p. 243-254.

——— 1973. Tertiary surfaces and deposits of the southern Rocky Mountains and their recognition: Geol. Soc. Amer. Abstracts with programs, v. 5, n. 6, p. 510-511.

Sylvester, G.H. 1973. Depositional environment of diamictites in the Headquarters Formation, Medicine Bow Mountains, southeastern Wyoming: M.A. thesis, Univ. of Wyo., Laramie, 84 p.

Trimble, D.E. and Carr, W.J. 1961. The Michaud delta and Bonneville river near Pocatello, Idaho: U.S. Geol Survey Prof. Paper 424-B, p. B-164-166.

Whitehead, A.N. 1925. *Science and the modern world:* Lowell lectures, MacMillan Co., reprinted (paperback) by Mentor Books 1948. 212 p.

Willman, H.B. and Frye, J.C. 1970. Pleistocene stratigraphy of Illinois: Ill. State Geol. Survey Bull. 94, 204 p.

CHAPTER 2

PROCESSES AND PATTERNS OF GLACIAL EROSION

by Geoffrey S. Boulton

ABSTRACT

Subglacial observations are presented of three important erosional processes, abrasion of the bed by debris embedded in the glacier sole, crushing of subglacial bedrock due to the pressure fluctuation produced by glacier flow over a bedrock hummock, and removal by plucking of the debris produced by the previous processes. The processes are then analyzed theoretically.

It is suggested that for any one ice velocity and increasing effective normal pressure at the glacier bed, that abrasion rates will rise to a maximum, thereafter falling rapidly until a critical pressure is attained beyond which the bed is not abraded and lodgment till accumulates. Normal abrasive processes are considered sufficient to explain p-forms such as *Sichelwannen, cavettos,* and glacial grooves. The progressive abrasion of a bedrock hummock is simulated, and for low pressures, high ice velocities or small bed roughness *roches moutonnees* or valley steps form with a down-glacier asymmetry, for high pressures, low velocities or high bed roughness, rock-cored drumlins form with an up-glacier asymmetry.

Analysis suggests that at constant ice velocity crushing on the lee flanks of bedrock hummocks becomes increasingly important under increasing ice thickness, rising to a maximum and then falling off rapidly. It is suggested that the plucking mechanism, a necessary complement to abrasion and crushing in eroding a rock bed, will be co-extensive with these processes. Plucking alone is necessary to erode unlithified sediments.

The theory also explains the forms of glaciated U-shaped valleys, *cirques,* valley steps, over deepened basins and fjords.

INTRODUCTION

Much of the work in the field of glacial erosion has concentrated on three themes: identification of the processes which might be responsible for such relatively small features as striae, friction cracks, *Sichelwannen,* and stoss and lee topography (Gilbert, 1906; Harris, 1943; Carol, 1947; Dahl, 1965; Gjessing, 1967); descriptions (numerical in recent years) of typical large scale erosional forms such as cirques, U-shaped valleys, overdeepened basins (Johnson, 1904; Lewis, 1960; Linton, 1963; Svensson, 1959; Doornkamp and King, 1971) and identification of patterns of erosion on a large scale (Clayton, 1965; White, 1972); and determination of the rates of erosional processes largely from sediment discharge in glacial streams (Thorarinsson, 1939; Østrem, Ziegler and Ekman, 1969). There have however, been surprisingly few attempts (e.g. Nye and Martin, 1967; Andrews, 1972) to produce more general, unifying theories which might interrelate these patterns of erosion on different scales, and above all to explain why a change occurs from erosion to deposition beneath glaciers, why different glaciers should erode at different rates, and why, as White (1972) has expressed it, glaciers should be "able or feckless erosive agents at different times and at different places".

This paper is an attempt to answer these questions, and to produce a more general theory of glacial erosion. It follows directly from a deposional and transportational theory (Boulton, 1974) and has the advantage that erosion, transport and deposition are analyzed as part of a unified continuum. It also produces specific, testable predictions.

The bed of a glacier may be composed of solid rock or fragmental materials. To erode the latter merely requires that the glacier should pick up fragments by some plucking process. These fragmental materials may be unlithified granular sediments, or equally, a highly jointed or fractured rock material. It has been argued by Boye (1950) and by Cailleux (1952) that erosion by a glacier is most effective when preglacial rocks have first been broken up by periglacial freeze-thaw activity.

A much-favoured plucking mechanism which might incorporate already fragmental materials, has been the freezing of meltwater in the low-pressure areas around the lee flanks of shattered bedrock obstructions (Carol, 1947). It has also been suggested (Weertman, 1961; Boulton, 1972) that, where a glacier is cold in its marginal area, net freezing of meltwater (derived from a temperate interior) to its sole would incorporate debris from subglacial sediments, and that subglacial materials which were themselves frozen could be incorporated in some bulk to produce very large erratics (Boulton, 1972).

However, previous discussions of plucking mechanisms have ignored the fact that such a mechanism is not enough to produce erosion of the bed. The force developed by the ice against the loose fragment must also be sufficient to overcome a frictional drag force between the fragment and subjacent bed materials. Thus, the existence of a plucking mechanism does not automatically imply erosion.

In addition to the plucking of already loosened rock fragments, it is clear that glaciers can break down hard, rigid, and unjointed rock masses. The scratching and grinding of a glacier bed by fragments in the glacier sole produces considerable abrasive smoothing of this surface (Forbes, 1842; Chamberlin, 1888; Demorest, 1939). It has also been strongly argued (Lewis, 1947) that large blocks of bedrock can be produced by glacial crushing of hitherto unloosened rock masses. Johnson (1904) suggested that sequential freezing and thawing beneath the margins of valley and corrie glaciers, where surface temperature fluctuations are still felt, might do this, and help explain such features as steep corrie headwalls. Lewis (1947) also argued that the step-shaped long profiles of many glaciated valleys may have developed beneath thin ice, where seasonal temperature fluctuations might still produce feeze-thaw fracturing and enlargement of rock steps. The importance of joints in weakening a rock so that it is susceptible to plucking has been well established (Matthes, 1930). Another crushing mechanism, suggested by Harland (1957), is that unloading of subglacial bedrock during glacier retreat might produce elastic rebound in them to the extent that fracturing occurs.

In this chapter, observations are presented, made both beneath and beyond the margins of modern glaciers, of a series of important erosional processes. Firstly the way in which bedrock is broken down by abrasion and by crushing is considered, and then the way in which materials loosened by these processes, and already fragmental materials, such as unlithified sediments, are plucked and transported away by the glacier. The processes are then modelled mathematically, and from a consideration of the glacier variables upon which they depend, it is shown that small and intermediate scale features such as *roches moutonnees,* and glacial grooves etc., can be synthesized and explained and that specific predictions about their forms can be tested. In order to establish large scale erosional patterns, several models are constructed of the way in which important glacier variables such as effective pressure at the bed and ice velocity, change over the area of the glacier. The predicted patterns of erosion are capable of explaining *cirques,* overdeepened basins, and U-shaped valleys. It is concluded that a relatively small number of erosional processes are capable of explaining most of the observed patterns of erosion.

ABRASION OF SUBGLACIAL BEDROCK

It is clear from subglacial observations that abrasion of subglacial bedrock by ice alone is of no consequence compared with that achieved by the debris particles with which most glacier soles are densely studded. Observations of the process of abrasion by such debris particles are first presented, and from these, a theory of abrasion is developed.

Observations

Figure 1 shows a clast of basalt held in the glacier sole, moving over the crest-line of a large basaltic *roche moutonnee,* and exposed in the wall of a tunnel 20 m below the surface of Breidamerkur jokull in southeast Iceland. A small cavity exists in the lee of the particle, and small particles have clearly been ploughed up in front of an obvious point of contact between particle and bed. The ice velocity over the bed was 9.9 m. yr^{-1} and the particle velocity 8.2 m. yr^{-1}.

The clast was taken from the ice and the surface which had been in contact with the bed inspected. Three points of contact with the bed were identified, of

Figure 1. A 30 cm basalt clast, held in basal ice and in traction over the glacier bed exposed in the wall of a subglacial tunnel. Ice movement right to left; a 30 cm cavity exists in the lee of the clast. Note the accumulation of smaller englacial particles behind clast, and ploughing of eroded debris before it.

which one was a heavily striated and smoothed facet, and two showed limited facetting and were relatively rough as if they had only recently made contact with the bed. The surface of the clast between these points of contact was covered by a thin and irregular layer of crushed debris which appeared to have been produced by grinding at the points of contact and laterally displaced into non-contact areas (Fig. 2). On the bed, areas of crushed debris can be seen to have been ploughed forward before the points of contact (Fig. 3). Crushed debris also forms a trail in the glacier sole behind large clasts moving over the bed, a relationship which can be clearly seen in the roofs of subglacial cavities. The striations ground into bedrock as a result of the passage of the most prominent point of contact of the clast may be traced for 3 m to a point where the up-glacier flank of the *roche moutonnee* dips up-glacier at 10°. This is probably the point at which the clast first made contact with the bed, as englacial movement vectors tend to descend towards the bed along the crestline of the *roche moutonnee.* The largest striation rapidly deepens to 3 mm, with a width of 4mm, but with increasing distance down-glacier it shallows to 1 mm, although becoming somewhat broader. This decrease in the depth of scratching is probably associated with the build up of crushed debris, which spreads the load at the clast/bed interface over a wider area. In general, the build up of such a layer of crushed debris coincides with a change in the nature of motion from a jerky stick-slip, to a relatively uniform sliding velocity. I suggest that stick-slip motion occurs when the sliding clast comes into contact with small rugosities on the bed which must be crushed before the clast can again move forward. When the particle has produced sufficient crushed debris, this is pushed before as a "bow-wave" which then provides a carpet over which the particle moves (Fig. 3). Presumably, when this carpet is exhausted, the clast will again come into contact with the bed, thereby, recutting a striation. Many large boulders in traction cut several striae, and these may disappear and reappear, perhaps as a result of this latter process. Striae may also terminate due to wearing away of asperities on the clast, but if the clast rotates new asperities will come into contact with the bed.

As a result of these observations, and those of several others (e.g. Chamberlin, 1888; Gilbert, 1906), I suggest two important abrasive processes, the first, and most important, the cutting of striae by asperities on clasts in traction over the bed, and the second, the abrasive polishing of the bed by small debris particles being sheared between the clast and the bed.

The abrasive power of a clast, and the form of the groove cut by it, also depend on the way in which ice flows around the clast. An increase in ice velocity or decrease in thickness, or an increase in clast size may lead to the development of a cavity in the lee of the clast. Where no cavity exists, the striae cut by the passage of the clast tend to be narrow and smooth-sided. Where, however, large cavities exist on the lee of a clast in traction, the grooves cut by the clast in bedrock are laterally more extensive than the area of contact, and fractures tend to extend beyond this area. Such grooves have also been observed beyond the margin of the glacier d'Argentiere where large boulders (diameters 3m) have merely been pushed over a bare rock surface, and have produced grooves up to 40 cm wide and 10 cm deep, having a very irregular floor and showing transverse fracturing. Similar grooves also occur where subglacial cavities exist, for here boulders may partially melt out of basal ice, or be partially expelled from it by locally high hydrostatic pressures (e.g. Vivian and Bocquet, 1973). When these boulders again come into contact with the glacier bed, a period of time elapses during which the boulder/bed contact is unconfined by any ice presence, and cutting of a groove more extensive than the contact area may result. I believe these over-sized and rough-walled grooves develop because the clast/bed interface is locally subjected to a very large pressure while the strength of the bedrock is merely its cohesive strength, no frictional strength is mobilized because of the absence of a confining pressure, thus, extensive brittle fractures are propagated.

Measured rates of abrasion

Direct measurements of the rate of erosion of a glacier bed have not hitherto been undertaken. Indirect estimates utilizing the sediment discharge produced by subglacial streams (e.g. Thorarinsson, 1939), and estimates of the volumes of rock

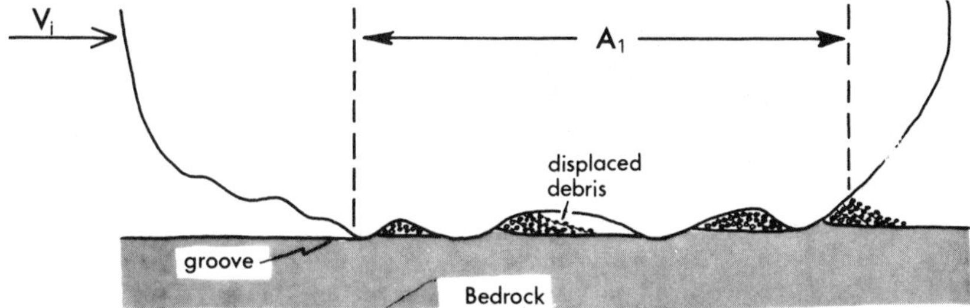

Figure 2. Contact between a clast in traction in basal ice and bedrock. Note *apparent* area of contact A_1 and the much smaller areas of *real* contact (A_r), both directly between clast and bed and indirectly via debris particles. Displaced debris fills gaps between clast and bed.

Figure 3. Debris produced by scratching of glacier bed which has been ploughed up before a clast in traction. Movement right to left. Two striae commence at top right hand corner.

Figure 4. Fragment of marble plate which was cemented to glacier bed and suffered abrasion for 30 days. Length of the fragment is 6.5 cm. Ice flow direction bottom to top. Note prominent line, oblique to flow direction, which separates abraded from un-abraded part. Ice rode over the plate edge (bottom of fragment) and left a cavity above un-abraded part. Lower part of fragment has lost 3 mm from its surface.

Figure 5. Part of aluminium plate attached to glacier bed and abraded for 30 days. Glacier movement was left to right. Note striation which terminates at middle top. Its depth varies before it terminates; the particle probably rotated over the plate and rotated into the ice at the point of termination. Photograph shows 4.5 cm in the direction of flow.

excavated from corries (e.g. Andrews, 1972) have yielded estimates ranging from 0.06 to 5 mm yr^{-1}, although these figures reflect all subglacial erosional processes.

In order to assess abrasion alone, rock and metal plates have been attached to bedrock beneath Breidamerkur jokull in Iceland and the Glacier d'Argentiere in the French Alps. This work will be reported in detail in a subsequent article, but some of the results are summarized in Table 1. The rock plates were each smoothly ground before being cemented to the glacier bed so that the pattern of abrasive wear could be readily established. Figure 4 shows part of a marble plate which has undergone up to 3 mm of abrasion beneath the Glacier d'Argentiere (4, in Table 1) and Figure 5 an aluminum plate into which a series of deep grooves have been cut. Where plates stand proud above the general level of bedrock, the edges may be worn down at a rate several times greater than in the middle, or where they lie flush with the surface (Fig. 4). The ice thicknesses shown in Table 1 cannot be used directly as an indication of ice pressure on the bed, as localities lie on near horizontal rock surfaces which occur immediately up-glacier of bedrock steps. The pressure at these points may be greater or less than the pressure that would be expected from the given ice thickness because of "bridging."

TABLE 1 Abrasion Rates

Locality	Average Abrasion rate		Ice thickness	Ice velocity	Comments
	Marble plate	Basalt plate			
Breidamerkurjökull 1	3 $mm.yr^{-1}$	1 $mm.\ yr^{-1}$	40 m	9.6 $m.yr^{-1}$	Ice thickness probably gives slight under-estimate of pressure. Average hardness of clasts in ice 6-7. Plates flush with rock.
Breidamerkurjokull 2	3.4 $mm.yr^{-1}$	0.9 $mm.yr^{-1}$	15 m	19.5 $m.yr^{-1}$	Ice thickness proabably gives reasonable estimate of pressure. Clast hardness 6-7. Plates flush with rock.
Breidamerkurjokull 3	3.75 $mm.yr^{-1}$		32 m	15.4 $m.yr^{-1}$	Ice thickness probably gives reasonable estimate of pressure. Clast hardness 6-7. Plates flush with rock.
Glacier d'Argentiere	Up to 36 mm. yr^{-1}		100 m	250 $m.yr^{-1}$	Ice thickness probably gives pressure over-estimate. Average hardness of clasts in ice 3-4. Some of jardmess 6-7. Plates project above rock surface, some areas receive greater than average pressure, some, less than average.

The error in measurement of average lowering of the surface of the plates to give abrasion rates is approximately 0.3 mm.

The purpose of Table 1 is simply to illustrate that abrasion rates under normal subglacial conditions can be relatively high, that they depend on rock type, ice thickness, and the concentration of debris in the glacier sole. The measured abrasion rates shown in Table 1 indicate that erosion by this process alone could excavate deep valleys and basins. It is worthy of note that the lowest figure in Table 1 is twice the world average for the erosion of lowland river basins, and that these measured rates would be adequate to produce by abrasion alone, most deep alpine glacial troughs given the lengths of time for which they have been occupied by ice during the Quaternary. It is quite clear that the abrading agent is not the ice itself, but the particles embedded in the glacier sole. In the many temperate glaciers that have been investigated, basal debris in transport tends to be restricted to the basal 10 cm or so, thickening exceptionally to 0.5-1 m. The concentration of debris in this zone varies considerably, from almost zero to 50-60% by volume.

Theory

In order to understand patterns of abrasion produced by glaciers, it is necessary to establish the controls on abrasion rates. The movement of particles over the bed has been analysed (Boulton, 1974) by adapting Weertman's (1957, 1962) theory of glacier sliding. The motion of a particle sliding over the bed is retarded because of a frictional drag against it, so that the particle velocity (V_p) is less than the glacier sliding velocity (V_i), and thus

$$V_p = V_i - V_r \tag{1}$$

The relative velocity (V_r) is produced by a net force between ice and particle which is equal to the frictional drag F between the particle and the bed. If a simple frictional law is assumed, and ignoring the weight of the particle, then the frictional drag is

$$F = (\varphi_i gh - \omega_p) A_1 \mu \tag{2}$$

where g is the gravitational acceleration, h the ice thickness, ω_p the value of any water pressure which might exist between the glacier and its bed, A_1 the apparent area of contact between particle and bed, φ_i the density of ice, and μ a coefficient of dynamic friction.

McCall (1960) has argued that the maximum vertical pressure of ice on a particle in traction over the bed will be given by the "yield strength" in compression of the ice, which he takes to be of the order of two bars. Thus for ice thicknesses greater than about 22 m, he suggests no increase in pressure above the limiting value of 2 bars, as ice would flow beneath the particle to maintain the pressure at that value. However subglacial observations beneath an ice thickness of 150 m show extensive areas of contact between particles and the bed, and that ice does not penetrate along the interface. Small cavities along the particle-bed interface are normally occupied by crushed debris and not ice. Thus, McCall's concept of a limiting value for $\varphi_i gh - \omega_p$ is rejected. Neither do I propose to use McCall's suggestion that clean ice will impose a shear stress against a tangential surface, but assume, as does Weertman (1957), the absence of such a stress in temperate ice.

The velocity of ice relative to the particle is composed of two components, one of plastic flow around the particle (V_1), and the other of mass transfer due to the complementary processes of pressure melting against the up-glacier flank and regelation against the down glacier flank (V_2). Assuming no cavitation around the particle, these two components are (Boulton, 1974)

$$V_1 = B\left(\frac{F}{A_2}\right)^n \left(\frac{z+y}{2}\right) \tag{3}$$

and

$$V_2 = \frac{CK}{L\varphi_i} \cdot \frac{F}{A_2 x} \tag{4}$$

where B and n are constants in Glen's flow law for ice, A_2 is the cross sectional area of the particle transverse to flow, x, y and z are axial lengths of the particle, C is a coefficient relating pressure and melting point of ice, L is the latent heat of ice, and K its thermal conductivity.

From (1) (2) (3) and (4), the velocity of the particle over the bed is

$$V_p = V_i - \left[B\left[(\varphi_i gh-\omega_p)\frac{A_1}{A_2} \cdot \mu\right]^n \frac{z+y}{2} + \left[\frac{CK}{L\varphi_i}(\varphi_i gh-\omega_p)\mu\frac{A_1}{A_2 x}\right]\right] \tag{5}$$

This equation also defines a boundary condition for abrasion, for when V_p=0, the particle ceases movement, and thus no longer abrades the bed.

The amount of abrasion produced by a single particle sliding over the bed may be theoretically calculated if some highly simplified generalizations are made about the processes involved. Tabor (1954) has shown that if a hard indenter is pressed into the surface of an otherwise brittle mineral, the high hydrostatic pressures developed around the contact are often sufficient to inhibit brittle fracture. Deformation in the mineral takes place largely by plastic flow, although the displaced material may subsequently fracture elastically. If these conditions hold it is possible to estimate the abrasive wear at the interface between two sliding surfaces.

Consider a particle such as that shown in Figure 2. An *apparent* area of contact (A_1) can be identified, but the *real* area of contact is very much smaller, contacts being made between small asperities on the surfaces. As debris is produced by the abrasive process, it tends to accumulate in the areas between points of contact. An individual fragment may become temporarily embedded in one of the sliding surfaces and thus abrade the other surface. After some distance of transport, when the sliding surface of a clast has been smoothed, this latter may be the most important process of abrasion of the bed.

Under the above conditions, the real area of contact (A_r),will be such that the average pressure across particle/bed junctions does not exceed the yield strength of the material. Its minimum value will be

$$A_r = \frac{NA_1}{p} \tag{6}$$

where N is the effective normal pressure ($\varphi_i gh-\omega_p$), and p is a plastic property of the rock known as its penetration hardness. It is commonly found that (6) gives the actual rather than minimum value of A_r. When a clast moves over the rock bed, asperities on the clast in transport must either penetrate the rock bed, or deform themselves, or both, in order that A_r should be of the magnitude specified by (6) (Fig. 2). Bed and particle may be of similar or different hardness, but nonetheless, it seems reasonable to assume that the volume of material displaced from the bedrock surface (A_b) will be

$$A_b = \frac{K_1 N A_1}{p} \tag{7}$$

where K_1 is a constant which depends on the relative hardness of the clast and bedrock, their detailed roughness, and the shapes of contact areas. If the particle moves over the bedrock surface, the rate of abrasion of the bed will be

$$A_b{}^* = \frac{K_1 N A_1 V_p}{p} \tag{8}$$

where V_p can be determined from equation (5).

The debris in traction over the glacier bed is not however of a single grain size, and thus both V_p and A_1 (and probably K_1) will vary considerably from grain to grain. Figure 6 shows the changing values of particle velocity (V_p) for debris particles of different size determined for a debris assemblage from Breidamerkurjokull in south-east Iceland and for an ice velocity of 10 m. yr $^{-1}$ and effective normal pressure of 10 bars. Large particles($> -7\phi$) and very small particles ($< -3\phi$) both have relatively small velocities over the glacier bed, while particles of intermediate grain size move relatively rapidly. This is a function of the changing tractive stresses on particles, which will be relatively small against large particles because of the efficacy of the plastic flow mechanism, small against small particles because of the pressure melting and regelation mechanism, and at a maximum against particles of intermediate grain size. A typical bimodal grain-size frequency distribution (c.f. Dreimanis and Vagners,1971) of a till from Breidamerkur jokull shows that the clast mode coincides with the relatively high values of particle velocity, a correlation I believe to occur in many tills. If we use measured values of A_1 from the same till, we can determine relative abrasion rates for the whole range of particle sizes which clearly indicate the predominant role in the abrasion process played by grain sizes in the clast mode.

Dreimanis and Vagners (1971) work on Pleistocene tills from Ontario, shows that progressive comminution of any one mineral species proceeds until a "terminal grade" is attained beyond which no further comminution occurs. The decreasing abrasion rate with grain size suggests that the terminal grade might occur when the real area of contact $A_r \geqslant \frac{NA_1}{p}$, in which case there will neither be failure in the particle or the bed.

This discussion assumed, however, that individual particles are free to move over the glacier bed without interference from others. In reality, this is not so. Particles which might otherwise lodge may be coerced into motion by other moving particles, or the motion of the latter might be blocked by more slowly moving particles. We clearly need to determine an effective velocity for the entire particle assemblage, which take particle interactions into account. By analogy with (1)

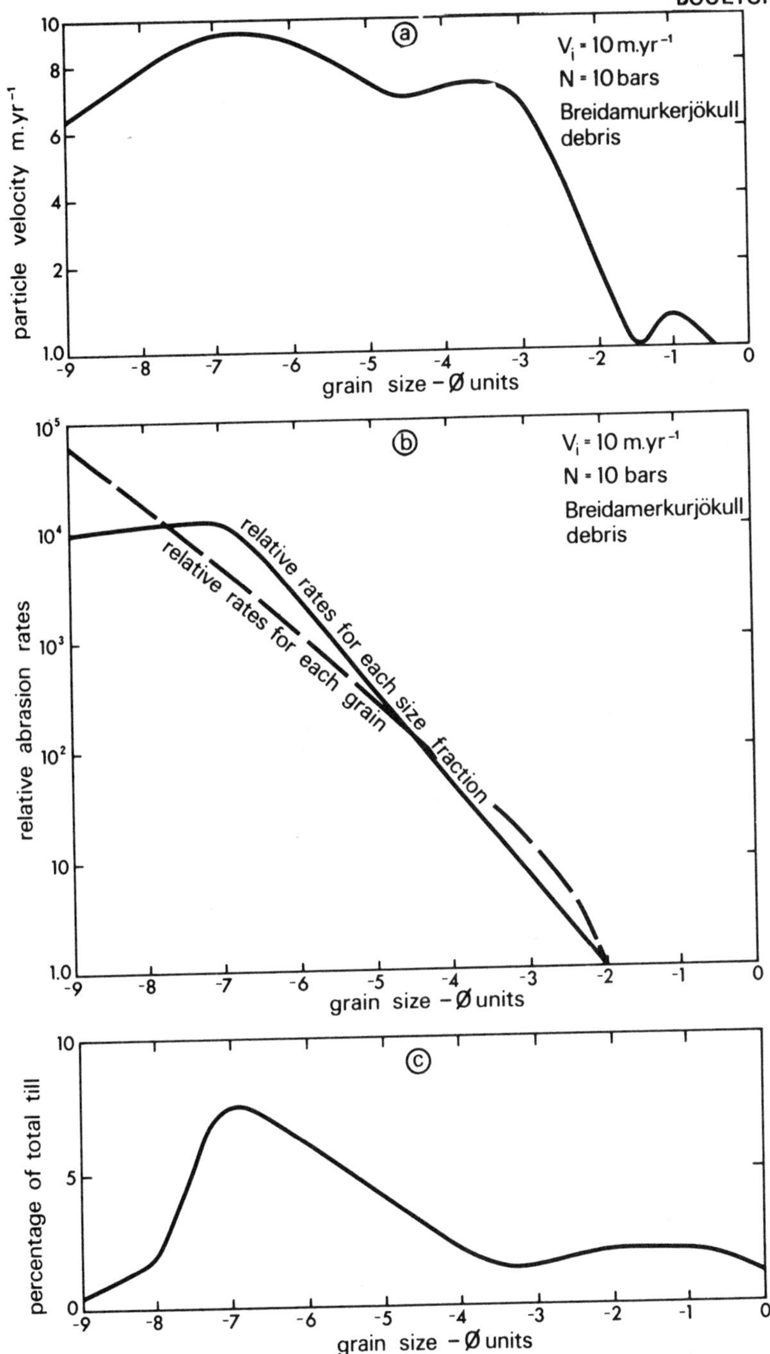

Figure 6. (a) Particle velocities for particles of different size and shape from Breidamerkurjokull Iceland.
(b) Relative abrasion rates for different grain sizes, and relative abrasion rates for different size fractions (grain abrasion rate x frequency).
(c) Grain size distribution between -9ϕ and 0ϕ.

$$V_p{}^1 = V_i - V_r{}^1 \tag{9}$$

Several ways in which $V_r{}^1$ might be determined have been discussed previously (Boulton, 1974). It has been suggested that

$$V_r{}^1 = \left(\frac{N}{L_c}\right)^{\frac{1}{m}} \tag{10}$$

where L_c is a *critical lodgement index* which depends on the grain size and shape distribution of the debris (Boulton, 1974) and seems to lie between 5 and 25 for a wide range of tills, and m is a constant (≈0.3). Thus from (8) and (10), for debris in transport

$$A_b{}^* = \frac{K_1 cN}{p} \left[V_i - \left(\frac{N}{L_c}\right)^{1/m}\right] \tag{11}$$

where c is the concentration of debris in traction over the bed in the plane of the glacier sole.

If we assume that there is no systematic change in c for different values of V_i and N, we can plot abrasion rates for a range of values of V_i and N. These are shown in Figure 7, using values of m and L_c of 0.3 and 9 respectively (taken from the Breidamerkerjokull till). K_1 could be determined experimentally, and for metals it rarely exceeds 0.03 (Rabinowicz, 1965). Abrasion rates $x\left(\frac{K_1 c}{p}\right)^{-1}$ have been plotted against normal pressure for several values of ice velocity. It can be seen that for any one ice velocity, and with increasing normal pressure, abrasion rates rise to a peak and then fall rapidly to zero. At pressures above this latter level, no

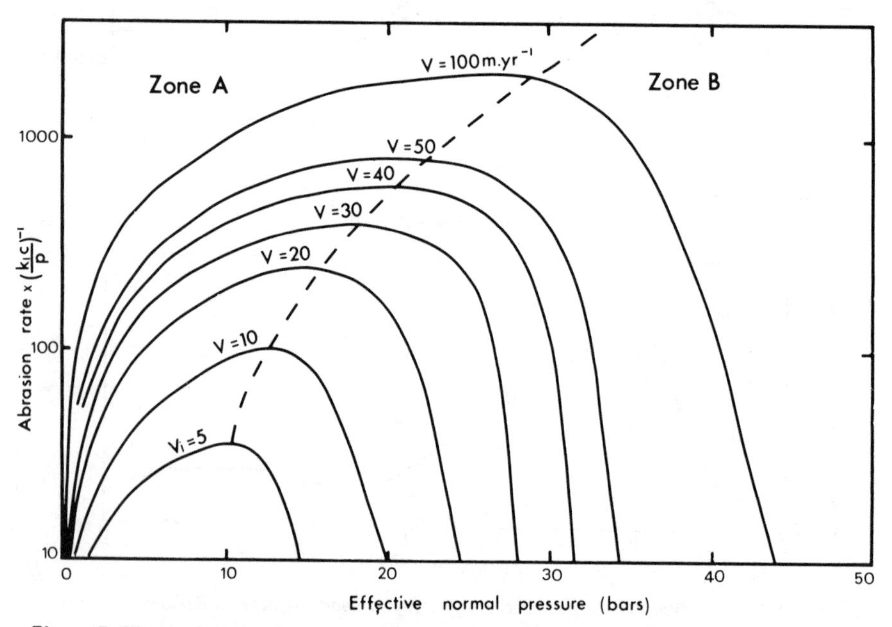

Figure 7. Theoretical abrasion rates plotted against effective normal pressure for different ice velocities. In zone A, abrasion rates increase with increasing pressure, in zone B, abrasion rates fall with increasing pressure, in zone C (to right of higher x-axis intercept for any one velocity), no abrasion occurs, debris in transport is deposited as lodgment till.

abrasion of the bed occurs, but debris hitherto in transport will be deposited as lodgment till. The horizontal axis gives critical conditions for lodgment. It is important to note that the continuum is not one of great abrasion → little abrasion → lodgment, but of little abrasion → great abrasion → little abrasion → lodgment.

The most effective means of testing and calibrating the theoretical abrasion rate curves is not to use laboratory experimental data, but subglacial measurements such as those presented in Table 1. Using examples (Table 1) from Breidamerkur jokull, and comparing them with the theoretical curves, we find that for the marble, $\frac{K_1 c}{p}$ would need to have a value of the order of 7.5×10^{-5} to give the observed values for examples 1 and 2 in mm. yr^{-1}. For the basalt, a value of the order of 2×10^{-5} would be necessary.

Tabor (1954) has demonstrated the very useful relationship

$$\text{Log } p = K_2 M \tag{12}$$

where M is the Mohs hardness number and K_2 is a constant, equal to log 1.6 for all hardness values less than 10.

The basalt has a Mohs hardness of 6-7 which gives an indentation hardness of approximately 900 kg. mm^{-2} and the marble a Mohs hardness of 3-4, giving an indentation hardness of 300 kg. mm^{-2}. In both cases, the concentration of debris in basal ice was approximately 50% by volume, this allows us to calculate values of K_1 (from 11) of 0.026 and 0.032 respectively. If we assume a value of K_1 of 0.032 for the third Icelandic locality, this predicts an abrasion rate of 3.6 mm.yr^{-1} compared with a measured rate of 3.75 mm.yr^{-1}, which is within the measurement error.

The role of subglacial water

In order to relate Equation 11 and Figure 7 to a real glacier, we need to know the value of effective normal pressure (N). If there were no water at the glacier bed this would be $N = \varphi_i gh$. However, water is continuously being produced at the bed of a temperate glacier, by frictional melting and by the geothermal heat flux. This water cannot build up subglacially if the glacier is to remain stable: it must be discharged.

If subglacial materials are permeable, the water may be discharged in these beds, and as the discharge must increase towards the glacier snout, the hydraulic gradient must also increase towards the snout. Water pressures at any point can be determined from the form of the piezometric surface if we know the glacier profile, and the permeability and the thickness of subglacial beds. Boulton, Dent and Morris (in press) have shown, that for a bed in which permeability is not strongly dependent on normal pressure,

$$N = \left[\frac{X^2 \omega}{2HbK_p} \left(\frac{x}{X}\right)^2 + \left[1-\left(\frac{x}{X}\right)^2\right]^{1/2} \right] p_o \tag{13}$$

where p_o is the ice pressure at the glacier source, X the half width of the ice cap and H its source thickness, ω is the rate of meltwater production at the bed, x is the distance from the glacier source to any point, b is the thickness of the permeable bed and K_p its permeability . Boulton (1974) has further discussed how a permeable bed of complex geology might effect subglacial water pressures.

The problem for an impermeable rock bed is more difficult, and has been discussed thoroughly by Weertman (1972). On such a bed, water may be discharged subglacially through tunnels, or via interconnected cavities and in a thin water layer between the glacier and its bed. In the latter case, and if the glacier rested on a smooth horizontal rock bed, the water produced basally would form a continuous layer in which the water pressure would be equal to the pressure of overlying ice, and N would thus be zero. Lliboutry (1969) has however pointed out, that such a water layer will be pierced by irregularities on the rock bed which will themselves bear part of the glacier load and thus the water pressure will be less than the ice pressure. Weertman (1972) has suggested that the distribution of load between these two can be approximated from the developments of his sliding theory by Kamb (1970) and Nye (1969, 1970), and that from Nye's (1970) work

$$N = \left(\frac{a}{\pi}\right)^{1/2} \eta V_i K_* \tag{14}$$

where a is a measure of bed roughness equal to $2\pi^3 r^2$ and r is the root mean square deviation of the bed from an average plane, divided by the distance over which this is averaged, η is a value of viscosity for ice equal to 1 bar year (the simplifiying assumption is made that ice is a Newtonian viscous solid) and K_* is a material constant equal to 0.1 cm^{-1}.

However, it is clear that this relationship must break down near the glacier snout, and I suggest the reason for this to be the existence of extensive cavities into which the subglacial water layer may drain. Thus, the water between the glacier and its bed is derived from a relatively small area of ice-bed contact and rapidly drains into cavities. As a first approximation, I shall assume that there is a linear increase in N with increasing ice thickness, from zero at the glacier snout, to the value given by Equation (14) at the ice thickness when cavities are no longer found. It could of course be that effective pressures might be higher in the zone of cavitation than in the zone of no cavitation where the ice is thicker. Where ice moves over major obstacles and irregularities on its bed, I shall assume, following Weertman (1972), that the water layer is squeezed out from the ice-bed contact, that the water pressure is negligible, and thus that the pressure developed between the ice and a bedrock obstruction is undiminished.

SMALL SCALE ABRASIONAL FEATURES

In this section, the observations and theory presented above are used to explain a series of typical small-scale features of glacial erosion, and to predict several forms which should be distinguishable by accurate measurement in the field.

Streamlined bedrock hummocks

One of the principal characteristics of glaciated bedrock surfaces is their high relief on a wide range of scales, and their relatively smoothly undulating character. In general the individual hummocks are streamlined, elongate parallel to ice flow, and may be asymmetric, with a steeper down-glacier flank showing abundant signs of plucking. They have been variously referred to as *roches moutonnees,* rock drumlins (Fairchild, 1907), ice moulded forms (cf. Linton, 1963), and whaleback and stoss and lee forms (Flint, 1971).

Using the theory developed above it is possible to predict theoretically how the form of a rock hummock on the bed of a glacier might develop as a result of the abrasion process. If it is assumed that the height of the hummock is small compared with the thickness of the glacier, glacier sliding velocity can be assumed constant over it. In addition, a constant hardness is assumed for the rock, and, for a first approximation, that debris concentration is constant along a flow line, although this latter might be an important source of error. The most important variable effecting the abrasion rate will be effective normal pressure (N) at the ice-bed interface. When a glacier moves over a horizontal bed, the effective normal pressure at the interface is simply ($\varphi_i gh - \omega_p$). However if it moves over a hummock, the pressure will fluctuate in the way shown in Figure 8. A higher than average pressure will be present on the up-glacier flank where the ice moves against the hummock (although at the foot of this flank there will be a local low pressure due to bridging), on its down-glacier flank, where ice moves away from it, there will be a lower than average pressure, and where ice moves against the foot of the down-glacier flank, there will be a subsidiary high pressure.

From Nye's (1970) glacier sliding theory, the value of this normal pressure fluctuation (ΔP) for a hummock of sinusoidal profile will be

$$\Delta P = [(a/\lambda)^2 2\pi^2]^{1/2} \eta V_i K_* \quad (15)$$

where a is the amplitude of the sinusoid and λ is its wavelength. Thus the value of normal pressure at any point will be $\varphi_i gh + \Delta P$ and for beds of low roughness a cavity will exist on the down-glacier flank of the hummock at all points where

$$\Delta P > \varphi_i gh - \omega_p \quad (16)$$

Equation 15 can be applied to a more complex bed approximated by the leading terms of the Fourier series (Nye, 1970).

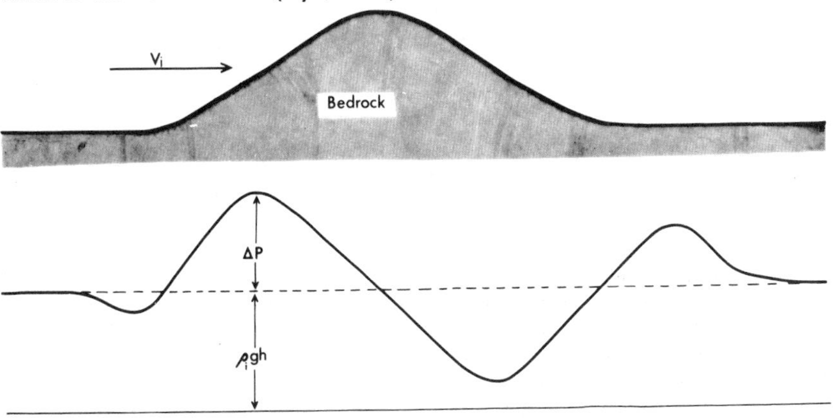

Figure 8. Schematic normal pressure distribution over glacier bed as ice flows over a bedrock hummock.

Thus, given the shape and size of an initial hummock, it is possible to predict from Equations 11 and 16 how this might evolve under given conditions of glacier velocity and thickness. Using an initial sinusoid superimposed on a horizontal bed, and using V_i=50 m.yr^{-1} and the initial value of ΔP=13 bars, the evolution of the bed form has been determined using two different values of pressure on the horizontal bed, of 7 bars and 24 bars. In the computation, the abrasion rate is determined for the bed form under both sets of conditions, and after a fixed time, the bed profile and normal pressure distribution are re-synthesized and the successive patterns of abrasive wear determined by the same method. Figure 9 shows the results, with the surface profile reproduced for every fourth step in the computation.

The two patterns of evolution are quite different. In Figure 9a, normal pressures

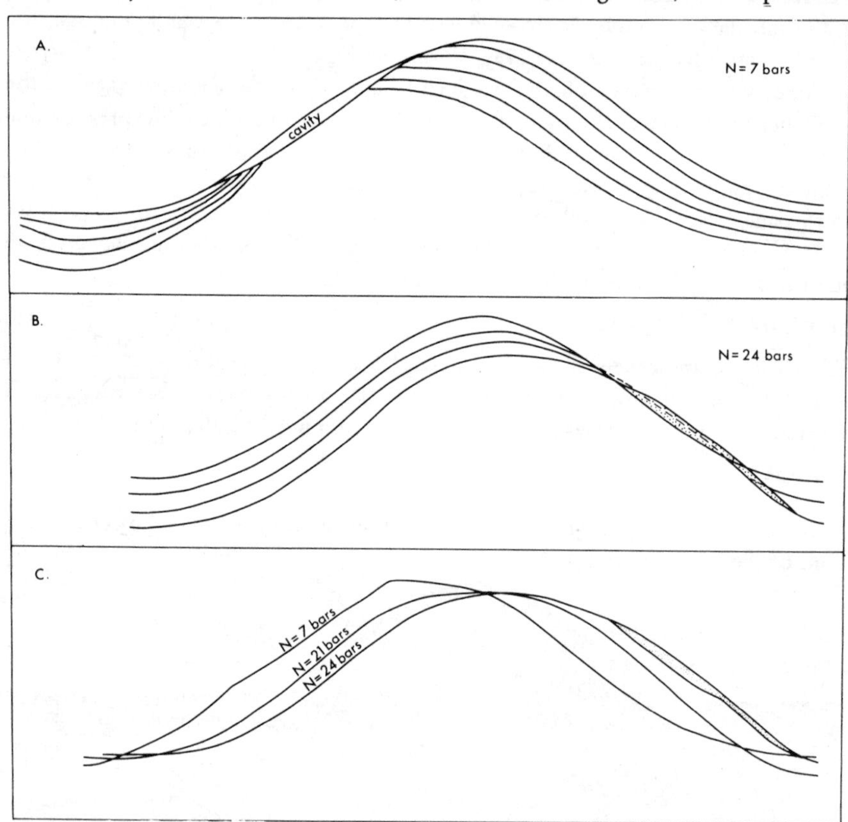

Figure 9. Patterns of abrasion of a sinusoidal bedrock hump for a constant velocity of 50 m.yr^{-1} for different values of ice overburden pressure. Shaded area represents lodgment till accumulation. In a, five stages of evolution are shown, and 3 stages in b. In c, the profiles of three humps are shown after 3 stages of evolution. Low overburden pressures produce a down-glacier asymmetry, and tends to the production of a stepped form with a basin on the down-glacier side, high overburden pressures produce lodgment on the up-glacier flank and an up-glacier asymmetry. Intervening pressures may produce symmetrical form.

at the ice-bed interface fall entirely within zone A of the abrasion curve in Figure 7. The rate of abrasion is thus high on the up-glacier flank of the hummock, low on its summit, zero on its lee flank, where cavitation occurs because $\Delta P > \varphi_i gh$, and relatively high at the foot of the lee flank. The result is to evolve a form in which the up-glacier flank is smoothed off to a low slope, although its steepest point migrates in a down-glacier direction, in which the position of the upper part of the lee flank remains stationary, and in which the foot of the lee flank is abraded so that this latter is made higher and steeper. Thus a large asymmetric step-like form may be produced. The hollowing out of a basin at the foot of the lee flank, produces, in effect, another hummock on its down-glacier side, and I would suggest that under stable conditions a series of smaller hummocks will be produced on the lee of a large one, with decreasing wavelengths depending on the original bump shape and ice velocity and thickness.

In Figure 9b, normal pressures at the ice-bed interface fall within zones B and C of the abrasion curve in Figure 7. Thus, lodgment occurs on the up-glacier flank of the obstruction (the computation of lodgment rates is explained in Boulton, in press) slight abrasion occurs on its summit, maximum abrasion on its down-glacier flank, and slight abrasion at the foot of this flank. The effect is to produce a form which migrates in an up-glacier direction with an asymmetry in the opposite sense to the first form, a steeper up-glacier rather than down-glacier flank. We have, in fact, produced a drumlin.

The contrasts are interesting. Typically, streamlined forms of eroded bedrock *(roches moutonnees)* have a down-glacier asymmetry, whereas streamlined depositional forms (drumlins) have an up-glacier asymmetry. The same effect could have been produced by a reduction in velocity, or a change in hummock shape. The latter explains why it might be possible to have both *roches moutonnees* and *drumlins* in the same area.

Lewis (1947) has drawn attention to the fact that the typical longitudinal asymmetry of glacially eroded bedrock noted here occurs on many scales, from small *roches moutonnees* a few meters across, to major valley steps several hundred meters high. The processes analyzed here can simulate the form of both, and help explain how the step and basin profile of a glaciated valley might evolve. It will be shown later however, that abrasion is not the only process capable of producing asymmetry, and it does not account for the irregular and fractured character of many steep down-glacier flanks, although absence of abrasion in a lee side cavity would avoid smoothing.

This analysis can be strictly applied only to two-dimensional forms, that is, forms, such as valley steps, whose width is greater than their length. Equidimensional forms, or those with a high length/width ratio are influenced strongly by the complex pattern of ice flow around them, although some general considerations noted above will apply. When ice flows over such an obstruction, normal pressures developed against the lateral flanks are generally larger than those developed by flow over the summit, because of the greater freedom the glacier has

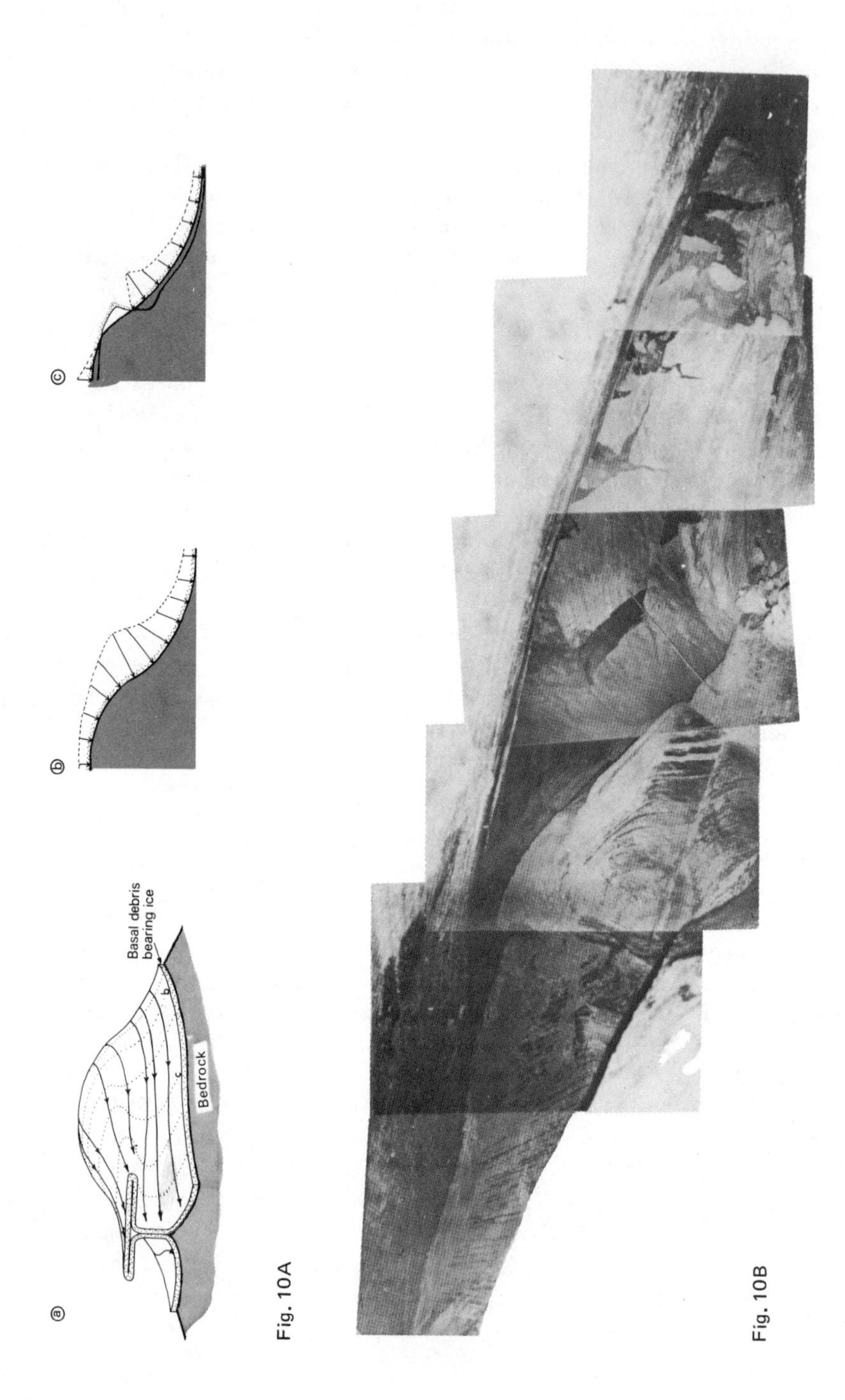

Fig. 10A

Fig. 10B

Figure 11. Ice flowing around the flank of hummock in Figure 10. Note clean ice above debris-rich sole, seen in fractures. The ribbing on the glacier sole is a reflection of rock profile at the point of *decollement* . The obvious large fluting in foreground emanates from a *cavetto* form.

to deform in a vertical direction towards its upper free surface, Because of this, flanking ice tends to flow into the lee side cavity more rapidly than overlying ice producing the pattern of flow shown in Figure 10. Such a flow pattern has been observed where there is a lee side cavity, but such a cavity is not essential. If there is no cavity, this flow pattern demands that at some point on the lee flank, ice moves at right angles away from the bed, producing a low pressure point,even if $\varphi_i gh \gg \Delta P$. The overfolding of basal ice in this position (Figs. 10 and 11) may translate debris hitherto embedded in the glacier sole into an englacial position, and it may be important in the plucking process on the lee flanks of bedrock hummocks.

Plastically-sculptured detailed forms

Many small scale patterns of glacial erosion have been grouped together as

Figure 10A. (a)Schematic diagram showing the flow of ice around a bedrock hummock. On the lee flank of the hummock, ice flow is directly away from bedrock, an ideal situation for plucking of that flank. This mode of flow takes debris in the glacier sole into an englacial position.

(b) Schematic representation of normal pressures along a transverse section at point b in (a).

(c) Schematic representation of normal pressures along a transverse section at point c in (a), and an impression of the probable pattern of surface lowering by abrasion.

Figure 10B. Cavity on lee flank of large bedrock hummock 150 m below the surface of the Glacier d'Argentiere. Ice flows over summit of hummock and meets ice flowing strongly around the flank which it folds over. Debris in glacier sole is folded into an englacial position. Length of metal bar is 2 m.

plastically-sculptured forms (p-forms), and have been recently reviewed by Dahl (1965), Gjessing (1966, 1967) and Embleton and King (1968).

Sichelwannen (sickle troughs) are troughs with two horns, the trough lying transverse and the horns parallel to ice movement and on the down-glacier side. They often occur in depressions, at the foot of up-glacier flanks of bedrock hummocks, and exceptionally on flat surfaces. Broad deep channels, known as *cavettos* may be as much as 1.5 m wide, 0.5 m deep and 10 m or more in length. They occur on the flanks of rock humps which lie parallel to ice movement, and may have an overhanging upper side with a sharp edge. A variety of other grooves and channels have been described; long straight grooves (≈20 cm wide) with smooth edges, which generally occur on extensive flat surfaces; and curved, winding channels. There appears to be little distinction in form between some small scale grooves and the giant glacial grooves described by Smith (1948) from Canada, which are as much as 100 m in width and 12 km in length. Dahl (1965) suggests that in Nordland (Norway) the frequency of p-forms is much greater in high plateau areas than in the fjord valleys.

Johnsson (1956) suggested that an ice/water paste extruded between the glacier and bedcould produce p-forms. Subsequently Gjessing (1966, 1967) suggested the agent of erosion to be fluid till, deforming between the glacier and its bed, whereas Dahl (1965) drawing on previous ideas of Lungner (1930) and Hjulstrom (1935), concludes they are produced by rapidly flowing subglacial meltwater streams, with cavitation erosion as an important mechanism. A common theme in each of these hypotheses is the necessity to find a process which can account for the association of many p-forms, which are clearly not normal fluviatile products, with such features as potholes which do appear to be fluviatile.

All the major forms referred to above have been observed subglacially on rock beds beneath Breidamerkur jokull and the Glacier d'Argentiere, where they appear to be produced simply by the abrasive processes referred to above. The wide spectrum of variation of the forms reported from previously glaciated areas can clearly not be matched exactly, as the subglacial areas of observation are much more limited. However, an understanding of some simple principles of the abrasion process enables one to comprehend how a large variety of forms might be produced in different situations.

Relatively narrow (≈10-20 cm) grooves, which may be very extensive and related to the passage of very large boulders have already been referred to. Subsequent smoothing by abrasion of the often-pitted floor of the groove would produce a long groove parallel to local striae, with a smooth cross profile.

Many p-forms are found on undulating bedrock surfaces, often related to *roches moutonnees.* If we consider the mode of flow around the bedrock hummock illustrated in Figures 10A and 10B, the pressure against the lateral flanks will be greater than that against the summit. A relatively high pressure exists midway down the lateral flank, which diminishes downwards and upwards. Towards the down-glacier end, folding of flanking ice and ice flowing over the summit of the hump will tend to produce low or zero pressures. Thus, a channel may be abraded down this flank above which there may be no abrasion, and which could develop an overhanging upper margin, a form very similar to *cavettos* (Fig. 10A). In general, I suggest that the margins of cavities may be identified by such lines.

Three forms analogous to what have elsewhere been called *Sichelwannen* have been observed near points of cavity closure. It has been observed at such points, that small obstructions on the bed cause basal ice in the immediate vicinity to flow readily around the obstruction (Fig. 12). As debris tends to be concentrated in the basal 10-20 cm or so, a stream of debris-rich ice flows around the flanks while the overlying ice which passes over the obstruction is relatively debris-poor. The flanking debris-rich ice appears to have ground grooves on the up-glacier flank and around the lateral flanks of the obstructions. I suggest that, as the points of cavity closure are also points of enhanced pressure against the bed, that small obstructions in the area of closure generate very considerable stress concentrations in the immediate vicinity. Basal debris-bearing ice thus tends to flow very readily, and the combination of the "debris streams" and high local normal pressures produce enhanced abrasion which leads to the typical *Sichelwannen* grooves (Fig. 12). In theory, these grooves could continue far down-glacier of the zone of stress concentration as the streaming debris-rich ice will tend maintain a zone with a high abrasion rate. I do not suggest that cavitation is necessary for the formation of *Sichelwannen* but that it is the most favorable circumstance, it produces very high local stress concentrations, and allows basal ice to deform readily. They could be produced on a hitherto horizontal surface if there is a local increase in rock hardness leading to the formation of an obstruction to flow. Alternatively, a boulder, lodged on the rock surface (Boulton, 1974) could cause their formation (cf. Hoppe, 1957).

The streaming of debris into debris rich bands separated by bands depleted in debris as a result of enhanced plastic flow around obstructions (Figs. 12a, 13) could clearly be responsible for many linear features, such as grooves and channels, because of the different abrasive power of the debris-rich and debris-poor streams. Indeed, one of the striking features of many glacial erosional forms is that of continuity (see Clayton and Moran, this volume).

There is a deeply ground groove 50 m in length which occurs on the down-glacier side of a col between two major bedrock hummocks beyond the margin of the Glaci d'Argentiere. Although its cross-profile is neither symmetrical or highly smoothed, there is great longitudinal smoothness and continuity. I suggest this form to be the result of preferential streaming of debris through the col from which it originates. Any small longitudinal rugosities will produce a locally high pressure and thus tend to be preferentially abraded, the form is thus smoothed out, and any longitudinal curves tend to have long wavelengths, although high small

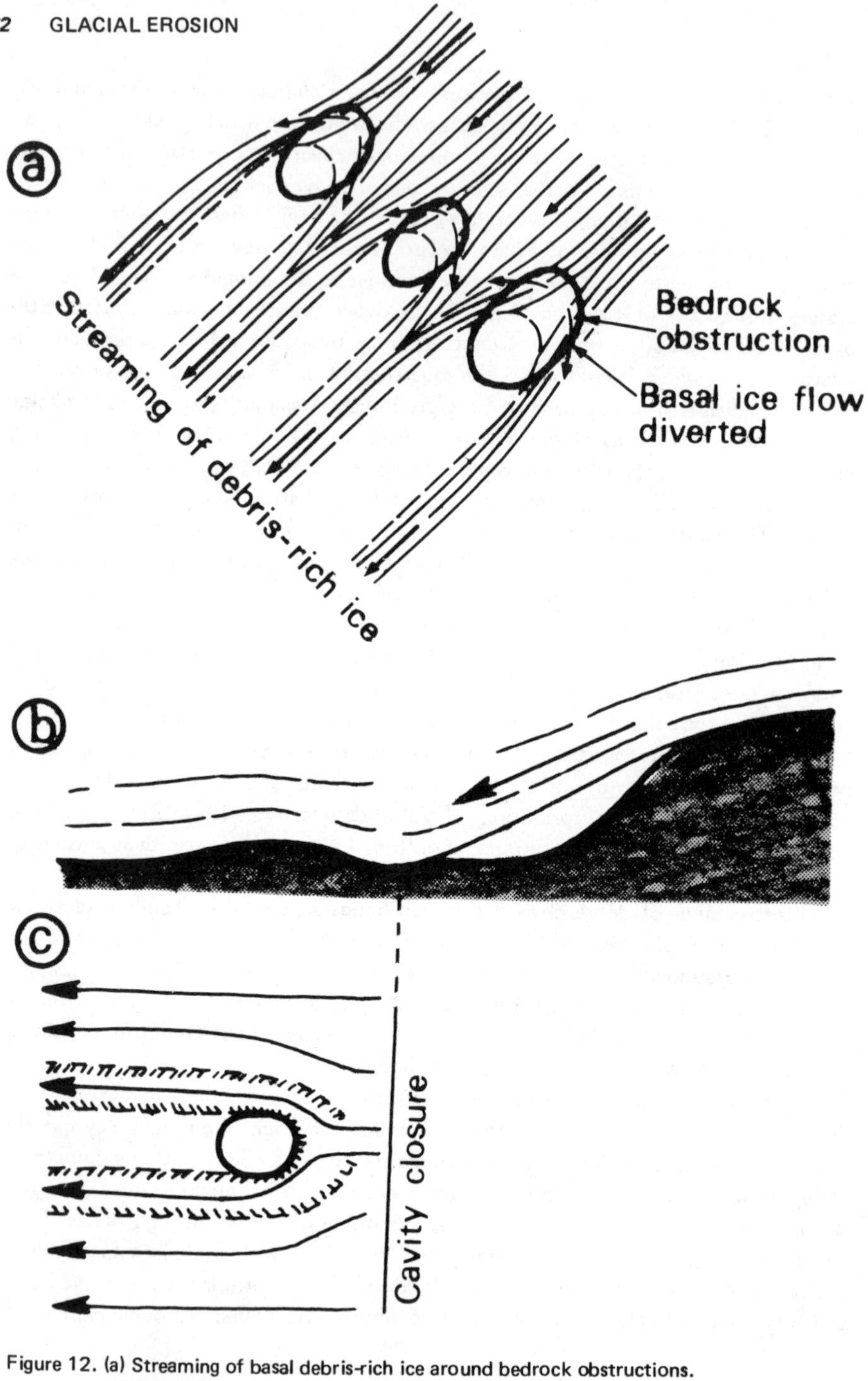

Figure 12. (a) Streaming of basal debris-rich ice around bedrock obstructions.
(b) Section showing the area of closure of a cavity.
(c) Plan of area of closure of a cavity, showing formation of *Sichelwannen* due to streaming of debris-rich basal ice around a bedrock obstruction.

scale roughness (<1 cm) may occur. Although "streaming" may initially cause a groove (or *cavetto*) to be a *locus* of enhanced abrasion, I suggest that it will not continue to be deepened. The increasing "bridging" support given to the ice on the flanks of the groove as it deepens will cause a decrease in normal pressure in the bed of the groove, and thus, although the debris concentration remains high, decreasing pressure will tend towards the establishment of an equilibrium form. I suggest this mechanism can also explain the giant grooves described from Canada by Smith (1948).

The observations of Dahl (1965) that many p-forms (*Sichelwannen* and *cavettos*) are much more frequent at higher altitudes in Nordland can be explained by the suggestion that their development is particularly favored by cavitation, for ice would be thinner than in the fjords and more susceptible to cavitation. The preferential occurrence of *Sichelwannen* between *roches moutonnees* and at foot of their up-glacier flanks is also explained by the tendency towards cavitation in such positions (Fig. 8).

The association of some forms with potholes does not necessarily imply that water plays a part in their formation. Vivian (1970) has pointed out the importance of glaciofluvial erosion beneath glaciers where cavities occur, and gives examples of potholes and other forms found in natural cavities. If ice flows over the irregular top of a pothole or fluvial depression, pressure fluctuations and "streaming" could produce p-forms emanating from them. It is found that for the Glacier d'Argentiere, ice velocities are high in summer, and cavitation is at a maximum as is subglacial stream flow. During winter, velocities decrease, and cavitation and streamflow are reduced (Vivian and Bocquet, 1973). Thus, areas of bed can be alternately affected by fluviatile and glacial processes, changes which can also occur on a longer time scale due to long term climatic changes. Thus I would suggest that the extreme p-forms, such as the most distinctive, *cavettos, Sichelwannen,* and troughs, are normal products of glacial abrasion, and that well-defined potholes are a product of fluvial erosion, but that a series of intermediate less easily classifiable forms exist, which have been affected by both processes.

SUBGLACIAL CRUSHING OF BEDROCK

Two processes have previously been proposed which might lead to subglacial fracturing of hitherto intact rocks to produce relatively large blocks. The first is that of freezing and thawing of meltwater, suggested by Carol (1947) to be the agent responsible for shattering the downstream face of a *roche moutonnee* some 50 m below the surface of the Grindelwald glacier, and by Johnson (1904) and Lewis (1940) to be important agents in shattering rock in the upper part of a *bergschrund* against a cirque headwall.

Whereas this process might be effective in fracturing rocks near the surface where extreme temperature fluctuations may occur, it seems unlikely to do this beneath the damping effect of a greater ice thickness. Ice masses do form in subglacial cavities, where water, hitherto under pressure between the glacier and its bed, freezes as it enters. However, unless rocks are highly permeable, this ice simply forms on their surface, it does not penetrate before freezing. Such a process may produce further loosening of already fractured materials (Fig. 20) but I would suggest that it does not itself produce fracturing.

A second process which clearly leads to fracturing, is that in which unloading of bedrock due to glacier thinning allows rebound and release of stored strain energy, (Lewis, 1954). The expected fractures are parallel to the glacier bed, and can be observed in many glaciated valleys.

As already noted, one of the most striking features of many forms of erosion, is their asymmetry. As suggested, this could arise by abrasion alone, but the existence of some other process is clearly indicated by the form of many of the steep down-glacier flanks of bedrock steps and *roches moutonnees*, which are often highly fractured and irregular (Fig. 14), and a plucking process has been invoked to account for them (Matthes, 1930). A plucking process alone may suffice if joint-bounded blocks alone are to be removed, but I suggest that some crushing process also operates to produce fractured masses in these positions, as some rocks showing highly fractured lee faces are otherwise unjointed, and in many cases fractures are unrelated to joint patterns (Fig. 14).

Theory

The normal pressure fluctuation at the ice/bed interface produced when a glacier moves over a bedrock obstruction in its path (Fig. 8), will induce a related pattern of stresses in the obstruction. Assuming an isotropic linear elastic response to stress in the bedrock, and given an approximate normal pressure variation based on Nye's (1969) analysis (Equation 15), Morland and Boulton (in press) have calculated the stresses and strains induced in a two-dimensional bedrock hummock of fixed shape ($a/\lambda = 1/4$) by glacier movement over its surface (Fig. 15).

For any given hummock shape and ice velocity, and in the absence of cavitation, the total normal pressure fluctuation is $2\Delta P$. Using a linearly viscous approximation for ice, and assuming a wavelength to amplitude ratio of 4/1, and wavelengths greater than 10 m,

$$\Delta P \approx \frac{10\eta V_i}{\lambda} \tag{17}$$

If however $\frac{\Delta P}{\varphi_i g h} > 1$ cavitation will occur, and the normal pressure fluctuation will be equal to $\Delta P + \varphi_i gh$. The shear stresses induced in the surface layers of bedrock are at a maximum where the normal pressure is least, on the down-glacier flank. Stresses in this position are greatest when the pressure fluctuation is $2\Delta P$, that is when no cavity exists. The relative magnitudes of shear stresses can be expressed as a fraction of ΔP (Fig. 15) for $\varphi_i gh = \frac{\Delta P}{2}$, and $\varphi_i gh \geqslant \Delta P$. If cavitation occurs, the maximum shear stress will occur immediately up-glacier of the point of closure, where there is a stress concentration in the rock due to the inhibition of strain by the high normal pressures beyond the point of closure.

Figure 13. Point of *decollement* of a subglacial cavity. The ice/bedrock contact occurs at 2/3 height of the photograph. A stream of debris-rich basal ice emanates from what was later exposed as a long groove cut in bedrock.

Figure 14. Boulder embedded in a lodgment till surface on the margin of Breidamerkurjokull. Ice flow was right to left. Present form probably due to erosion in that position. Note smoothed abraded up-glacier flank, and shattered down-glacier flank.

In order to predict loci of failure, a Coulomb failure criterion is used such that the shear strength is given by

$$|\tau| = \tau_o + N \tan \phi \qquad (18)$$

where τ_o is a cohesive strength and $\tan \phi$ a coefficient of internal friction. The shear strength of the rock will be a minimum where effective normal pressure (N) is

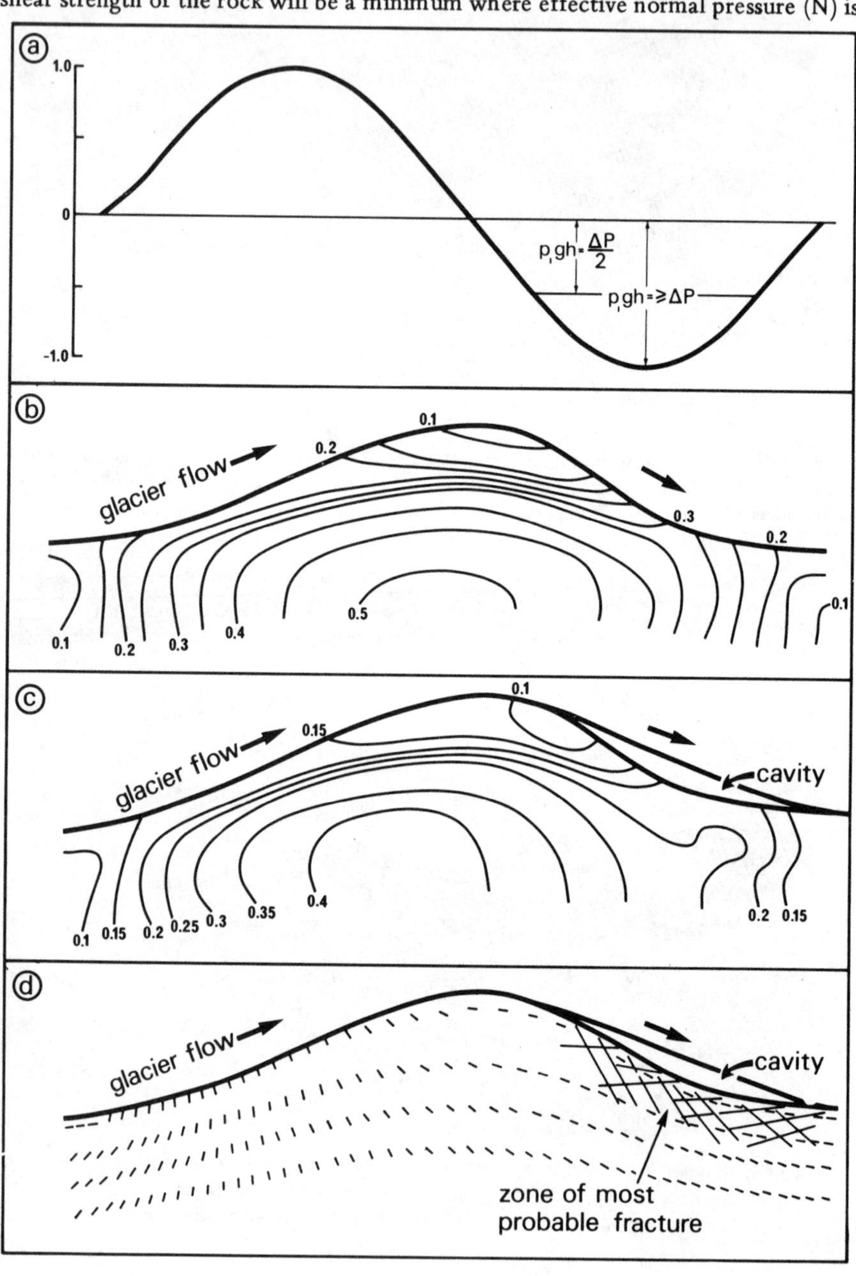

zero; no frictional component of strength will be mobilized and $|\tau| = \tau_o$ Thus the strength of near surface rocks will have a minimum value in the downglacier position where stresses are at a maximum, suggesting the greatest tendency towards failure in that position. For given values of hummock wavelengths and ice velocity, the changing strength of the rock can be plotted against the stresses induced in it for changing values of ice overburden pressure. This relationship is shown in Figure 16 using a value of $\tau_o = 5$ bars (typical of many siltstones and unfractured sandstones) and $\tan \phi = 0.7$, and with the assumption that the rock is impermeable. As ice thickness increases and the size of cavity decreases, stresses rise to a maximum value at the point at which $\Delta P = \varphi_i gh$ and the cavity just closes. For greater ice thicknesses, the stresses remain constant, as they depend merely on ΔP. In the zone of cavitation, the rock strength beneath the cavity will be constant and equal to τ_o, no frictional strength will be mobilized. After cavity closure, under greater ice thicknesses, rock strength will increase due to mobilization of internal friction.

It is useful to measure the tendency towards failure by a safety factor given by $\frac{\text{strength}}{\text{stress}}$, such that failure will occur when this is less than 1. From Figure 16 and Equation 18 it can be seen that

for cavitation

$$\text{safety factor} = \frac{\tau_o}{\left(\frac{1.25\eta V_i}{\lambda} + 0.188\,\varphi_i gh\right)} \tag{19}$$

for no cavitation

$$\text{safety factor} = \frac{\tau_o + \left(\varphi_i gh - \frac{10\eta V_i}{\lambda}\right)\tan\phi}{\frac{3.13\eta V_i}{\lambda}} \tag{20}$$

The failure tendency for bedrock form of different wavelength and constant shape (in this case $a/\lambda = 1/4$) can now be determined for different values of ice thickness and velocity (Fig. 17). Safety factors for $\tau_o = 5$ bars fall to a minimum at the point at which the lee side cavity just closes. For ice velocities of the order of 3 $m.yr^{-1}$ and less (Fig. 17a) no failure will occur at any ice thickness (the analysis

Figure 15. Stresses induced in elastic bedrock hummocks due to ice flow over them.
(a) Normal pressure fluctuation at the ice bed interface due to ice flow over a hummock. The full sinusoidal variation occurs when the ice pressure $\varphi_i gh = \Delta P$. A cavity will occur, below which normal pressure is zero, if $\varphi_i gh \ngtr \Delta P$. The pressure fluctuation for $\varphi_i gh = \frac{\Delta P}{2}$ is also shown.
(b) The distribution of stresses in bedrock for $\varphi_i gh \geqslant \Delta P$. The stresses are given as a fraction of $\varphi_i gh$. Note that maximum stresses in surface rocks occur on the down-glacier flank. This coincides with the zone of minimum normal pressure, where minimum frictional strength is mobilized. It is thus the most likely zone of fracture.
(c) The distribution of stresses for $\varphi_i gh = \frac{\Delta P}{2}$. Maximum stresses, and therefore greatest tendency to failure, occur just up-glacier of the point of cavity closure.
(d) Principal stress axes for c. Assuming an angle of friction of $\phi = 35^o$, the most probable directions of shear failure have been included.

breaks down for $\lambda < 10$ m). For $V_i = 31.5$ m.yr^{-1}, failure will occur in all forms where $\leqslant 50$ m for ice thickness between 130 m and 230 m. For $V_i = 315$ m.yr^{-1}, failure will occur in all forms where $\lambda \leqslant 500$ m for similar ice thicknesses. Failure will always take place more readily for smaller wavelengths, and in the latter case will still occur in $\lambda = 10$ forms if $\tau_0 = 125$ bars, a cohesive strength typical of that of many hard unfractured rocks. Thus, the hardest unfractured rocks may be crushed by this process if ice velocities are high.

In a population of glacially eroded hummocks of given shape but varying wavelength, there should be a progressive decrease in the intensity of lee side fracturing as wavelength increases. Similarly, for hummocks of constant wavelength but varying amplitude, intensity of fracturing should decrease with decreasing amplitude. Different shapes can also be analyzed. Bedrock forms in which $a/\lambda > 1/4$ will fracture more readily, and those in which $a/\lambda < 1/4$ with greater difficulty.

It should be stressed that the above analysis refers to a two-dimensional rock bed in which individual hummocks are elongated transverse to ice flow. The analysis of three-dimensional hummocks, in which the width is of the same order or smaller than length, would pose considerable problems. I suggest that in these latter cases, the tendency towards fracture on the lee flanks would be enhanced because of the high normal pressures on lateral flanks and the tendency for maintenance of a low pressure point on the lee flank at the point of stagnation which occurs when flanking ice flows into the lee side position.

Structural anisotropy, including planes of jointing, may considerably weaken subglacial bedrock, as could fatigue induced by ice velocity and thickness

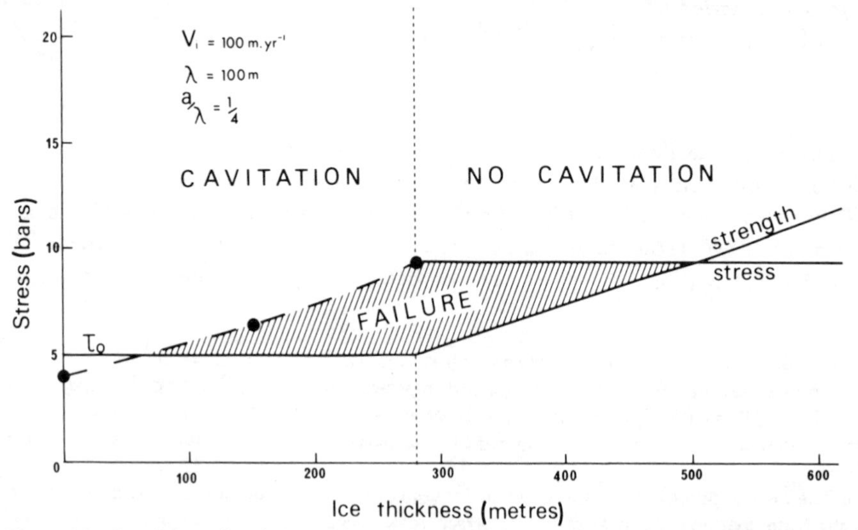

Figure 16. Conditions for failure of a rock in which $\tau_0 = 5$ bars, forming a hummock of $\lambda = 100$ m and $a/\lambda = 1/4$. In the zone of cavitation rock strength is constant while stress rises for an increasing ice thickness, where no cavitation occurs, stress is constant, and strength rises for an increasing ice thickness. The tendency towards failure is thus greatest at point where cavities are just suppressed. This figure effectively shows the ease of quarrying under conditions of constant velocity and increasing ice thickness.

fluctuations. As Matthes (1930) concluded from his study of glacial erosion in the Yosemite valley, "joint structure plays a very important part in glacial quarrying," and even the hardest rocks, with their bulk cohesive strength reduced by jointing, would be prone to failure under the stresses considered here. If subglacial rocks are permeable, then the horizontal axis in Figures 16 and 17 should refer, not to ice thickness, but to effective pressure ($\varphi_i gh - \omega_p$).

Morland and Boulton (in press) have also determined the most probable directions of principal stresses induced in a bedrock hummock (Fig. 15). From these, the directions of shear failure can be determined. They will be offset by angles of $\phi/2$ from the principal stress axis. Typical values of ϕ are 35° for sandstones, and 45° for granites. The lee side position is also one in which tensile fracture can occur, and is most likely to occur along planes parallel to the surface.

The forms of many bedrock humps and steps have been described in the literature (Johnson, 1904; Matthes, 1930; Demorest, 1939; Lewis, 1947; Linton, 1963) and a common form is found in all these descriptions, of a smoothed and striated up-glacier face, produced by abrasion, and a steep and fractured down-glacier face, attributed to "plucking". As Matthes (1930) has observed, joints are of fundamental importance in the plucking process. However, rock must still be fractured before it can be removed, for joint planes rarely cut up a hard rock into loose blocks which need merely be "plucked". The process analyzed here is capable of crushing intact weak rocks, and producing loose blocks by further fracturing of already jointed hard rocks. It also explains the location of crushing, and predicts optimum conditions for its development. Steep down-glacier slopes provide an ideal "crushing block" beneath the glacier, and will presumably retreat as do many other scarp features in the natural environment.

The effect of individual boulders on the crushing process is also in places important. Boulders in the glacier sole above a cavity, which had partially melted out of basal ice or were being expelled from it, produce considerable abrasion and crushing of bedrock when they again come into contact with it at the point of cavity closure becasue of very high contact pressures. Spalling immediately above a bedrock step may also be produced by this process, which may fracture large boulders or rock humps beneath the glacier sole.

PLUCKING

To produce erosion of a hard glacier bed, the fragmental debris produced by abrasion and crushing must be subject to a complementary plucking process. There are two requirements for this process: that the moving ice should exert some tractive force against the loosened fragments, and that this force should be sufficient to overcome the resistence to movement offered by frictional drag of the fragment against subjacent materials. To produce erosion of an already fragmental bed, such as a bed of unlithified sediments, plucking alone is required.

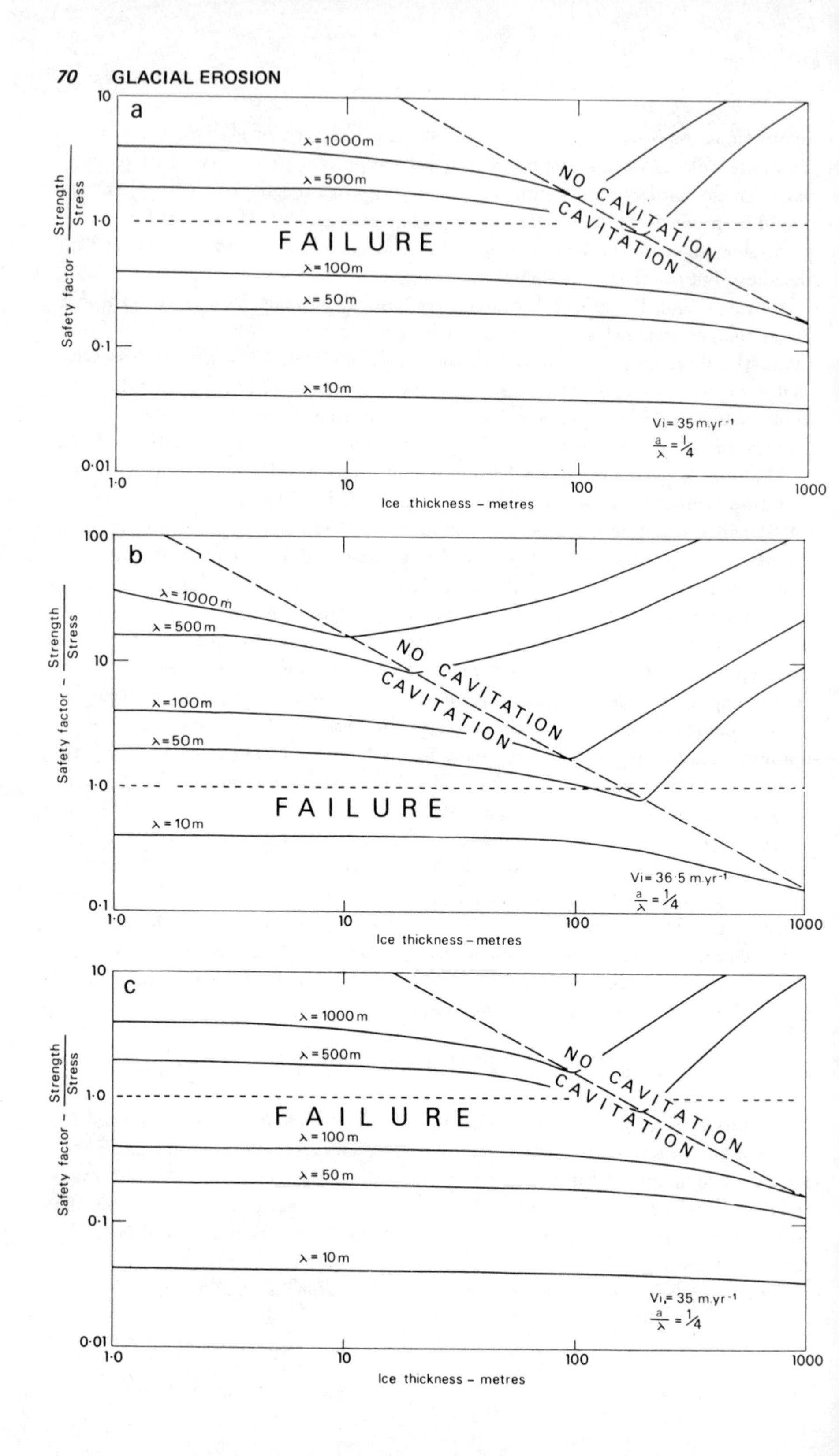
a
Safety factor - Strength/Stress
λ = 1000 m
λ = 500 m
NO CAVITATION
CAVITATION
FAILURE
λ = 100 m
λ = 50 m
λ = 10 m
Vi = 35 m.yr⁻¹
a/λ = 1/4
Ice thickness - metres
b
Safety factor - Strength/Stress
λ = 1000 m
λ = 500 m
NO CAVITATION
CAVITATION
λ = 100 m
λ = 50 m
FAILURE
λ = 10 m
Vi = 36·5 m.yr⁻¹
a/λ = 1/4
Ice thickness - metres
c
Safety factor - Strength/Stress
λ = 1000 m
λ = 500 m
NO CAVITATION
CAVITATION
FAILURE
λ = 100 m
λ = 50 m
λ = 10 m
Vi = 35 m.yr⁻¹
a/λ = 1/4
Ice thickness - metres

Observations

The two processes of plastic flow of ice, and regelation of meltwater enable loosened fragments to be tightly gripped by the moving glacier. It can be shown that plastic flow is more effective in achieving this for large particles and regelation for small particles.

An example of plucking due largely to plastic flow was observed beneath the Glacier d'Argentiere. A series of boulders were found lying on the rock floor of a cavity beneath the glacier, a cavity which is prone to periodic closure as a result of ice thickness and velocity fluctuations. During closure of the cavity, the glacier sole pressed down against the boulders, and the force developed against them caused icy to flow around the boulders so as to encase them almost completely. During this process, the boulders began to move with the ice and were transported away.

The process of plucking by regelation appears to work in a similar fashion. A rock plate was cemented to a near-horizontal surface immediately up-glacier of a rock step beneath the Glacier d'Argentiere. The bedrock surface at this point was revealed by a tunnel cut into the ice, which after construction was allowed to close due to ice movement. The basal ice flowing over this point was observed to transport, from time to time, relatively large boulders up to 50 cm in diameter, which tended to have cavities in their lee. Some 30 days after its emplacement, and shortly after the passage of one such boulder, fragments of one plate were observed in transport in the basal ice where it flowed over the lip of the rock step to form the roof of a cavity (Fig. 18 and 19). The larger fragments ($\approx$ 3-4 cm) were observed to have been partially pressed into the glacier sole, which had presumably yielded by flow. They were also fringed by mats of spicular ice crystals which completely encased the smaller fragments. I suggest that after the initial fragmentation of the rock plate due to the very large contact stresses imposed by the large boulder, the ice flowed plastically around the larger fragments, and that regelation ice also formed around these in addition to completely encasing smaller fragments after pressure-melting against their up-glacier flanks. Moving of ice around the fragments in this way, generated a tractive force capable of coercing them into motion against the resistance offered by the underlying cement. The broad groove in the glacier sole down-glacier of the fragments reflected the

Figure 17. Safety factors for rock hummocks (τ_0 = 5 bars, tan ϕ = 0.7) of different wavelength and constant shape (a/λ = 1/4), for different conditions of ice velocity and thickness. A safety factor (strength/stress)$\leqslant$1 indicates that failure will occur. The diagram can be considered an indication of the effectiveness of erosion by quarrying. The zone in which cavitation is just suppressed will be one of greatest quarrying potential.

Figure 18. Point of *decollement* of a subglacial cavity. Groove in glacier sole is formed by flow of ice around a rock plate attached to bedrock. The plate was then crushed by a large boulder in traction over the bed and individual fragments have been plucked from the cement by flow and regelation of ice around them. After removal of the fragments, it can be seen that the groove largely ceased to exist.

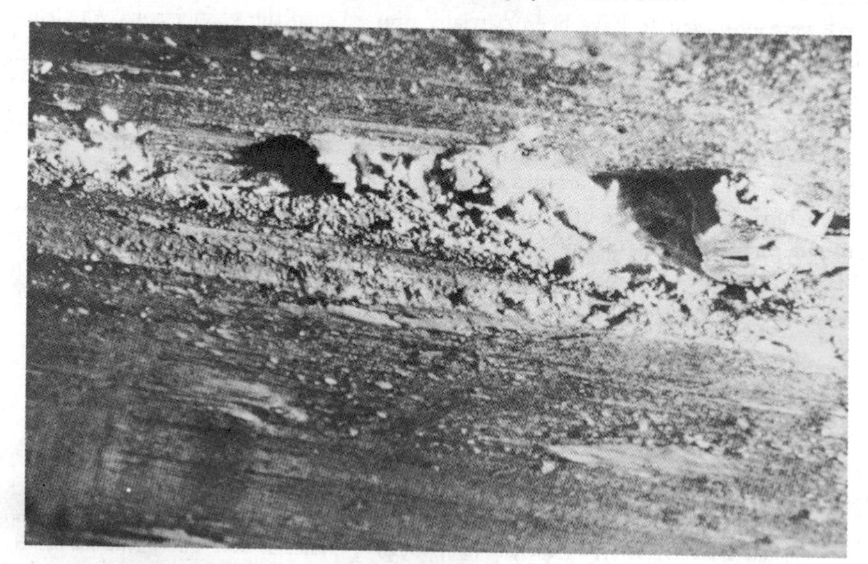

Figure 19. Plucked fragments of plate (Fig. 18) contained in basal ice. Ice flow is right to left. Note cavity in lee of largest fragment.

movement of the ice around the plate before it was crushed. After the crushing episode, most of the plate was removed by plucking, although two very narrow grooves remained which reflected the two fragments left cemented to bedrock. These observations are relevant to the removal by ice of the debris produced by abrasion, the plucking of already fragmental rock masses, and erosion by plucking of unlithified subglacial sediments.

Observations were also made on the down-glacier flanks of bedrock obstructions of removal of fractured materials. From that position Figure 20a shows the upper part of the steep down-glacier flank of a large roche moutonnee which was exposed in a small cavity no more than 2 m in length. This flank is irregularly fractured along a series of planes which cut across the dominant vertical foliation in the rock. It is not possible to tell if the joints are tectonic or glacially induced. Immediately below the glacier sole in Figure 20a, a steeply inclined fresh joint plane is heavily ribbed by a series of criss-crossing ice crystals. The joint has clearly formed very recently. A meter and a half below the location of Figure 20a, a large block, which had fallen from the joint-bound step shown in that figure, lies on an ice surface. As shown in Figure 20b, this ice appears to be stagnant, lying in the cavity between glacier sole and rock. Such stagnant masses tend to be temporary, and as ice velocity and thickness changes cause changes in flow over the bed, they tend to be drawn into the flow of the glacier. I suggest that the large block was not fractured from the bed by freezing of ice, but that this merely dislodged it by its growth in a pre-existing fracture. The block then fell onto the stagnant ice mass, and will probably be drawn at some later time into the flow of the glacier. The unsteady flow of ice in a lee side "eddy" could clearly be an important agent in removing fractured blocks from this position.

Discussion

The above observations show how ice may 'grip' loosened fragments and thus coerce them into motion. In order to do this, it must also overcome the frictional drag between the fragments and the glacier bed. In the case of debris produced by the abrasion process, I have already defined boundary conditions for this process, and determined when the frictional drag against particles is sufficient to inhibit their movement by the glacier. Thus "plucking" of the products of abrasion is co-extensive with the abrasive process itself.

Plucking will take place readily on the lee side of rock humps when cavitation conditions hold. When cavitation is inhibited, the normal pressure which generates frictional resistance to plucking is

$$N = \varphi_i g h - \frac{10 \eta V_i}{\lambda} \qquad (21)$$

The plucking force will depend very much on the way in which ice flows around loose fragments, and is thus difficult to assess. However, frictional resistance will rise very rapidly after the closure of cavities, and thus I shall assume that plucking

Figure 20. (a) Point of *decollement* at head of a small subglacial cavity. Striated glacier sole can be seen at top left. Below lies a near vertical joint plane 30 cm high covered by a mat of spicular ice crystals. Below the vertical joint lies another gently sloping joint. Growth of ice crystals together with the shear stress imposed by the moving ice have removed a joint block.

(b) The removed joint block lying some 1.5 m below location of (a). It has fallen into an 'eddy' of ice in the cavity, and will probably be removed when this ice is pulled into the main glacier flow.

will remove all fractured material from these positions for ice thicknesses just greater than those for closure of cavities. If we consider a bump with a width similar or smaller than its length, rapid flow into the lee position from lateral flanks will tend to produce a low-pressure 'stagnation point' where ice movement is directly away from the rock surface, and where plucked fragments will tend to be carried into an englacial position. Plucking should take place readily in this position under the greatest ice thickness.

Glacial erosion of unlithified sediments merely requires the existence of a plucking process. The incorporation of such materials by temperate glacier ice has been discussed by Boulton (1974), and it was concluded this would occur when the internal frictional resistance of the sediment was overcome by the shear stress imposed by the moving glacier, that is when

$$\tau_g = \tau_o + (\varphi_i gh - \omega_p) \tan \phi \qquad (22)$$

where τ_g is the shear stress at the glacier bed. If we assume τ_g to be approximately constant, with a value of about 1 bar, then for any one sediment, erosion will only occur for relatively small values of effective normal pressure $(\varphi_i gh - \omega_p)$.

LARGE SCALE PATTERNS OF EROSION

It has been suggested that the plucking process, a necessary complement to abrasion and crushing in eroding a glacier bed, is approximately co-extensive with them on a bedrock surface, and that it will carry away any material produced by them. Large scale patterns of erosion can thus be reconstructed by consideration of the variation in the controls on these processes in real glaciers. Their most important characteristic is that for a constant ice velocity, their importance rises with increased pressure, to an optimum, beyond which it declines rapidly, until, at high pressures, both processes are suppressed (Figs. 7 and 17). In examining whether the theory can also explain large scale patterns of erosion, we need to determine where conditions for abrasion and crushing are satisfied, and the relative rates at which these processes operate.

Cirque Glaciers

There is relatively little information about the dynamics of cirque glaciers apart from the study undertaken by Lewis and his collaborators (Lewis, 1960). It is thus convenient to use Vesl-Skautbreen, the glacier they studied, as a model for others. Because of their small size, cirque glaciers tend to be thin, slowly moving ice masses. Vesl-Skautbreen had a maximum ice thickness of about 75 m, and a measured slip velocity beneath the equilibrium line of 2.33 $m.yr^{-1}$. It has often been suggested that glacial cirques develop from a nivation hollow in which nival processes have produced a dish-shape (e.g. Flint, 1971). Growth of the snow-patch to form a glacier might still allow sapping processes in a *bergschrund* to push-back and steepen the headwall (Lewis, 1940; McCall, 1960), but these cannot explain the over-deepened rock basin characteristic of many cirques. From Figure 17, which shows erosion by quarrying (crushing and plucking) for an ice velocity of 3.15 $m.yr^{-1}$, it can be seen that even for a relatively rough bed on which $a/\lambda = 1/4$, only the very weakest rocks would fail, and then only beneath ice thicknesses of the order of 100 m. Thus quarrying of large blocks is only likely in a cirque if they are already completely loose.

Although large blocks are unlikely to be produced by quarrying once loose material has been initially swept from the cirque floor, it has been shown (Johnson, 1906; McCall, 1960) that large blocks sapped from the headwall or falling into a bergschrund may enter the glacier sole to become abrasive tools. McCall's observations suggest that both ice thickness and velocity rise to a maximum beneath the equilibrium line, and thus (as cavitation is likely to be important beneath the entirety of such a glacier) I would expect from Figure 7 that maximum abrasion rates will occur here. This will lead to a preferential lowering of the cirque floor and the formation of an overdeepened basin (Fig. 21). Excessive over-deepening will be avoided by cirque retreat, but several other processes could also inhibit this, such as stagnation of basal ice or a progressive decrease in velocity as the basin deepens.

Therefore, I conclude that in a small cirque, abrasion will be the dominant erosive process and that deposition is unlikely to occur, the rock floor will tend to be smooth and swept clear of till.

Valley Glaciers

Glaciers channelled in valleys exhibit a wide range of velocities and thicknesses. The greatest (non-surging) velocities are shown by those which channel the ice

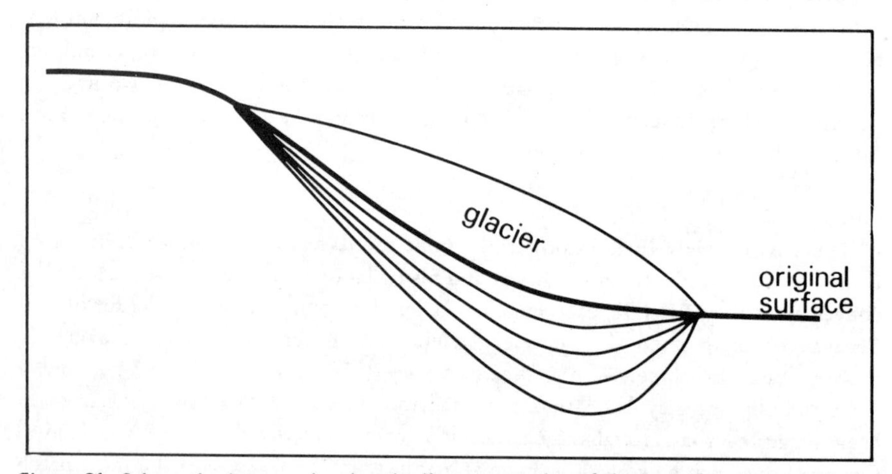

Figure 21. Schematic diagram showing the likely deepening of the bed of a cirque glacier by abrasion under steady-state conditions. Highest velocities and thicknesses beneath the equilibrium line are likely to concentrate the greatest abrasion rates in that position.

discharge from wide areas of major ice caps and ice sheets. Jakobshavn Isbrae in west Greenland is one such case, with a velocity of the order of 5 $km.yr^{-1}$ (Bauer, 1961), and the Byrd glacier in Antarctica has a velocity of 0.84 $km.yr^{-1}$ (Swithinbank, 1963). Glaciers of this type will clearly be very effective agents of erosion, both by quarrying and abrasion.

For valley glaciers which transmit discharges from smaller basins of accumulation, Andrews (1972) has suggested that a glacier's velocity through the equilibrium line is proportional to its "ablation gradient" (a measure of "energy of glacierization", Meier, 1961) which is greatest in maritime areas and lower latitudes. For such glaciers, ice thicknesses up to 400 m are common with sliding velocities from 1 to 800 $m.yr^{-1}$. Some illustrative measured values are: Mer de Glace, (accumulation area), 35-800 $m.yr^{-1}$ for ice thicknesses of 100 m – 30 m (Reynaud, 1973); Glacier d'Argentiere (ablation area), 150 $m.yr^{-1}$ and 700 $m.yr^{-1}$ for ice thicknesses of 150 m. and 40 m respectively (Vivian, 1970; and personal observation); Athabasca glacier, 3 $m.yr^{-1}$ and 300 m thickness (Raymond, 1971), and 5 $m.yr^{-1}$ and 200 m thickness (Savage and Patterson, 1963); Blue Glacier, 7.5 $m.yr^{-1}$ and 26 m thickness (Kamb and LaChapelle, 1964); Breidamerkur jokull, 6-20 $m.yr^{-1}$ and 40 m thickness (personal observation).

For such a range of velocities and ice thicknesses it can be seen from Figure 17 that except for those glaciers where the ice thickness is great and sliding velocity small (e.g. Athabasca glacier), subglacial cavitation will be an important process. This is true at the accessible sites beneath theGlacier d'Argentiere (Vivian, 1970; Vivian and Bocquet, 1973) where beneath 150 m of ice and with a sliding velocity of 150 $m.yr^{-1}$, cavitation is widespread and water drains freely in subglacial channels. Thus, I shall assume for all these glaciers that the effective normal pressure against the bed is dependent on ice thickness; it becomes possible to explain the typical U-shaped cross profile of a glacial valley. Consider a V-shaped valley occupied by ice for the first time. The normal pressure of ice against the valley walls will rise linearly with depth. The maximum pressure in the valley bottom may fall in any of the three critical regions A, B or C in Figure 7. If it falls in region A, the lower part of the valley profile will be abraded at a greater rate than its upper part; if in B, the maximum abrasion rate will occur at some point low on the valley side; and if in C, deposition will occur in the valley bottom, and the greatest abrasion rate will occur at some point on the valley side. Figure 22 shows schematically how all of these combinations will lead to the development of a U-form.

The quarrying process will also show a similar tendency to produce U-forms. High on the lateral flanks, only the most acute bumps of short wavelength will be crushed, but deeper in the valley, forms which are less acute and have longer wavelengths will suffer quarrying. If cavitation is entirely suppressed in the valley bottom, only acute forms of short wavelength will continue to be quarried (Fig. 17).

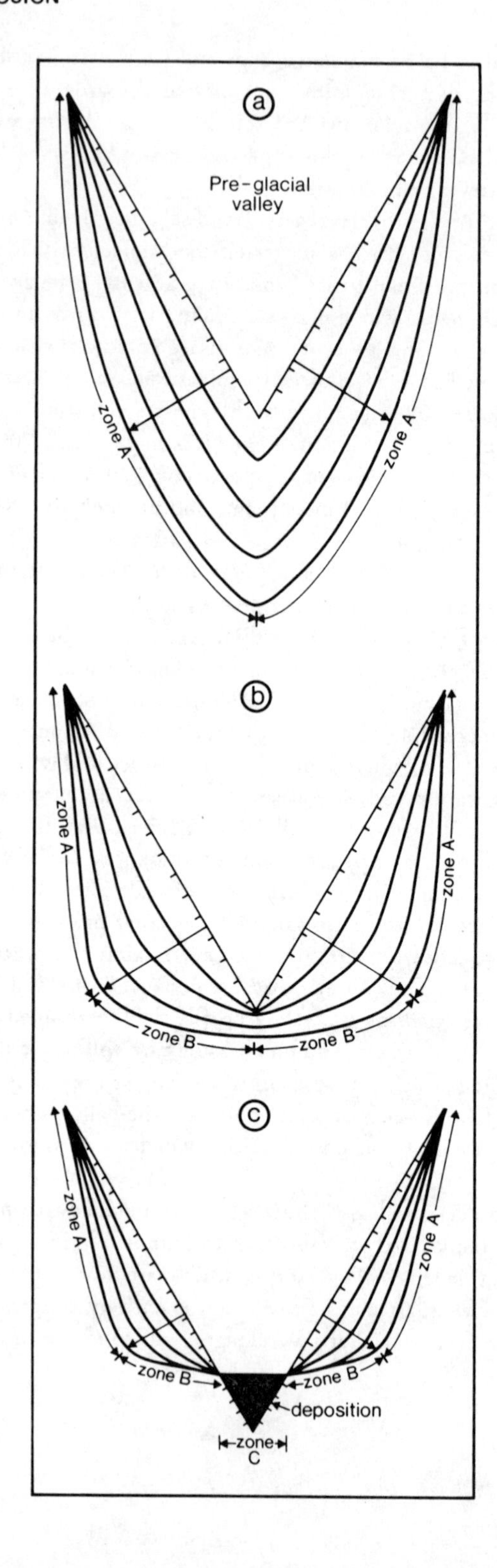
a
Pre-glacial valley
zone A
zone A
b
zone A
zone A
zone B
zone B
c
zone A
zone A
zone B
zone B
deposition
zone C

One of the characteristics of rock floors of glaciated valleys is their relatively high relief. Lewis (1947) has stressed the importance of asymmetric forms which grade from small *roches moutonnees* to major valley steps. The way in which valley steps might evolve from relatively high points on the valley floor is illustrated by Figure 9a. In this case, maximum pressures occur in region A of Figure 7, and abrasion rates are highest against the up-glacier flank and at the foot of the down-glacier flank. This will lead to the planing away of the up-glacier flank, and the formation of a basin at the foot of the down-glacier flank, the form of a typical valley step. The steepening of a downglacier flank by abrasion in this way will tend to develop conditions under which quarrying can occur, and many field examples tend to suggest that both processes operate hand in hand.

Individual bedforms beneath a glacier may retreat due to quarrying, or as a result of heavy lee flank abrasion (region A, Fig. 7), and light up-glacier flank abrasion or even deposition (regions B or C). If abrasion is concentrated on the up-glacier flank and the foot of the down-glacier flank, the step which may form will not necessarily retreat, unless lee-side quarrying occurs.

Another characteristic of glaciated valleys is the frequent occurrence of overdeepened basins. A tendency for local overdeepening is clearly intrinsic in both abrasion and crushing processes (Figs. 7 and 17), for the importance of both tends to increase to a maximum with increasing ice thickness and subsequently to decline. The very high velocities characteristic of some outlet glaciers of major ice caps and ice sheets may maintain subglacial erosion beneath very considerable ice thicknesses. I suggest that the deep erosion and basin formation under considerable ice thicknesses inferred for many Norwegian fjords during the Pleistocene ice ages (Gregory, 1913; Ahlmann, 1941; Strom, 1948) is a reflection of the high velocities characteristic of ice streams which drain large areas of ice caps and ice sheets.

Widespread deposition of till may also occur beneath valley glaciers. According to this theory, this will occur when ice velocities are small, when thickness is large or when the bed is rough. It will tend to be localized around high pressure bedrock humps in drumlinoid forms. Slowly moving glaciers, such as many studied in Iceland (Okko, 1955), tend to deposit extensive lodgment till, whereas more rapidly moving glaciers, such as many in western Norway and the Alps, tend to sweep their beds clear of till.

Ice Caps and Ice Sheets

Ice caps and ice sheets have two important characters which in this theory are considered to be important controls on glacier erosion, a systematic increase in

Figure 22. Patterns of abrasive erosion of an originally V-shaped valley. If these are continued under uniform conditions, each would lead to a characteristic U-form.
(a) Pressures fall entirely within zone A, Figure 7, and greatest erosion thus occurs in the valley bottom.
(b) Pressure in valley bottom falls in zone B, Figure 7, locating greatest erosion low on the valley sides.
(c) Pressure in valley bottom falls in zone C, Figure 7, deposition therefore occurs there, with greatest erosion on the valley sides.

thickness from snout to source and low velocities over the bed. Because of these, cavitation will be suppressed only a little distance from the margin. Thus, if the glacier has a rock bed, effective normal pressure will be largely controlled by bed roughness, apart from in a narrow marginal zone. However, the only ice sheet beds accessible to inspection are those which were occupied by large Pleistocene ice sheets, and in these cases, the major part of the bed is composed of permeable deformable sediment, *not* of rigid impermeable rock as is often the case for valley glaciers. The sediment is often till, and may have been deposited largely because of low basal ice velocities. A water–pressure model for such a glacier bed has been developed by Boulton, Dent and Morris (in press), and Boulton (1974). For a realistic bed, composed of sediments of both high and low transmissibility (permeability x thickness = coefficient of transmissibility), the latter, by requiring a high hydraulic gradient to maintain the water discharge through them, will act as dams to water flow and maintain high water pressure in the up-glacier sediments even though the piezometric gradient may be low (Fig. 23).

If the glacier bed is made entirely of loose, deformable sediments, plucking only is required to erode it, which will only occur where ice pressures are low in the marginal zone, or where high water pressures produce low effective pressures in internal zones. Till will be deposited in areas where $N/V_i^m \geqslant L_c$ and will be eroded where $\frac{\tau_g - \tau_o}{\tan \phi} \geqslant$ N and abrasion of the bed will occur in intervening zones. If we assume a fluctuating value of N depending on the character of bed sediments, then the excavation of basins will be the principal erosive effect, basins whose distribution on an ice-sheet wide scale reflects the bed geology. These erosional basins may occur in areas of predominant deposition where effective pressures are reduced or ice velocities increased so that $N/V_i^m < L_c$, or in areas of erosion where velocities or pressures are locally increased (Fig. 7 and 17). A general trend would also be the development of sub-marginal basins.

For an ice sheet with a rock bed the important erosive processes are abrasion, which forms a continuum with deposition of lodgment till, and quarrying (quarrying= crushing + plucking) of rock masses. These two processes are relatively independent. Quarrying depends largely on ice velocity and thickness changes and rock strength, whereas abrasion/lodgment depends largely on changes in rock hardness, ice velocity and effective normal pressure at the glacier bed, and this latter depends upon bed roughness and permeability. Thus, the geology of the bed is of fundamental importance in controlling erosion, a clear implication of studies of the beds of Pleistocene ice sheets, such as those of White (1972) in North America and Rudberg (1973) in Sweden.

Assuming a constant ice velocity, increasing ice thickness will tend to enhance the rate of quarrying to a maximum, beyond which it will decline and give way to a zone of lodgment till accumulation. Local decreases in rock strength, or increases in joint frequency may locally enhance quarrying (Rudberg, 1973), which may of course also occur in a zone of lodgment till accumulation. One of the effects of this

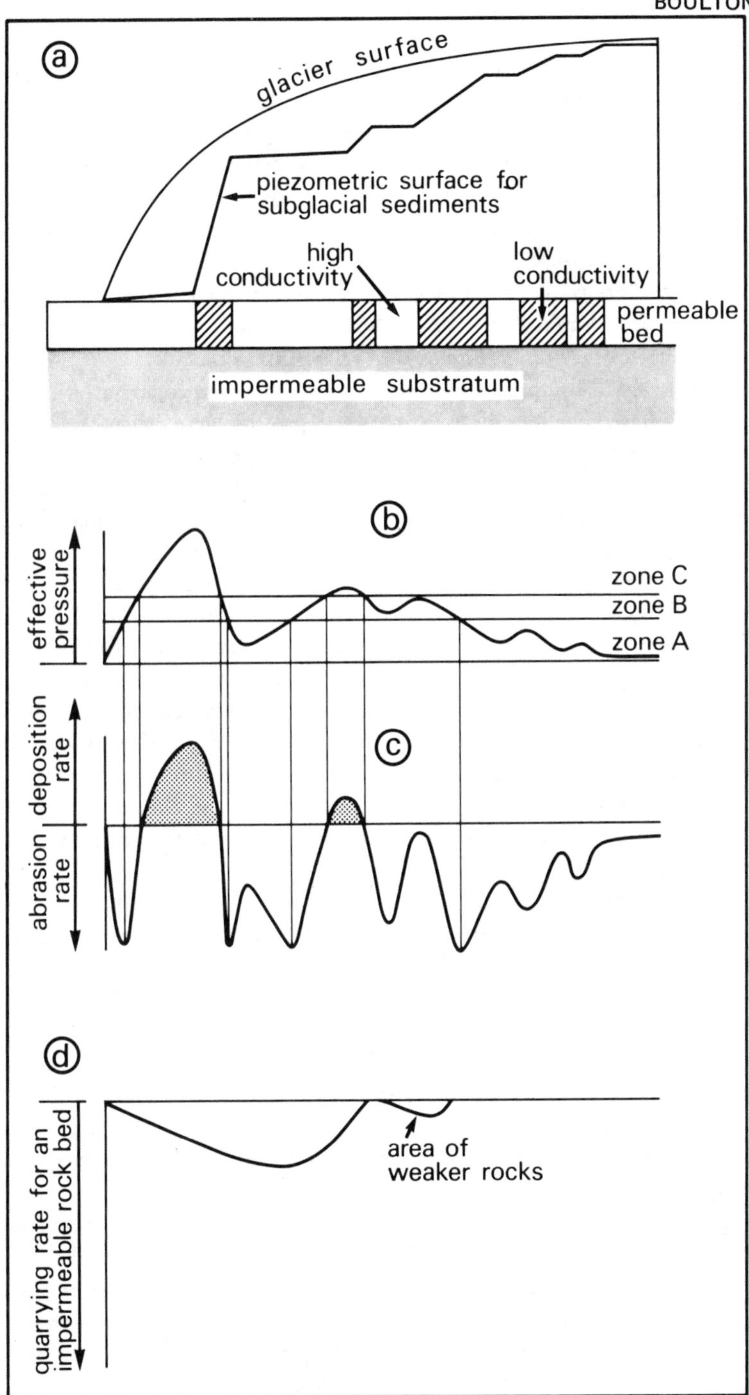

Figure 23. Schematic diagram showing the controls on erosion beneath a large Quaternary ice sheet.

pattern of quarrying will be to produce sub-marginal basins, or local basins where rocks are weak (Rudberg, 1973; Linton, 1957), and it may well increase local relief.

The intensity of the abrasive process will largely be dependent, not on ice thickness, but on bed geology and roughness. Abrasion may thus still occur in the internal parts of ice sheets where effective pressures are relatively low, eventhough quarrying is completely suppressed by large ice thicknesses. The effective pressure may be increased because of relief enhancement due to plucking, but in general depositional and abrasional processes will be controlled largely by bed geology and ice velocity, and should show no systematic glacier-wide variations related to thickness changes, other than those in turn related to velocity.

It is interesting to compare these predictions with broad zonal patterns of glacial erosion which have been identified in areas occupied by large Pleistocene ice sheets. For instance, Clayton (1965) identified in the Finger Lakes region of New York State, a series of zones in which the intensity of glacial erosion increased from south to north (although there is some dispute about the significance of some of the evidence, Coates, 1966). Bedrock hummocks are also described as more strongly streamlined to the north, indicating an increase in intensity of abrasion. Thus at high points on the glacier bed, the normal pressure would be directly related to ice thickness, as any water would be squeezed away from these zones (Weertman, 1972), and increased streamlining of such forms is to be expected with increased distance from the ice margin. However, immediately north of the Finger Lakes region, there is a drumlin field in the area south of Lake Ontario, which fits well with a pattern of increased effective pressure to the north, which would firstly cause a decline (zone B, Fig. 7) in the abrasion rate, and then the deposition of lodgment till (zone C, Fig. 7). It is presumed that the quarrying process was also suppressed in this area by large ice thickness.

Under the source areas of ice sheets and ice caps, horizontal ice velocities will tend to zero and thus erosive processes should be entirely suppressed. Such glacial protection could adequately explain such features as the presence of deeply weathered soils on potential ice cap source areas as the Cairngorm *plateaux* (Linton, 1959).

In this chapter, little account has been taken of glacier variation nor the cumulative effects of multi-cyclic Pleistocene glaciation. Steady conditions have been assumed throughout. Small features, up to the size of *roches moutonnees*, may well owe their form to relatively stable conditions, and it is these forms which will best provide a test for the theory.

If typical forms do exist on a larger scale a non-steady state theory must be sought to account for their detailed shape, hardly a tractable problem in our present state of knowledge. For instance, it can be seen from Figure 22 that a U-shaped cross profile will tend to develop in a glaciated valley no matter what the ice thickness or velocity, and thus any one cross profile is likely to reflect a series of events rather than approximating exactly to any one of the forms in Figure 22.

It has been argued that over-deepened basins beneath ice caps may be produced both by abrasional and quarrying process. In the former case, the location will be governed largely by bed geology, independent to a certain extent of glacier fluctuation. Quarrying is however dependent on ice thickness and velocity, and is thus only likely to systematically produce basins if the glacier is stable for a long period or repeatedly occupies the same area.

At the present time, we know relatively little about the lengths of time for which certain ancient morainic lines were occupied during the Pleistocene, and thus the pattern and tempo of glacial occupation is difficult to elucidate. However, the southern limits of successive glaciations in Europe and North America were very similar on a continent-wide scale, and it may be that many erosional patterns (such as those described by White, 1972) are the product of systematic re-occupation of the same areas during successive glacial episodes, rather than a direct reflection of the distribution of subglacial processes.

Non-temperate ice

It has been assumed in the preceding analysis, that ice in contact with the glacier bed is at the melting point. Many glaciers in high latitudes are however composed of ice below the melting point, although we believe that only in a few of these is the ice of the glacier sole below the melting point in other than the terminal zone. It has also been suggested (Weertman, 1961; Boulton, 1972) that an appropriate model for large Pleistocene ice sheets which existed in latitudes now temperate would be one in which the outer terminal zone is composed of cold ice to which meltwater derived from basal melting in a temperate interior freezes (see also Clayton and Moran, this volume). In this outer terminal zone, two sub-zones have been suggested, the outermost, in which subglacial sediments are frozen to some depth, and an inner zone in which the rate of freezing of internally produced meltwater is sufficient to maintain the glacier sole at the melting point. Although I would suggest there is no sliding of cold ice directly over its bed because of the relatively high adhesive strength in shear of an ice/rock interface (e.g. Jellinek, 1959; Goldthwait, 1960), a high shear strain rate is likely in the ice immediately above a stationary boundary layer.

How would the erosional theory presented here be effected if the glacier-bed interface were frozen?

The crushing process discussed here would not be fundamentally changed, except that the pressure fluctuation, ΔP would not be determined from a sliding velocity but from the shear strain rate in the basal ice. If we assume a stationary boundary layer, then it is difficult to envisage an englacial particle coming into contact with the glacier bed, and there is some experimental evidence to support this contention (Nye, personal communication). However, plucking of part of the bed material might leave the plucked fragment in contact with the bed so that it ploughed through a stationary boundary layer and thus abraded the bed. Thus, although abrasion might be reduced in importance, I still expect it to occur.

Lodgment is likely to take place much less readily. Firstly, particles may infrequently make contact with the glacier bed because of the existence of a stationary boundary layer, secondly, the regelation process will be suppressed and thus small particles will be retarded much less easily, and thirdly, adhesive forces at particles boundaries will allow moving ice to exert a tangential shear stress against them, a process rejected for temperate ice (Weertman, 1957).

The existence of a significant adhesive force between the glacier and its bed should enhance the plucking process considerably, and Boulton (1972) has suggested that very large erratics might be incorporated because of this. Nonetheless it is still necessary that the resistance of subjacent materials should be overcome, and if these are frozen, this resistance will be much increased.

Glacier power and erosion potential

Andrews (1972) has made the reasonable suggestion that the erosive power of a glacier can be estimated from a determination of the power expended in sliding over its bed. Although clearly a potentially rewarding line of attack, Andrews' formulation is a little over-simplified and I suggest it leads to conceptual errors. He defines total power (ω_T) as $\omega_T = \tau_g \bar{U}$, where $\bar{U}$ is the average glacier velocity. The effective power is then some fraction of ω_T, depending on the relationship between glacier slip velocity and internal flow velocity. Implicitly this assumes that erosion is done by the glacier ice, whereas it is largely produced by debris particles.

I re-define the problem in different terms to take the abrading debris into account. The average shear stress at the glacier bed is $\langle\tau\rangle = \varphi_i g h \sin\alpha$, where α is the slope of the glacier surface. The drag at the glacier sole must balance this, and is composed of drag generated by movement of basal ice around fixed obstructions on the bed, and drag of the ice against debris particles which are themselves in frictional contact with the rock bed. Thus

$$\langle\tau\rangle = \frac{2\eta V_i K^2{}_*}{\pi} + \bar{F} \tag{23}$$

$\left(\frac{2\eta V_i K^2{}_*}{\pi}\right)$ gives the **drag against** fixed obstacles from Nye's (1970) theory. Thus the nature of the glacier bed must be taken into account. In addition, the power expended by movement of ice around debris particles is greater if the particles are lodged. The erosive power can be better estimated from the transporting power (P_t) of the glacier, discussed by Boulton (1974), and will be $V_p{}^1\bar{F}$, which is, from (9) and (10),

$$P_t = \left[V_i - \left(\frac{N}{L_c}\right)^{1/m}\right]\bar{F} \tag{24}$$

Therefore erosive power is zero for lodgment conditions, although it should be recognized that this estimate applies only to abrasion and not to quarrying.

ACKNOWLEDGEMENTS

Much of this work was supported by generous grants from the Royal Society and NERC. The author is grateful to Dr. R. Vivian and others from the Institut de Geographie Alpine, Grenoble, and to Dr. E.M. Morris, for their assistance in the field and valuable discussions; to Electricite de France, Electricite d'Emosson, and Pegaz et Pugeat for permission and assistance in working in the galleries beneath the Glacier d'Argentiere; to Flosi, Halvdan and Sigidur Bjornsson of Kvisker and to many UEA students for their help in Iceland; to Sue Middlege for her typing, David Mew for his drawing, and Denise and Kate for their forbearance.

REFERENCES

Ahlmann, H.W. 1941. The main morphological features of north-east Greenland: Geogr. Annlr., 23, 148-82.

Andrews, J.T. 1972. Glacier power, mass balance, velocities and erosion potential: Z. Geomorph. Supplementband, 13, 1-17.

Bauer, A. 1961. Influence de la dynamique des fleuves de glace sure celle de l'inflandsis du Groenland. Union Geodesique et Geophysique Internationale, Association Internationale d'Hydrologie Scientifique, Commission des Neiges et Glaces, Assemblee Generale de Helsinki, July 25—Aug. 6, 1960, 578-84.

Boye, M. 1950. Glaciaire et periglaciaire de l'Ata Sund nord-oriental (Groenland): Paris.

Boulton, G.S. 1972. The role of thermal regime in glacial sedimentation: Inst. Br. Geogr. Spec. Pub., 4, 1-19.

—————— 1974. Processes and patterns of subglacial sedimentation: a theoretical approach: in Wright and Moseley (eds.). *Ice Ages Past and Present.*

Boulton, G.S., Dent, D.L. and Morris, E.M. (in press). Subglacial shearing and crushing, and the role of water pressures in tills from south-east Iceland: Geogr. Annlr.

Cailleux, A. 1952. Polissage et surcreusement glaciaires dans l'hypothese de Boye: Rev. Geomorph. Dyn., 3, 247-57.

Carol, H. 1947. The formation of roches moutonnees: J. Glaciol., 1, 58-9.

Chamberlin, T.C. 1888. The rock scorings of the great ice invasions: U.S. Geol. Surv. 7th A. Rep., 155-248.

Clayton, K.M. 1965. Glacial erosion in the Finger Lakes Region, New York State: Zeit. f. Geomorph., 9, 50-62.

Coates, D.R. 1966. Discussion of K.M. Clayton. "Glacial erosion in the Finger Lakes Region, New York State": Zeit. f. Geomorph., 10, 469-474.

Dahl, R. 1965. Plastically sculptured detailed forms on rock surfaces in northern Nordland, Norway: Geogr. Annlr., 47, 83-140.

Demorest, M. 1938. Ice flowage as revealed by glacial striae: J. Geol. 46, 700-25.

Doornkamp, J.C. and King, C.A.M. 1971. *Numerical Analysis in Geomorphology:* Arnold, London.

Dreimanis, J.A. and Vagners, U.J. 1971. Bimodal distribution of rock and mineral fragments in basal tills: in Goldthwait, ed., *Till: a Symposium.* Ohio State University Press.

Embleton, C. and King, C.A.M. 1968. *Glacial and periglacial Geomorphology.* Arnold, London, 608 p.

Fairchild, H.L. 1907. Drumlins of central western New York State: Bull. N.Y. St. Mus., 111, 391-443.

Flint, R.F. 1971. *Glacial and Quaternary Geology*. Wiley, New York, 892 p.

Forbes, J.D. 1843. *Travels through the Alps of Savoy:* Edinburgh.

Gilbert, G.K. 1906. Crescentic gouges on glaciated surfaces: Geol. Soc. Am. Bull., 17, 303-13.

Gjessing, J. 1967. Potholes in connection with plastic scouring forms: Geogr. Annlr., 49, 178-87.

------ 1966. On plastic scouring and subglacial erosion: Norsk Geogr. Tidsskr., 20, 1-37.

Goldthwait, R.P. 1960. Study of ice cliff in Nunatarssuag, Greenland: Tech. Rep. Snow Ice Permafrost Res. Establ:, 39, 1-103.

Gregory, J.W. 1913. *The nature and origin of fjords:* London.

Harland, W.B. 1957. Exfoliation joints and ice action: J. Glaciol., 3, 8-10.

Harris, S.E. 1943. Friction cracks and the direction of glacial movement: J. Glaciol., 51, 244-58.

Hjulstrom, F. 1935. *Studies of the morphological activity of rivers as illustrated by the River Fyris:* Uppsala.

Hoppe, G. 1957. Frostfenomen och glacialmorfologi i sodra Sverige: Ymer, Stockholm.

Jellinek, H.H.G. 1959. Adhesive properties of ice: J. Colloid Sci., 14, 268-80.

Johnson, W.D. 1904. The profile of maturity in alpine glacial erosion: J. Geol., 12, 569-78.

Johnsson, G. 1956. Glacialmorfologiska studies i sodra Sverige med sarskild hansyn till glaciala riktningselement och periglaciala frostfenomen: Medd. fr. Lunds Univ. Geogr. Inst., Avh. 30.

Kamb, B. 1970. Sliding motion of glaciers: theory and observation: Rev. Geophys. Space Phys., 8, 673.

Kamb, B. and LaChapelle, E. 1964. Direct observation of the mechanism of glacier sliding over bedrock: J. Glaciol., 38, 159-72.

Lewis, W.V. 1940. The function of meltwater in cirque formation: Geogr. Rev., 30, 64-83.

------ 1947. Valley steps and glacial valley erosion: Trans. Inst. Br. Geogr., 14, 19-44.

------ 1954. Pressure release and glacial erosion: J. Glaciol., 2, 417-22.

Lewis, W.V., ed. 1960. Norwegian cirque glaciers: R. Geogr. Soc. Res. Ser., 4.

Linton, D.L. 1957. Radiating valleys in glaciated lands: Tijd. Kon. Ned. Aard. Genoot., 74, 297-312.

------ 1959. Morphological contrasts of Eastern and Western Scotland: in Miller, R. and Watson, J.W., ed.- *Geographical Essays in Memory of Alan G. Ogilivie.*

------ 1963. The forms of glacial erosion: Trans. Inst. Br. Geogr., 33, 1-28.

Lungner, E. 1930. Spaltentektonik und Morphologie der schwedischen Skaggerrack-Kuste: Bull. Geol. Instn. Univ. Uppsala, 21, 1-478.

Matthes, F.E. 1930. Geologic history of the Yosemite Valley: U.S. Geol. Surv. Prof. Pap., 160, 137 p.

McCall, J.G. 1960. The flow characteristics of a cirque glacier and their effect on glacier structure and cirque formation: in Lewis, ed., *Norwegian cirque glaciers,* R. Geogr. Soc. Res. Ser. 4, 39-62.

Meier, M.F. 1961. Mass budget of South Cascade Glacier, 1957-1960: U.S. Geol. Surv. Prof. Pap., 424-B, 206-11.

Morland, L.W. and Boulton, G.S. (in press). Stress in an elastic hump: the effects of glacier flow over elastic bedrock.

Nye, J.F. 1969. A calculation on the sliding of ice over a wavy surface using a Newtonian viscous approximation: Proc. Roy. Soc. London, 311A, 445-67..

------ 1970. Glacier sliding without cavitation in a linear viscous approximation: Proc. Roy. Soc. London, 315A, 381-403.

Nye, J.F. and Martin, P.C.S. 1967. Glacial erosion: Int. Ass. Sci. Hydrol., Comm. Snow and Ice, Berne, 78-83.

Okko, V. 1955. Glacial drift in Iceland, its origin and morphology: Bull. Comm. Geol. Finl., 170, 1-133.

Østrem, G., Ziegler, T. and Ekman, S.R. 1969. Slamtransportundersokelser i Norsk Bre-Elver 1969: Vass. Hydrol. Avdeling, Oslo, Rapp. Nr. 60/70, 68 p.

Rabinowicz, E. 1965. *Friction and wear of materials:* Wiley, New York.

Raymond, C.F. 1971. Flow in a transverse section of the Athabasca Glacier, Alberta, Canada: J. Glaciol., 10, 55-84.

Reynaud, L. 1973. Flow of a valley glacier with a solid friction law: J. Glaciol., 12, 251-8.

Rudberg, S. 1973. Glacial erosion forms of medium size—a discussion based on four Swedish case studies: Zeit. f. Geomorph., 17, 33-48.

Savage, J.C. and Patterson, W.S.B. 1963. Borehole measurements in the Athabasca Glacier: J. Geophys. Res., 68, 4521-36.

Smith, H.T.U. 1948. Giant glacial grooves in Northwest Canada: Am. J. Sci., 246, 503-14.

Strom, K.M. 1948. The geomorphology of Norway: Geogr. J., 112, 19-27.

Swithinbank, C.W. 1963. Ice movement of valley glaciers flowing into the Ross Ice Shelf, Antarctica: Science, 141, 523-4.

Tabor, D. 1954. Moh's hardness scale--a physical interpretation: Proc. Phys. Soc. B., 67, 249-257.

------ 1956. The physical meaning of indentation and scratch hardness: British Journal of Applied Physics, 7, 159-66.

Thorarinsson, S. 1939. Observations on the drainage and rates of denudation in the Hoffellsjokull district: Geog. Annlr., 21, 189-215.

Vivian, R. 1970. Hydrologie et erosion sous-glaciaires: Rev. Geogr. Alp., 58, 241-64.

Vivian, R. and Bocquet, G. 1973. Subglacial cavitation phenomena under the Glacier d'Argentiere, Mont Blanc, France: J. Glaciol. 12, 439-59.

Weertman, J. 1957. On the sliding of glaciers: J. Glaciol., 3, 33-8.

------, 1961. Mechanism for the formation of inner moraines found near the edge of cold ice caps and ice sheets: J. Glaciol., 3, 965-78.

------ 1964. The theory of glacier sliding: J. Glaciol., 5, 287-303.

------ 1972. General theory of water flow at the base of a glacier or ice sheet: Reviews of Geophysics and Space Physics, 10, 287-333.

White, W.A. 1972. Deep erosion by continental ice sheets: Geol. Soc. Am. Bull., 83, 1037-56.

CHAPTER 3

A GLACIAL PROCESS-FORM MODEL

Lee Clayton
and
Stephen R. Moran

ABSTRACT

The glacial geomorphology of the interior of North America can be interpreted using a glacial process-form model that requires a quarrying zone near the glacial margin (the glacier is frozen to its bed in most places, excess pore-water pressure is dissipated by marginal seepage, and the glacial flow is upward) and an abrasion zone behind the margin (the bed is thawed and the glacier is sliding, excess pore-water pressure is not dissipated, and the glacial flow is horizontal or downward). Most erosion takes place in the quarrying zone, producing well-homogenized englacial sediment, most of which becomes superglacial mudflows as the glacier melts.

The resulting glacial landscape involves four elements: preadvance, subglacial, superglacial, and postglacial. First, subglacial erosion may replace the preadvance morphology with either longitudinal shear marks (in places in the abrasion zone where the shear strength is high as a result of dissipation of excess pore-water pressure) or transverse compressional features (in places in the quarrying zone where the excess pore-water pressure can not be dissipated). As the englacial and superglacial sediment thicken, they obliterate the underlying preadvance or subglacial elements. Washboard moraine may result where thin englacial sediment is concentrated along transverse shear zones. Topography with hummocks 200 m in diameter will result if the englacial sediment is isotropically distributed; the hummocks result from superglacial mudflows down the side of glacial sinkholes. Finally, postglacial erosion or deposition gradually modifies the superglacial, subglacial, or preadvance elements.

Glacial landforms result from the interaction of several continuously varying processes and can be grouped into four genetically significant morphological suites.

INTRODUCTION

In this chapter we outline a glacial process-form model. The model indicates our present understanding of how glaciers erode, transport, and deposit sediment, and how these processes produce glacial landforms. This model is rapidly evolving as different aspects are tested and the resulting modifications fed back into the model. We point out, directly or by implication, parts of the model that we have not tested or cannot test because we lack the needed information; we hope geologists working in other areas will be able to test various aspects of our model.

We write from a midcontinent point of view. Clayton has worked with mountain glaciation in New Zealand and Alaska, but most of his experience has been with the morphology of several counties in North Dakota at scales ranging from 1:20,000 to 1:125,000. Moran has mapped the glacial geology and stratigraphy of several areas in Ohio, Illinois, and Saskatchewan, and most of his recent work has been with glacial geology and lithostratigraphy in North Dakota and northwestern Minnesota. The model we present here was developed because of a need to explain the Late Wisconsinan glacial landforms in an area (1) that has low relief, (2) that is underlain by Cretaceous and Paleocene offshore, shoreline, or fluvial clay, silt, and sand, (3) that is blanketed by 0 to 200 m of glacial sediment with about equal parts sand, silt, and clay, a large amount of which is montmorillonite, (4) where the glacial landforms are fresh and well preserved, (5) where the distribution of the surface sediment is known in some detail (every county in North Dakota has been mapped at a scale of at least 1:125,000), and (6) where the lithostratigraphy of the near-surface glacial sediment is known in considerable detail.

GLACIAL PROCESSES

This section contains discussions of glacial erosion, transport, and deposition plus background discussions of the effects of the movement of heat, groundwater, and ice.

Heat Flow

Weertman (1961) and Boulton (1972) discussed the importance of the thermal regime in determining many of the internal processes of continental glaciers and the resulting deposits and landforms. They suggested that glaciers in high latitudes are frozen to their beds near the margin but have liquid water at their base some distance back from the margin. Recent studies in the Midwest have shown that permafrost existed beyond the Wisconsinan glacier. Wright (1971, p. 324) concluded that permafrost existed in northeastern Minnesota until at least 11,500 B.P. because of the presence of tundra vegetation. Bickley (1972a, 1972b) described extensive fossil ice-wedge polygons (15 to 90 m across) on Early Wisconsinan drift and older surfaces in south-central North Dakota just beyond boundary of Late Wisconsinan drift, but no evidence of permafrost on the Late Wisconsinan drift was seen. On this basis we feel justified in suggesting that the Pleistocene glaciers advanced over terrain containing permafrost; the marginal zones of these glaciers were frozen to their beds.

Before proceeding farther with the glacier model it is necessary to review heat transfer. The transfer of heat by conduction can be described by the equation $Q=kIA$, where Q is the amount of heat, k is the conductive constant, which is an intrinsic property of the materials through which the heat is transferred, I is the thermal gradient, and A is the area (Welty and others, 1969, p. 212-220).

In the following discussion we consider the same area *(A)* in all cases. We assume that the conductive properties of the glacier and bed *(k)* remain constant through the range of debris and ice mixtures considered. The thermal gradient *(I)* is therefore the only variable producing changes in heat transfer. The thermal gradient between two points is a function of the temperature difference *(T)* and the distance *(l)* between the points: $I = T/l$. That is, heat flow through a glacier is controlled by the thickness of the glacier and by the difference between the basal and surface temperatures. For example, where the distance between the two points is increased (the glacier is thickened) and the top and bottom temperatures are constant, the thermal gradient *is decreased.* This causes a decrease in the amount of heat that can be transferred through the glacier. Similarly, where the temperature above the glacier is lowered while the temperature at its base is constant and the thickness of the glacier is constant, the thermal gradient *is increased.* This causes an increase in the amount of heat that can be transferred through the glacier.

Thermal zones. With this basic idea about heat transfer in mind, we now look at the thermal regime of a continental glacier. Both Weertman (1961) and Boulton (1972) used a similar approach in their discussions of thermal regime. We consider three zones (Fig. 1A): (1) the proglacial permafrost zone, (2) the frozen-bed zone, and (3) the thawed-bed zone.

(1) In the proglacieal zone, the cold climate has caused the formation of a layer of permafrost. The cold climate has existed long enough that the thickness of the permafrost layer has reached equilibrium. All of the geothermal heat that is supplied to the base of the layer of permafrost is conducted upward to the atmosphere.

(2) In the frozen-bed zone, the geothermal heat continues to be supplied to the base of the permafrost layer. The presence of overlying glacial ice increases the factor l in the heat-flow equation, causing a decrease in the thermal gradient, a decrease in the amount of heat that can be transferred upward, an increase in the heat remaining at the base, and a decrease in the thickness of the permafrost. The permafrost thins toward the center of the glacier as the thickness of the ice increases; the rate of thinning is irregular because of differences in permeability of the sediment.

(3) In the thawed-bed zone the glacier is thick; the thermal gradient is low, little heat can be transferred upward, and much heat, both geothermal and frictional, is available at the base of the glacier. As a result, permafrost is generally absent, and the excess heat melts the base of the glacier or is transferred toward the margin of the glacier along with the meltwater in the form of subglacial streams or groundwater.

During the initial advance of a major glaciation, such as the Late Wisconsinan advance of the Laurentide ice sheet about 20,000 B.P., all three thermal zones were probably present. The width of the frozen-bed zone was controlled by the climate, the surface profile of the glacier, and the rate of advance: the colder the climate, the wider the zone; the steeper the ice front, the narrower the zone; the more rapid the advance, the wider the zone. It seems likely that the thawed-bed zone extended outward to the margin of the glacier during periods of glacial retreat, but a frozen-bed zone was probably reestablished during temporary readvances, even though little or no permafrost developed in the proglacial zone.

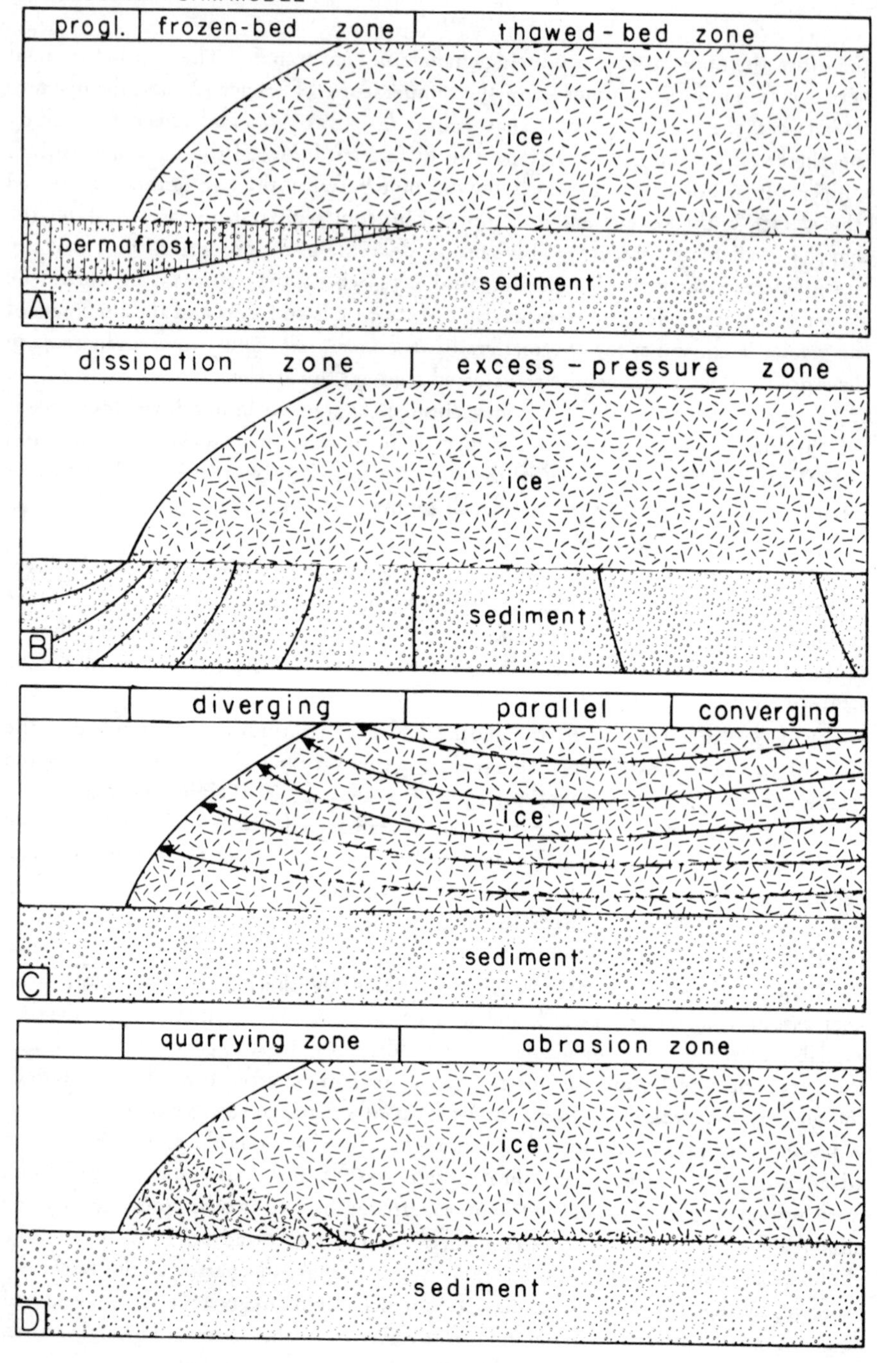

Figure 1.
Thermal zones (A), groundwater-pressure zones showing isopotential lines, with the potential decreasing to the left (B), glacial-flow zones (C), and erosion zones (D).

Groundwater

Groundwater is important in glacial erosion because of its influence on the shear strength of subglacial sediment. Shear strength *(s)* is related to the intergranular pressure (effective stress) (σ): $s = c + \sigma \tan \phi$, where c is cohesion and ϕ is the angle of internal friction (Terzaghi, 1950). Interangular pressure is controlled only by pore pressure *(u)* where the total pressure is constant: $\sigma = p - u$, where p is the total pressure (the thickness of the material over the shear plane times the average, bulk, weight density of the material). That is, shear strength is controlled by the balance between interangular pressure and pore pressure. In any saturated, unconsolidated sediment, the grain-to-grain structure is much more compressible than the pore water. Therefore, any load applied to a saturated sediment is initially transferred entirely to the pore water, causing excess pore-water pressure. This excess pore-water pressure begins to dissipate immediately by causing the intergranular water to flow along the path of least resistance toward an area of lower potential. As the excess pore-water pressure is dissipated the load is transferred from the pore-water to the granular matrix of the sediment in the form of increased intergranular pressure. The shear strength gradually increases as the excess pore-water pressure is dissipated and the load is transferred to the sediment structure. Before loading, $s_0 = c + (p_0 - u_0) \tan \phi$. On applying the load, $s_1 = c + [(p_0 + p_e) - (u_0 + u_e)] \tan \phi$, where p_e is the part of the total overburden pressure due to the weight of the glacier, u_e is the excess pore-water pressure caused by the glacial load, and $p_e = u_e$; it can be seen that $s_0 = s_1$. As the excess pore-water pressure dissipates, part of the weight of the glacier is transferred to the granular matrix of the sediment, and $s_2 = c + [(p_0 + p_e) - (u_0 + u_e - \Delta u)] \tan \phi$, where $\Delta u = \Delta \sigma$. Dissipation of the excess pressure continues until the entire weight of the glacier has been transferred to the sediment structure, and $s_3 = c + [(p_0 + p_e) - u_0] \tan \phi$, $\Delta u = \Delta o = p_e = u_e$. It can be seen that s_3, the final shear strength of the sediment, is greater than s_0; the shear strength of the sediment has been increased.

The rate at which the excess pore-water pressure can be dissipated and transferred to the sediment structure is related to the rate at which groundwater can flow out of the sediment. Where the potential difference causing the water to flow is constant, the permeability of the sediment and the length of the flow path control the rate of groundwater flow. The most rapid dissipation of excess pressure occurs where the potental decreases over a short flow path in coarse-grained sediment, and the least rapid dissipation of excess pressure occurs where the potential decreases over a long flow path through fine-grained sediment.

Differences in compressibility are also important in determining the rate at which the sediment structure responds to an applied load. Coarse-grained sediment is much less compressible than fine-grained sediment, so the intergranular pressure increases sooner in coarse-grained sediment than in fine-grained sediment.

Where drainage outlets are available so that groundwater flow can occur, the combined effect of lower compressibility and higher permeability of coarse-grained sediment result in a much more rapid transfer of an applied load from the pore water to the sediment structure than in fine-grained sediment. As a result, coarse-grained sediment tends to become more rapidly strengthened by glacial overriding than fine-grained sediment.

Where a buried aquifer extends from back under a glacier, and ends near the margin of the glacier, the fine-grained sediment overlying the aquifer can experience extremely high excess pore-water pressure leading to a severe reduction in shear strength. In addition to undissipated excess pressure resulting from the load of the ice over the site, pressure is transferred from farther upglacier through the aquifer. This results in locally intense erosion (discussed below in the section on glacial erosion).

Groundwater pressure zones. We consider two groundwater pressure zones (Fig. 1B): (1) a zone of persistent excess pore-water pressure and (2) a zone of potential dissipation of excess pore-water pressure. The boundary between these two zones is not sharply defined. Its position is controlled by the permeability and availability of drainage outlets. Where the sediment has low permeability or no drainage outlets are available, the boundary between the zones can be close to the glacial margin. Where the sediment is highly permeable and drainage outlets are available, the boundary is much farther behind the margin.

(1) The zone of persistent excess pore-water pressure approximately corresponds to the thawed-bed zone. As the glacier advances, the distance to drainage outlets increases and the length of groundwater flow paths increases. The rate of dissipation of any residual excess pore-water pressure decreases. Under these conditions, fairly high excess pore-water pressure may persist almost indefinitely in fine-grained undrained sediment.

(2) The zone of potential dissipation of excess pore-water pressure corresponds approximately to the frozen-bed zone. Here the flow-path from the sediment beneath the permafrost wedge to drainage outlets beyond the margin of the glacier is short, and in permeable sediment the excess pore-water pressure caused by glacial loading is dissipated rapidly. In less permeable sediment the potential for dissipation is great, but where the glacier advances rapidly, the rate of increase of the flow-path length can quickly exceed the rate of dissipation. As described above, where buried aquifers that terminate near the margin transfer higher pore-water pressure from farther up glacier into this zone, the excess pore-water pressure can become abnormally high.

Glacial Flow

Glacial movement can occur in two ways: by basal sliding or by flow. In incompressible flow, velocity (v) is related to the rate of discharge (Q) and the cross-sectional area (A) between flow lines: $v = Q/A$. If we assume that climate and other factors affecting glacial regime are steady and discharge is constant at a point, changes in velocity along a flow line are related only to changes in distance between flow lines (Albertson and others, 1960, p. 86-101). That is, in a glacier with steady flow, deceleration along a flow line occurs where the flow lines diverge, uniform flow occurs where they are parallel, and acceleration occurs where they converge. The distinction between diverging and converging flow is important, because sediment can be carried up into the glacier only in areas of diverging flow, where ice at the bed is forced over more slowly moving ice farther down glacier.

The location of the diverging-flow zone, the parallel-flow zone, and the converging-flow zone is related to several factors (Fig. 1C). In the accumulation area of a glacier with uniform width and flat bed, the flow must change from downward to horizontal, resulting in converging flow; in the wastage area the flow must change from horizontal to upward, resulting in diverging flow. Where the bed is not flat, converging flow occurs where the bed steepens downstream, and diverging flow occurs where the bed flattens or reverses in gradient. Where the glacier is not of uniform width, converging flow occurs where the channel narrows downstream, and diverging flow occurs where the channel widens.

Glacial Erosion

Glacial erosion has traditionally been considered to occur by one of two processes: (1) abrasion and (2) quarrying. On the basis of our understanding of the thermal regime, the groundwater regime, and the glacial-flow regime, we suggest that glacial erosion is basically different under the central part of the glacier--the abrasion zone--and under the margin of the glacier--the quarrying zone (Fig. 1D).

Abrasion zone. The abrasion zone is in the central or submarginal part of a glacier. It corresponds to the thawed-bed zone, the zone of excess groundwater pressure, and the zone where the glacier flow lines are converging or parallel. Because the base of the glacier is not frozen to its bed in most places, basal sliding occurs (Fig. 2A), and abrasion by debris in the base of the glacier is the principal mode of erosion. Quarrying also occurs in the abrasion zone wherever irregularities in the bed are high enough and have discontinuities oriented in the right direction and where the glacier is frozen to its bed. The bed is frozen here as a result of refreezing because of pressure differences around knobs or because of the persistence of isolated patches of permafrost. This quarrying produces the debris that serves as tools to abrade the bed in the abrasion zone.

As described above, the permeability, compressibility, and degree of drainage of the subglacial sediment control the amount of abrasion that occurs. Where the bed has low permeability, is highly compressible (fine-grained sediment), and has no drainage outlet, the subglacial water supports much of the weight of the glacier; the excess pore-water pressure is high. Therefore the intergranular pressure (effective stress) in the subglacial sediment is low and the shear strength remains low. Where the bed is flat, little force is available to cause abrasion, and little erosion takes place. Where the bed has irregularities, they may bear most of any existing intergranular pressure and undergo considerable erosion, which tends to flatten the bed. The intergranular pressure under glaciers was probably commonly low, because Harrison (1958) and others have shown that subglacial sediment is commonly much less compacted (less consolidated) than it would be if it were freely drained and the full weight of the glacier rested on the sediment. Where the bed is highly permeable and is only slightly compressible (coarse-grained sediment) or where a drainage outlet is available, the excess pore-water pressure is dissipated. Therefore the weight of the glacier is supported by the sediment, producing high intergranular pressure and high shear strength. A large amount of force is available for abrasion, and considerable erosion takes place.

Quarrying zone. The quarrying zone is generally in themarginal zone of a glacier. It corresponds to the frozen-bed zone, the zone of dissipation of groundwater pressure, and the zone where the glacial flow is diverging (Fig. 1). The glacier is frozen to the subglacial sediment, so blocks of sediment can be quarried from the bed, depending on the orientation of discontinuities in the sediment. The quarried masses tend to be moved up into the glacier because of the diverging flow: the ice and its load move upward over slower moving ice farther downglacier.

The excess groundwater pressure determines the size of the quarried masses. Where the excess pressure is dissipated by drainage of groundwater from beneath the glacier, erosion is confined to piece-meal incorporation of small masses at the ice-sediment interface (a in Fig. 2B). Where drainage of groundwater is reduced

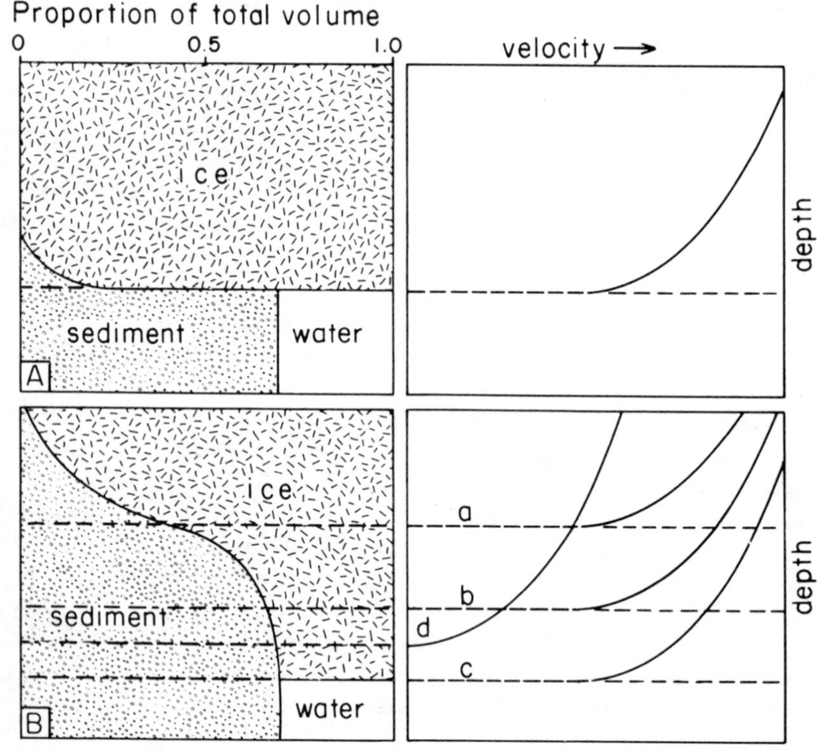

Figure 2.
Material distribution (left) and velocity profile (right) in abrasion zone (A) and quarrying zone (B). The horizontal dashed lines are the depth to the "bottom of the glacier" at times a, b, c, or d.

either by decreased permeability of the bed or by the absence of drainage outlets beyond the limit of the glacier, the dissipation of excess pore-water pressure is decreased. The strength of the sediment beneath the marginal zone of the glacier is correspondingly reduced in response to the resulting increase in excess pore-water pressure. The lower the strength of the sediment beneath the glacier, the greater the tendency to incorporate larger blocks of the bed (b and c in Fig. 2B). Where a buried aquifer that extends from back under the glacier to the marginal area pinches out beneath the frozen-bed zone, very high excess pore-water pressure can occur, and very large blocks of the bed can be incorporated into the glacier (c in Fig. 2B).

It is likely that the greatest amount of erosion took place at the upglacier edge of the quarrying zone, because there the velocity of the ice was greatest and the thin edge of the permafrost wedge could be most easily sheared off.

In the abrasion zone, the base of the glacier is at the sharp contact between ice and sediment (Fig. 2A). However, in the quarrying zone, the amount of sediment in the ice gradually increases downward. The shear plane marking the "base of the glacier" may move up and down with time (a to c in Fig. 2B). It is at "a" much of the time, at "b" part of the time, and sometimes as deep as "c." In some situations the glacier may move entirely by flowing, with no basal shear plane (d in Fig. 2B); in this case, the "base of the glacier" is the point in the sediment-ice column where movement stops.

To summarize our ideas about glacial erosion, there are two erosion zones in a glacier: the abrasion zone and the quarrying zone. In each of these zones, the nature and degree of glacial erosion is determined by the temperature at the base of the glacier, the permeability and compressibility of the subglacial sediment, and the direction of the vertical component of ice flow. Where the bed has low permeability and high compressibility and lacks a drainage outlet in the abrasion zone, much of the weight of the glacier rests on the pore water, and little erosion takes place; where the bed is highly permeable and only slightly compressible and has drainage outlets, much of the weight of the glacier rests on the subglacial sediment, and considerable abrasion occurs. In the quarrying zone, especially at its up-glacier edge, small blocks of sediment are frozen onto the sole of the glacier and moved upward over the slower moving down-glacier ice; where drainage is blocked, causing excess pore-water pressure, large masses of sediment can be moved up into the ice.

Glacial Transportation

In the abrasion zone (Fig. 1D), only a small amount of sediment is eroded by the glacier. The abrasion zone is characterized by nonturbulent flow with flow lines descending toward the bed or parallel to the bed; therefore sediment can not be carried up into the glacier, but is transported only at the very base. The small amount of sediment that is present tends to be deposited as the basal ice melts.

However, where there are irregularities on the bed, sediment may be carried higher into the glacier. The higher and more abrupt the knob, the higher the sediment will be lifted up into the glacier. Descending flow lines tend to return the sediment to the bed downice from the irregularity. So the sediment supplied to the glacier in the abrasion zone moves up and down while passing irregularities in the bed, but it generally stays near the base of the glacier (Fig. 1D and 2A).

In the quarrying zone, especially at its upglacier edge, a much larger amount of sediment is supplied to the glacier. The quarrying zone is characterized by decelerating flow, and as a result flow lines diverge upward away from the bed, lifting the material supplied from the abrasion zone plus the material originating in the quarrying zone higher and higher up into the ice as the terminus of the glacier is approached (Fig. 1D and 2B).

Glacier Deposition

In the following discussion the word "deposition" is used to indicate that the material has come to rest and the intergranular ice and the underlying ice have melted. That is, sediment may be under a glacier and have stopped moving: but if it contains intergranular ice, it has not been deposited.

Sediment can be deposited from a glacier in four different ways (Boulton, 1971). It may melt out from actively moving ice. It may melt out from beneath inactive ice. It may melt out from the top of the glacier and then undergo mass movement before it comes to rest on the solid ground as the underlying ice melts out. And it may melt out from the top of the glacier and then be let straight down without mass movement when the underlying ice melts.

Subglacial deposition from active ice. Subglacial deposition from actively moving ice must be restricted to the thawed-bed zone; as the base of the glacier thaws, any enclosed sediment is deposited. Because only small amounts of englacial debris are available in the thawed-bed zone, glacial sediment of all kinds deposited there is generally thin (typically 1 m or less in North Dakota), and probably only a small amount of it was deposited directly from active ice. The sediment of the last advance in most areas is much less compacted than sediment of earlier advances, suggesting that the sediment of the last advance never had the full weight of the glacier on it after it became entirely free of intergranular ice.

Subglacial deposition from inactive ice. Most subglacial deposition probably involves the melting out of sediment from inactive ice. The ice may be inactive because the glacier has stagnated, because a slice of basal ice has been trapped in an irregularity in the bed of an active glacier, or because a thin basal zone of an active glacier has become so choked with sediment that movement is no longer possible. In any case, glacial sediment deposited beneath inactive ice should inherit the fabric of the englacial sediment, but with decreased plunge angles as a result of compaction during removal of the intergranular ice (Harrison, 1957).

Superglacial deposition with mass movement. Englacial sediment is high in the ice in the diverging-flow zone, so superglacial sediment rapidly accumulates on the marginal zone of a glacier as the surface melts downward. As the superglacial sediment accumulates, it controls the rate of melting. Where it is less than a few millimeters thick it speeds up melting by absorbing solar heat. Where it is thicker than a few millimeters it insulates the ice. Where the superglacial sediment is irregular in thickness, insulation is irregular, and melting is irregular, producing an irregular topography on the ice. Where meltwater is produced faster than it can drain away, the superglacial sediment becomes waterlogged. This promotes mass movement of the superglacial sediment down the surface irregularities and destroys any fabric that the superglacial sediment may have inherited from the englacial sediment. After a large amount of superglacial mass movement, the last ice melts out, letting the superglacial mudflows down onto solid ground.

Superglacial deposition without mass movement. Where the superglacial sediment is coarse enough or where it is thick enough that the underlying ice melts slowly, all of the meltwater produced will be able to drain away through the superglacial sediment fast enough that it is no longer waterlogged. As a result, mass movement stops, and a layer of melt-out sediment accumulates above the ice and below any mudflow sediment. Like subglacial sediment deposited from inactive ice, it retains the fabric it inherited from the englacial sediment.

Recognition of mode of deposition. Superglacial mudflows play a basic role in our geomorphic model. We find no way of explaining most glacial topography without them. Although we find little evidence for other modes of glacial deposition, we do not rule them out. Boulton (1971) found no conclusive way to differentiate the four kinds of sediment once they have been deposited from modern Spitzbergen glaciers. This agrees with our observations on glacial sediment in North America: glacial sediment deposited during the last advance generally lacks any obvious subdivision. If subglacial sediment was deposited in North Dakota, it is not noticeably different from the superglacial sediment, contrary to the popular opinion that superglacial sediment is coarser grained.

GLACIAL GEOMORPHOLOGY

With a glacial-process model in mind, we may now evaluate the factors that determine the final form of a glacial landscape. If we ignore the effects of fluvial and lacustrine processes, glacial landscapes can be considered to be made up of four superimposed elements. (1) The preadvance element involves the landscape existing before the most recent glacial advance. (2) The subglacial element involves subglacial erosion or deposition on the preadvance surface. (3) The superglacial element involves variations in thickness of collapsed superglacial sediment. (4) The postglacial element involves landforms cut into or built up on the glacial surfaces by postglacial erosion or deposition.

Any glaciated landscape contains these four elements. The differences among various glacial landscapes are in part the result of the degree to which each element has eradicated or masked the preceding elements. First, the degree to which the preadvance element can be seen in the final landscape depends on the amount of subglacial erosion and deposition. The degree to which the subglacial element and the preadvance element can be seen in the final landscape depends next on the amount of masking by superglacial deposition. Finally, the degree to which all three of these elements can be seen in the final landscape is determined by the amount of postglacial erosion or deposition.

Preadvance Element

White (this volume) shows that the effects of the preadvance landscape have been commonly overlooked in the past. The landscape in most glaciated parts of North Dakota partially retains at least the large-scale elements of the preadvance landscape; the Red River valley, for example, has persisted as a valley all through late Cenozoic time. In many areas, smaller landforms, like meltwater channels, are still clearly visible, even though they have been glaciated at least once. In some areas, even small-scale features, such as fluvial terraces or morainic hummocks, were not obliterated by the last glaciation. Where the preadvance element is conspicuous, it is an indication that subglacial smoothing has not been severe and glacial sediment is not thick.

Subglacial Element

If a glacier is eroding at all, the preadvance element is gradually destroyed and replaced by either of two subglacial landforms: *longitudinal shear marks* or *transverse compressional features.*

Longitudinal shear marks. Longitudinal shear marks are the most widespread of the two subglacial landforms. They consist of ridges and troughs elongated and streamlined parallel to the direction of ice movement. They range in length from a few millimeters to many kilometers. Their shape ranges from short, squat forms with length-to-width ratios of less than 2:1 to long, narrow forms with length-to-width ratios of more than 1000:1 (Gravenor, 1953).

Longitudinal shear marks, in general, are given a variety of names, depending on the setting. On fault planes they are called "slickensides" or "fault-plane grooves." Those formed on a waxed floor beneath a sliding box may be called "scratches" or "drag marks." Those formed by a knife on peanut butter may be called "smear marks." Glacial shear marks also have a variety of names, depending on size, shape, and composition, including "drumlins," "rock drumlins," "roches moutonnees," "crag and tail," "flutings," "fluted ground moraine," "glacial grooves," and "striations." We believe these all result from the same process, subglacial shearing. The details of this process are discussed by Muller (this volume); we will mention only a few of the more obvious aspects here.

The size of shear marks is related to the size of the inhomogenities in the shearing material (small serrations on a knife produce small smear marks, small chunks of peanuts in the peanut butter result in small smear marks) and to the size of preexisting irregularities (raisins in peanut-buttered muffins might result in raisin-sized smear marks).

The shape of shear marks may be related to the duration and intensity of shearing. The longer ones may be in more easily eroded material or may have been sheared for greater lengths of time. Shape may also be influenced by the homogeneity of the material (smear marks in crunchy peanut butter have a lower length-to-width ratio than in creamy peanut butter).

Field evidence from North Dakota supports the conclusion that erosion rather than deposition is dominant. In every place where evidence is available, the bulk of large subglacial shear marks is composed of preexisting highly permeable or slightly compressible sediment, either sand and gravel, as in McHenry Country (Fig. 3; Moran, in preparation) and Rolette County (Deal, 1972, p. 63), or highly fractured siliceous shale of Cretaceous age, in Cavalier County (Arndt, in press). In McHenry County, truncation of dune cross bedding in sand by the veneer of glacial sediment suggests that the sand, which forms the core of the shear marks, is a remnant of a once more extensive outwash plain. Large shear marks are most generally composed of highly permeable and slightly compressible sediment, probably because this favored dissipation of excess pore pressure and the increase of intergranular pressure. This provided a large amount of force for abrasion, as on tectonic thrust planes (glacial sediment in the abrasion zone is analogous to fault gouge).

In every longitudinal shear mark that we have looked at, the glacial sediment of the last advance is so thin that it contributes little or nothing to the volume of the landform. Where the surface layer is thicker, it masks the shear marks; landforms composed of let-down glacial sediment are draped over the shear marks. Where the surface layer of glacial sediment is thicker than the height of the shear marks, they are completely buried and can be identified only by subsurface stratigraphic studies. We suspect that drumlins are sometimes considered to be molded from the glacial sediment of the last advance rather than eroded from preadvance sediment because exposures are not deep enough to expose the nonglacial core of the drumlin or the glacial stratigraphy has not been worked out in enough detail and the thinness of the surface layer of glacial sediment is not recognized (see White, this volume).

Whatever the details of their origin, we believe that "drumlins" and related features are longitudinal shear marks eroded anywhere the subglacial material was unfrozen so the glacier could slide over its bed. They form in the thawed-bed zone (abrasion zone) some distance back from the terminus, behind the frozen-bed zone (quarrying zone) (Fig. 1). They are conspicuous in these areas not only because they form there, but also because the glacial sediment of the last advance is thin (the material eroded to produce the shear marks has been moved farther downglacier), and they have not been buried.

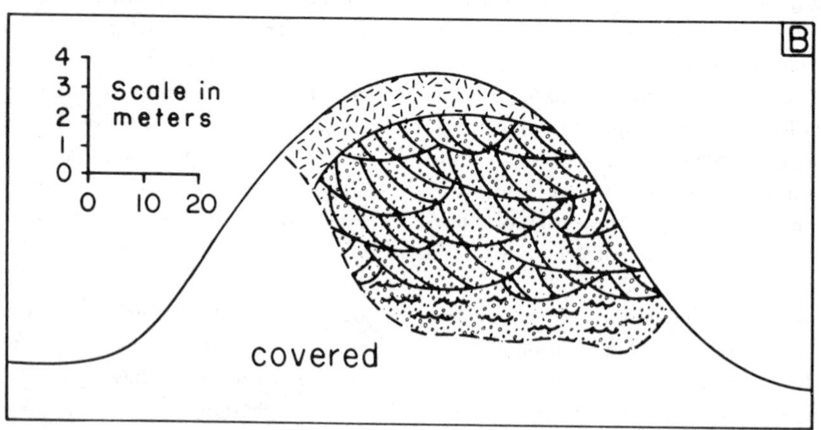

Figure 3.
Longitudinal shear marks. A. Air photo, southern McHenry County, N. D. (U.S.D.A. BAI-7BB-190 and 192; area shown is 4.1 km long). B. Cross section through a shear ridge showing glacial sediment over dune cross-bedded fluvial sand over ripple cross-bedded fluvial sand (0.4 km southeast of Verendrye, southern McHenry County, N. D.).

Transverse compressional features. Nearer the terminus, the glacier was frozen to its bed, and a different kind of subglacial feature, compressional folds and thrusts in the subglacial sediment, resulted. Vertical displacement is typically tens of meters, and the individual folds or thurst masses are commonly 200 m across. The folds are generally overturned; their axial plane and the thrust faults dip upglacier 30° to 60°. The strike is parallel to the ice margin. In map view these forms are concave up glacier, typically with a radius of curvature of 5 km in North Dakota (Fig. 4).

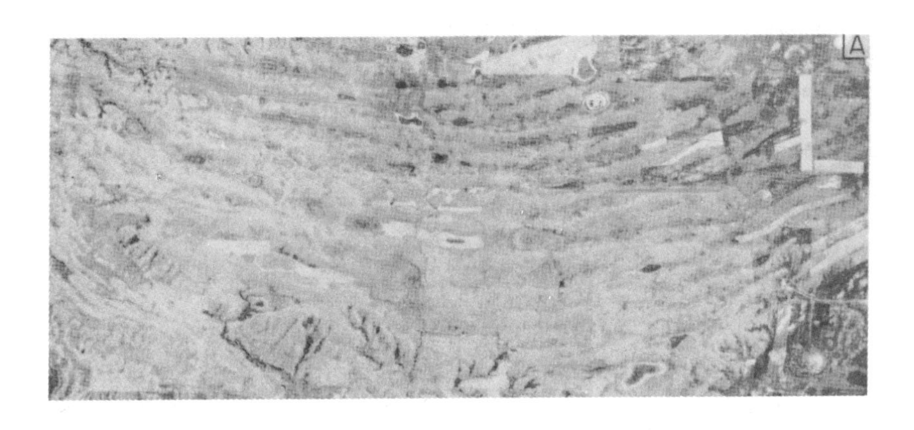

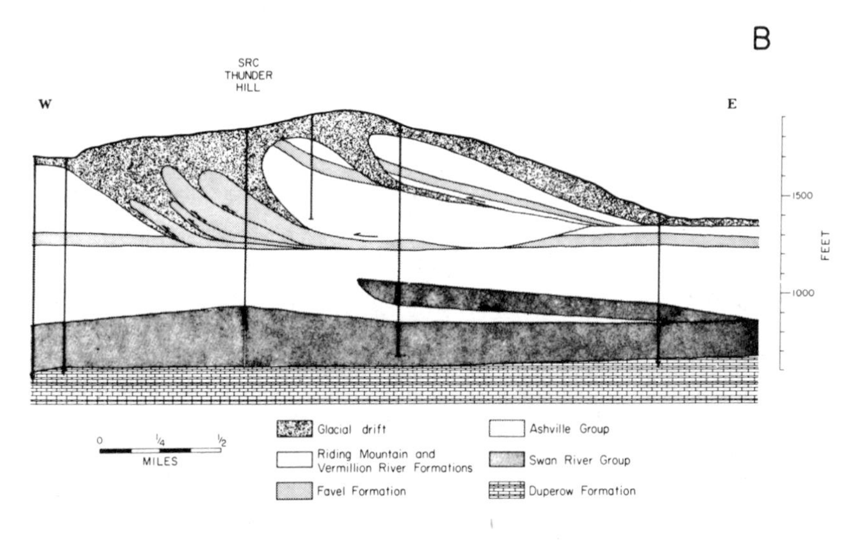

Figure 4.
Transverse compressional features. A. Air photo of Prophets Mountains, Sheridan County, N. D. (U.S.D.A. BAN-2AA-154 and BAN-3AA-8; area shown is 6.4 km long). B. Cross section of Thunder Hill, eastern Sask. (from Moran, 1969, Figure 17).

In some places the thrust masses are more or less equidimensional and untilted, forming prominent hills as much as 5 km across and 50 m high. Upglacier from the hill is typically a depression of the same size and shape as the hill; the thrust mass came out of the depression (Bluemle, 1970). In the past these hills have commonly been misidentified as "kames" or in-place outliers of bedrock, depending on their composition. These hills are especially common in the "Martin Moraine" of Pierce, Sheridan, and McHenry Counties and just south of Devils Lake and Stump Lake in Nelson, Ramsey, and Benson Counties, North Dakota.

Many fold and thrust masses are composed of fluvial gravel deposited before the last glaciation. Some are composed of pre-Pleistocene rock. Sibley Buttes, in Kidder County, are made of thrust Cretaceous sandstone (the beds retain their continuity in a transverse direction for 10 km). Antelope Hills in Pierce County, are composed of folded and thrust sand and shale of Cretaceous and Paleocene age. Where the thrust masses are composed of unbedded material, like preexisting glacial sediment, they can generally be recognized only by their surface form; the crest of each fold or thrust mass corresponds to a ridge.

Many are buried under a thick layer of glacial sediment. They can be recognized only by subsurface studies. It is a common occurrence in parts of North Dakota and Saskatchewan to drill through masses of bedrock 50 m thick and then back into glacial sediment.

Compressional folds and thrusts form in the marginal part of a glacier, where the glacier is decelerating, it is frozen to its bed, and there is excess groundwater pressure. Deceleration is necessary to cause the compression and to cause upward shearing of the material. A frozen bed is necessary so that the subglacial material can be incorporated into the glacier; the effective base of the glacier, as a flowing mass, is beneath the ice-sediment interface. Excess groundwater pressure is needed to reduce the shear strength of the rock, helping the glacier to move the sediment (Moran, 1971; Bluemle, 1970). Compressional features are most commonly found in North Dakota and Saskatchewan associated with buried valley-fill deposits because of the extremely high excess pore-water pressure that developed over buried aquifers. They also occur where the glacier advanced against an upland under which the groundwater flow system hindered the dissipation of the excess pore-water pressure.

Superglacial Element

Glacial sediment of the last advance was not an important part of the subglacial morphological element discussed in the previous section. Subglacial landforms are composed largely of sediment deposited before the last glacial advance. Where the surface layer of glacial sediment is 1 m or less thick, it only slightly modifies the preadvance or subglacial elements. Where the glacial sediment becomes thicker, it begins to mask the underlying landforms. Where the thickness of the surface layer of glacial sediment is greater than the preadvance or subglacial relief, it completely obliterates the underlying landforms. In this section we move on to those aspects of glacial geomorphology that are directly related to variations in the thickness of the glacial sediment.

There are two significant ways that this variation in thickness might come about. (1) The original distribution of the englacial sediment may be nonuniform. (2) Uniformly distributed englacial sediment may be redistributed and deposited nonuniformly by superglacial mass movement.

Englacial nonuniformity. The landform that is commonly called "washboard moraine" or "minor moraines" in the midcontinent area (Gravenor and Kupsch, 1959) is the only widespread landform whose detailed morphology need be explained in terms of the nonuniformity of englacial sediment. Washboard moraine is a series of transverse ridges and trenches spaced about 200 m apart and with local relief of 1 to 3 m. The ridges and trenches are generally gently curved, with a radius of curvature of 10 to 20 km, concave upglacier (Fig. 5). These features apparently form as the result of greater concentrations of glacial sediment along periodically spaced transverse zones of shearing.

They occur in glacial sediment that is 1 to 3 m thick. Where the glacial sediment is thinner, the relief is generally too low to produce a distinctive landform; instead any preadvance or subglacial landform shows through. As the glacial sediment becomes thicker, subglacial landforms become buried and the washboard moraine may become conspicuous.

In many places where the glacial sediment is 1 to 3 m thick and in most places where it is greater than 3 m, transverse lineations are lacking, for one of three possible reasons. (1) Upward sloping shear planes are absent where the glacial flow lines are parallel; this is most likely at the up-glacier edge of the quarrying zone. (2) Where the englacial sediment is concentrated near the base of the glacier, it has little ice under it, so it undergoes little disruption when the ice melts. The shear-plane masses are let down intact to form washboard moraine. However, if the same amount of englacial sediment is lifted higher in the glacier, it undergoes repeated mass movement, resulting in disruption of the englacial sediment when the underlying ice melts, preventing washboard moraine from forming. (3) The transverse shear zones may be close enough together to produce a uniform blanket of superglacial sediment when the englacial sediment in the overlapping shear zones melts out.

Superglacial nonuniformity. Even if the englacial sediment arrives uniformly at the surface of the glacier, it soon becomes nonuniformly distributed. As described above in the section on glacial deposition, any slight differences in the insulating ability of superglacial sediment will cause differential melting, which results in topographic irregularities on the ice, which promotes mass movement and a nonuniform distribution of the superglacial sediment.

As this process continues, the surface irregularities undergo inversion of topography (Fig. 6). Ice that has thinner superglacial sediment, and so is less well

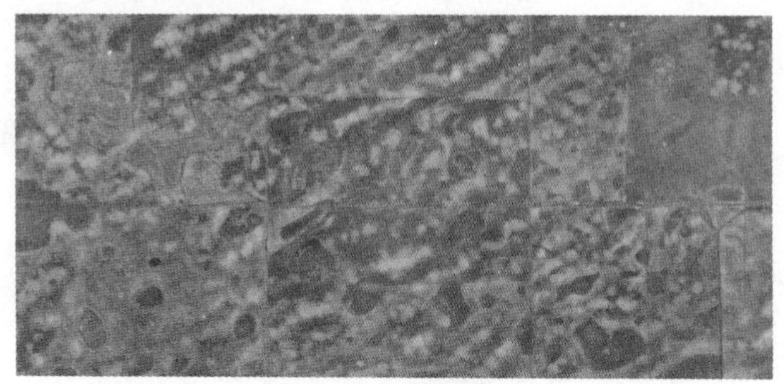

Figure 5.
Air photo of washboard moraine in Nelson County, N. D. (U.S.D.A. CWM-3W-108).

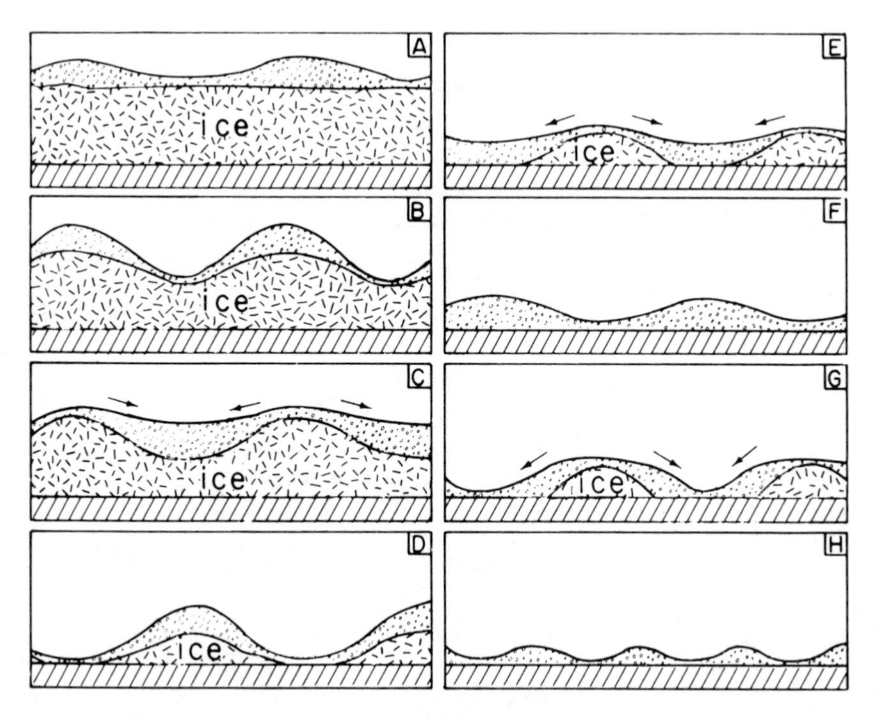

Figure 6.
Glacial sinkholes and the formation of hummocks. A. Nonuniformity of superglacial sediment. B. Nonuniformity of insulation and melting. C. Mass movement. B, C, D. Inversion of topography. E, F. Formation of hummocks. G, H. Formation of circular disintegration ridges.

insulated, melts faster to form depressions. Areas of thicker superglacial sediment are higher, and mudflows occur when the slopes become steep enough. These mudflows tend to fill up the depressions, which then have thicker superglacial sediment than the high areas; the ice under them melts more slowly, and they then become the high areas. The topography has become inverted.

This superglacial topography is similar to solution topography on limestone. The sinkholes are circular, closely spaced, have a maximum diameter of about 200 m, and have a maximum depth of roughly 60 m on modern glaciers (Fig. 7A) (Clayton, 1964). Their depth is limited by the minumum depth at which ice becomes flowable enough to start to close the bottom of the sinkhole. The slope of the sides of fully developed sinkholes is maintained at an angle of about 30° on modern glaciers by a balance between two factors. (1) Mass movement results from the steepening effect of thick sediment and slow melting on the tops of the slopes and thin sediment and fast melting on the bottoms of the slopes (where waves in ponds in the sinkholes tend to wash sediment from the base of the slopes). (2) Mass movement in turn causes a flattening effect as the result of thin sediment and fast melting at the tops of slopes and thick sediment and slow melting at the bottom of slopes. Combining the maximum depth of fully developed sinkholes with the consistent angle of slope on their sides gives a diameter of about 200 m, which is the same as that of fully developed sinkholes on modern glaciers.

Collapse topography. As the inversion of topography on the glacier approaches the final phase of its last cycle, the bottoms of the sinkholes melt through to solid ground, resulting in circular holes in the glacier. Material flowing down the sides of these holes may completely fill in the hole, resulting in a hill of material occupying the position of the former sinkhole when all of the ice finally melted (Fig. 6F); these hills are commonly called "hummocks" and the resulting topography (Fig. 7B) is commonly called "moraine" (or "hummocky moraine," "stagnation moraine," "dead-ice moraine," "end moraine," or "ground moraine"). If the amount of material flowing into the hole is not enough to completely fill it (Fig. 6G), the material forms a doughnut-shaped ridge at the base of the sides of the hole (Fig. 6H and 7C); these ridges are called "circular disintegration ridges" (Gravenor and Kupsch, 1959) or "doughnuts." If, in the final stages of topographic inversion, thick deposits of material in the bottom of sinkholes cause them to invert into ice-cored cones, the material may flow down the sides of the cones, producing, when all of the ice has melted, doughnut-shaped ridges, also called "circular disintegration ridges." (Any ridges formed by material moving down ice slopes and collecting at their base are called "disintegration ridges." They may be any shape, from circular to straight, depending on the shape of the ice slope; Parizek, 1969).

The maximum diameter of primary hummocks in collapse topography is about 200 m (Fig. 7B). The diameter of circular disintegration ridges is about 200 m (Fig. 7C). This corresponds with the diameter of primary sinkholes on the Martin River Glacier in Alaska (Fig. 7A), which tends to confirm our interpretation that hummocks and circular disintegration ridges form in sinkholes and that "morainic topography" is the result of the collapse of superglacial sediment.

Differences in the steepness of hillslopes in collapse topography are primarily the result of differences in the viscosity of superglacial sediment. If we restrict our attention to Late Wisconsinan glacial sediment in central North America, which all has about the same grain size and clay mineralogy, the variable that produced the largest differences in flowability was water content. Water content is related to the rate of ice melting, which is related primarily to the thickness of the insulating cover of superglacial sediment. That is, thicker superglacial sediment has less water in it, is more viscous, and produces hummocks with steeper sides. As a result, local relief in collapse topography is related to the thickness of superglacial sediment, because, if the diameter of primary hummocks stays constant at 200 m, the local relief is only a function of slope angle.

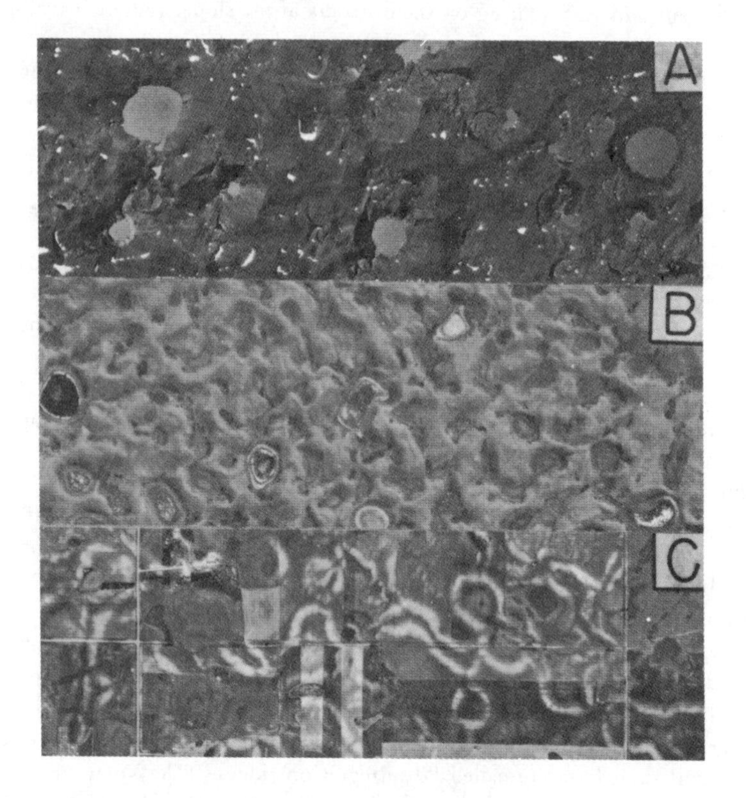

Figure 7.
A. Glacial sinkholes, Martin River Glacier, south-central Alaska (U.S.F.S. EEV-15-82). B. High-relief collapsed glacial sediment with round hummocks, Mountrail County, N. D. (U.S.D.A. BAL-6V-90). C. Medium-relief collapsed glacial sediment with circular disintegration ridges, Mountrail County, N. D. (U.S.D.A. BAL-4V-108). Each area shown is 2.4 km long.

Slope angles and local relief are easy enough to measure, but the thickness of the superglacial sediment can no longer be directly measured because we know of no easy way to distinguish superglacial sediment from subglacial sediment. However, the total thickness of subglacial sediment and superglacial sediment can be measured in most outcrops or drill holes because the glacial sediment deposited as a result of the last glaciation is distinctive in much of North Dakota (Ulmer and Sackreiter, 1973; Salomon, in press; Harris, 1973; Harris, Moran, and Clayton, 1974; Bickley, 1972a). We have suggested above, however, that subglacial sediment is negligible and most of the sediment deposited during the last advance is made up of superglacial sediment. If this is true, the total thickness of the glacial sediment of any advance should correlate closely with the thickness of superglacial sediment and with the slope angles or local relief. A few unsystematically collected observations in North Dakota suggest that this predicted correlation is accurate (Fig. 8A).

Postglacial Element

A final set of factors that influences glacial morphology includes postglacial processes. All glacial morphology has been at least slightly altered by creep or slopewash on the tops and sides of hummocks and by deposition of the eroded material in the intervening depressions. Dissection may also completely remove the glacial topography. In about 10% of the part of North Dakota that was glaciated in Late Wisconsinan time, postglacial fluvial or lacustrine erosion or deposition has completely eliminated the Late Wisconsinan glacial morphology. In the part of North Dakota that was glaciated in Early Wisconsinan time or earlier (Clayton, 1966), about 90% of the glacial morphology has been completely eroded away.

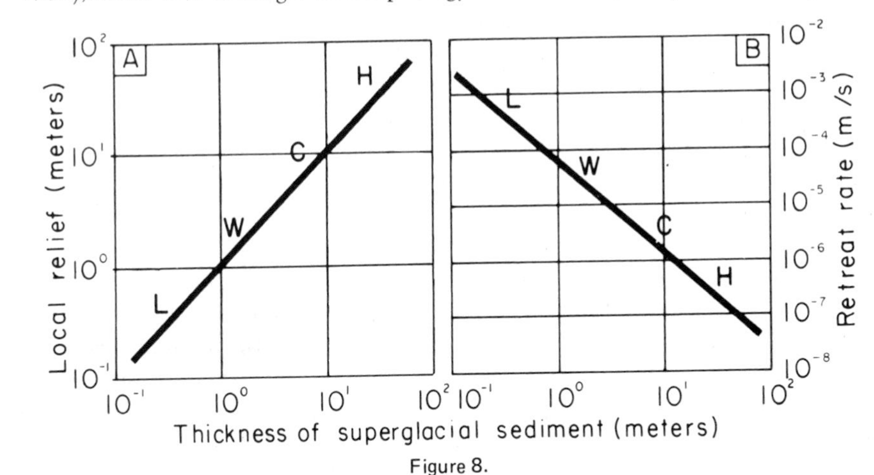

Figure 8.

A. Predicted relationship between thickness of glacial sediment and local relief of collapse hummocks. B. Predicted relationship between the rate of retreat of the active-ice terminus of the Late Wisconsinan glacier in the Upper Midwest and the thickness of the sediment deposited by the glacier. Typical location of longitudinal shear marks (L), washboard moraine (W), circular disintegration ridges (C), and round hummocks (H) are shown on the plots.

Regional Aspects

The previous sections dealt with the local aspects of glacial morphology that affect the form of individual hills and depressions. In this section we will discuss broader aspects that determine the distribution of landforms over areas the size of a county or a state. The individual landforms can be grouped together into associations or suites: the fringe suite, the marginal suite, transitional suite, and the inner suite (Fig. 9).

Fringe suite. The outermost fringe of a layer of glacial sediment is commonly thin (1 to 4 m). Relief is low (1 to 4 m) and slopes are gentle (1° to 4°); the landscape is undulating. Meltwater channels and outwash plains are common; in many places this suite has been completely washed away by meltwater. In many places this suite is absent because the glacier never advanced beyond the position of the marginal suite. In North Dakota this suite is widest where the land in front of the glacier sloped away from the ice and narrowest where it sloped toward the ice. The thinness of the glacial sediment in this zone may be the result of the small amount of time the glacier stood at this position.

Marginal suite. Here the glacial sediment is thicker than in any of the other suites, generally 4 to 30 m. Local relief is high (4 to 30 m) and slopes are steep (4° to 15°). This suite is characterized by rolling to hilly collapse topography. Circular disintegration ridges may be abundant. Meltwater channels, eskers, washboard moraine, and longitudinal subglacial marks are commonly absent. Transverse shear masses most commonly occur in this zone, generally over buried aquifers. Collapsed superglacial fluvial sediment and ice-walled-lake plains are common. Where this suite is narrow, it is sometimes called an "end moraine." The marginal suite occurs where the glacial sediment is thickest, so it occurs just downglacier from the site of greatest erosion. That is, this suite corresponds to the mean position of the quarrying zone, which roughly coincided with the frozen-bed zone, the zone of dissipation of excess pore-water pressure, and the zone where glacial flow lines diverged.

Transitional suite. Behind the marginal suite, the glacial sediment becomes gradually thinner, about 4 m to 1 m, up glacier. Relief is low (4 to 1 m) and slopes are gentle (4° to 1°); the topography is undulating. Circular disintegration ridges are characteristic of the downglacier part of this suite, and washboard moraine is characteristic of the upglacier part of it. Eskers are most abundant in this suite. Partly buried longitudinal subglacial shear marks may be present in the up-glacier part of this suite.

This suite probably corresponds to the zone of fluctuation of the up-glacier edge of the quarrying zone. Glacial erosion may have been great at times, but much of the eroded sediment was transported down-glacier to the marginal zone. The

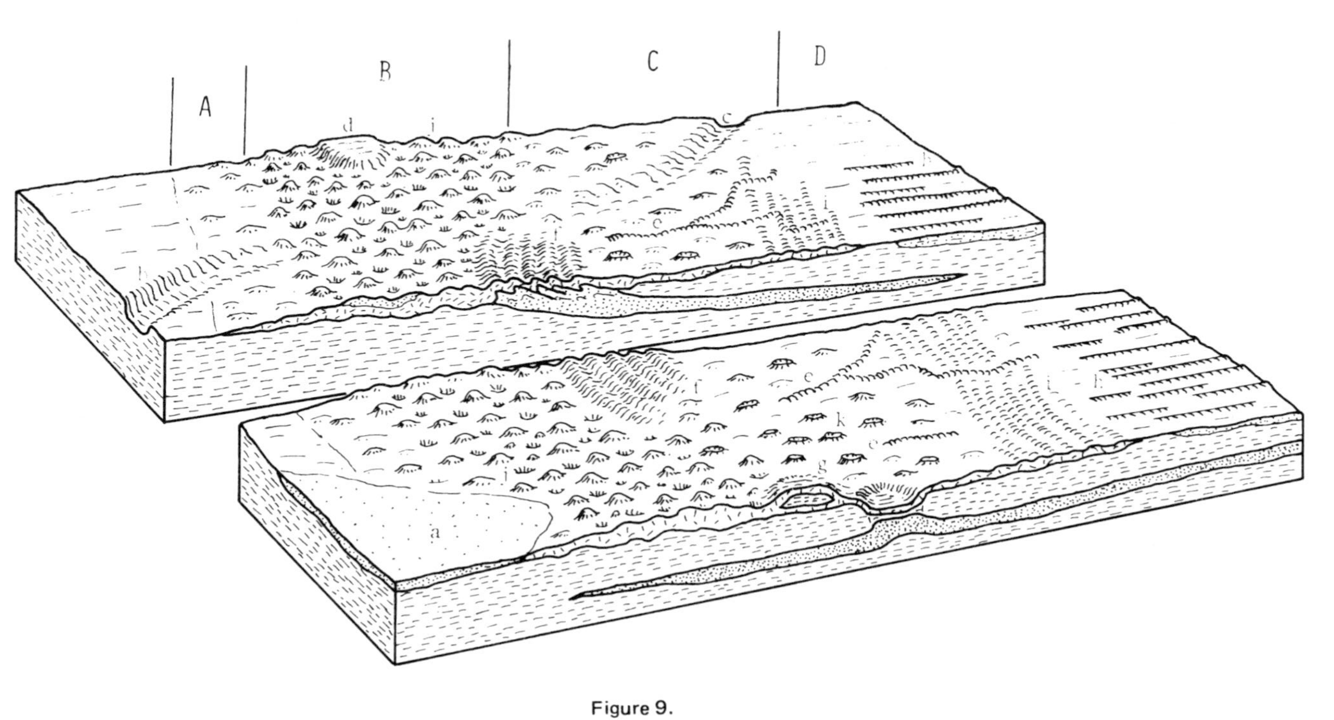

Figure 9.

A typical sequence of glacial suites. A. Fringe suite. B. Marginal suite. C. Transitional suite. D. Inner suite. a: fluvial plain. b: meltwater channel. c: partly buried meltwater channel. d: ice-walled-lake plain. e: eskers. f: transverse compressional features. g: equidimensional thrust mass. h: longitudinal shear marks. i: washboard moraine. j: simple hummocks. k: circular disintegration ridges. The area shown is perhaps 50 km wide.

longitudinal shear marks in the up-glacier part of this suite formed where the abrasion zone entered this area.

Inner suite. The inner suite is characterized by thin glacial sediment; the sediment of the last advance is commonly less than 1 m thick. As a result, primary depositional glacial relief is nearly absent. The topography is of two types. (1) Longitudinal shear marks occur where abrasive erosion was intense because of high intergranular pressure at the bed and low pore-water pressure resulting from coarse subglacial sediment or a drainage outlet. (2) Slightly modified landforms originating before the last advance occur where abrasive erosion was negligible because of low intergranular pressures at the bed and high pore-water pressure resulting from fine-grained subglacial sediment or the absence of a drainage outlet.

Sequences of suites. Generally these four suites occur in a definite sequence of parallel zones, as implied by their names (Fig. 9). In some places individual suites may occupy only narrow zones or they may be missing from the sequence. In other places the suites may be intermixed in a complex pattern as a result of local complications in the subglacial topography or stratigraphy. Each sequence was controlled by the position of the abrasion and quarrying zones through time; the contact between the abrasion and quarry zones was controlled by the climate and other factors that affect the glacial regime, the proglacial and subglacial topography, and the subglacial stratigraphy.

We therefore distinguish between sequences that have only local significance and those that have regional significance. Climatically controlled regional sequences can be correlated from region to region as far as the climatic changes affected the glacier. They occur in areas where there is no topographic or other obvious local control.

Local sequences, such as topographically-controlled sequences, cannot be correlated beyond the region of local control. For example, the Missouri Coteau in South Dakota, North Dakota, and Saskatchewan is the marginal suite of a topographic sequence. The Coteau Slope to the southwest contains the fringe suite, and a transitional suite and inner suite occur to the northeast of the Missouri Coteau in many areas. The entire Coteau sequence is topographically rather than climatically controlled. It occurs along the northeast edge of the Missouri Plateau and is crossed at an angle by climatically controlled Late Wisconsinan ice-marginal positions.

The regional effects of topography and glacial regime are to some degree mutually complementary. Once a glacier terminus has arrived at a long-term stationary position, a marginal accumulation begins to build up. This intensifies the decelerating flow in the ice, which promotes increased marginal accumulation. The presence of this barrier in turn helps to stabilize the glacial terminus at that position.

Retreat rates. The thickness of glacial sediment in a climatically controlled sequence should be systematically related to the rate of retreat of the active-ice terminus; the inactive-ice terminus lags far behind (Moran and Clayton, 1972). The most rapid retreat rates resulted from large-scale stagnation. As shown in Figure 8B (based on only a few known retreat rates from the Upper Midwest) landforms with the thinnest glacial sediment, such as "drumlins," should be better indicators of large-scale stagnation than those with thicker sediment such as "stagnation moraine" and "dead-ice moraine.'"

A Test of the Model

A good test of this model should be to try it out by mapping an area, creating a workable map legend, and writing a report on the map area using this approach. The following section illustrates this approach by giving a sample explanation for a map of the glacial geology of North Dakota.

The scale of the map will to a large degree control the format of the legend. For example, on small-scale maps, such as 1:2,000,000 or 1:5,000,000, probably only the glacial suites or sequences of suites can be shown. On large-scale maps, such as 1:20,000, 1:50,000, or 1:100,000, individual hummocks or disintegration ridges can be shown. We here concentrate on medium-scale maps, such as 1:200,000, 1:500,000, or 1:1,000,000.

We deal only with the glacial part of the legend. We ignore lacustrine and fluvial landforms, such as ice-walled-lake plains or eskers, that are associated with glacial deposits. The criteria to be used in the legend must first of all be mappable. So we will have to ignore some interesting genetic aspects of glacial geomorphology, because we have inadequate information. Most of the appropriate criteria can be expressed in the form of a continually varying property--a "continuum" (Clayton, 1972).

Obliteration-of-preadvance-element continuum. The obliteration of the preadvance morphologic element by subglacial erosion can be expressed more-or-less quantitatively in terms of the proportion of the preadvance element that has been obliterated. For purposed of our legend, we will divide this continuum into three categories: none (0 to 0.2), partial (0.2 to 0.8), and complete (0.8 to 1.0). We have found that this is most easily judged on air photos; the presence of obvious preadvance river channels and other nonglacial features indicates little or no obliteration. The presence of longitudinal shear marks indicates nearly complete obliteration.

Obliteration-of-subglacial-element continuum. The subglacial element is obliterated by the superglacial element. We again use three categories: none (0 to 0.2), partial (0.2 to 0.8), and nearly complete (0.8 to 1.0). This can often be judged on air photos; longitudinal shear marks indicate little or no obliteration, collapse hummocks superimposed on shear marks indicate partial obliteration, and hummocks alone indicate complete obliteration. In many areas, especially where there has been little obliteration of the preadvance element, and the preadvance element consists of glacial morphology of a previous glaciation, it is necessary to know the thickness of the glacial sediment of the last advance before this judgment can be made.

Thickness-of-superglacial-sediment continuum. We have encountered no obvious problems in assuming that the entire thickness of sediment deposited by a glacier is made up of superglacial sediment. The best way to determine the thickness is in outcrop or in a drill hole, providing the glacial stratigraphy of the area has been worked out. For reasons discussed in the next paragraph, we divide this continuum into four categories: none (0 to 1 m), thin (1 to 4 m), medium (4 to 8 m), and thick (8 to about 30 m).

Fluidity-of-superglacial-sediment continuum. Fluidity is controlled by grain size, clay mineralogy, and water content. We assume that grain size and clay mineralogy were the same at the time of deposition as today. If we restrict our attention to typical glacial sediment in areas such as North Dakota, we have only one variable, water content, because most has roughly the same grain size and clay mineralogy.

Water content and fluidity are interpreted from the steepness of hillslopes in the present collapse topography. We assume that water content was controlled by the rate of melting and therefore by the thickness of the superglacial sediment, and we assume that the slope of the side of hummocks was controlled by the viscosity of the sediment. Using the field observation that the thickness of the glacial sediment is generally nearly equal to the local relief in North Dakota and that the largest collapse hummocks average about 200 m across, we arrive at the approximate geometric relationship illustrated in Figure 10: $65 \tan \theta = d$; θ is numerically the same as d if θ is the maximum slope angle in degrees and d is the average thickness of sediment or local relief in meters. We have decided to divide the slope continuum

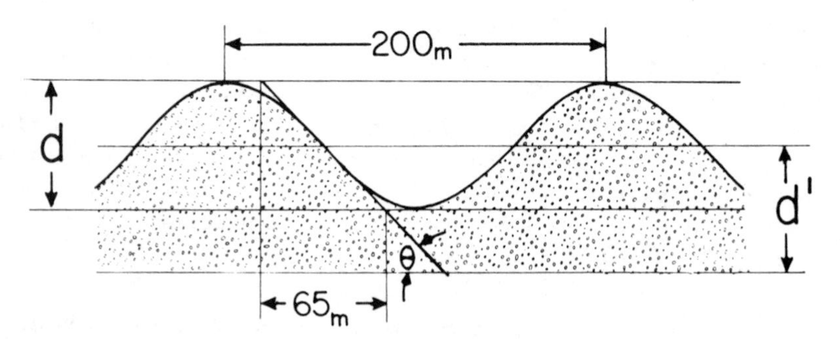

Figure 10.
Relationship between local relief *(d)* and average thickness of glacial sediment *(d')*, which are about equal, the average distance between hummocks (200 m), the average horizontal distance (65 m) between the intersection of the tangent to the slope and the elevation of the top and bottom of the hummock, and the maximum slope (θ): $65 \tan \theta = d$, where θ is almost the same as d if θ is measured in degrees and d is measured in meters.

into the following four categories, because they correspond to values easily obtainable from 1:125,000 soil maps that are available for the entire state: level (0° to 1°), undulating (1° to 4°), rolling (4° to 8°), hilly (greater than 8°). As shown by the equation, these four categories closely correspond to the following four local-relief categories: flat (0 to 1 m), low relief (1 to 4 m), medium relief (4 to 8 m), and high relief (greater than 8 m). And these categories correspond to the four thickness categories given in the previous paragraph. In summary, we conclude that the four descriptive slope-angle categories (level, undulating, rolling, and hilly) correspond to four genetic fluidity categories (very highly fluid, highly fluid, medium fluid, and slightly fluid).

Longitudinal-nonuniformity-of-englacial-sediment continuum. If the englacial debris was isotropically (but not necessarily homogeneously) distributed in map view, the resulting collapse topography is isotropic in map view. But if there is a longitudinal bunching of the englacial sediment, transverse lineations, or washboard moraine, will result. The degree of nonuniformity could be quantified using the degree of linearity. However, for the purposes of this legend, we prefer to break this continuum into only two categories, with the dividing line at the point where the lineations become conspicuous on air photos: isotropic topography and topography with transverse lineations.

Degree-of-compression continuum. This continuum could be quantified using the degree of longitudinal shortening. However, for this legend we divide the continuum into two categories at the point where the lineations become conspicuous on air photos: isotropic topography and topography with transverse lineations. These lineations involve compression of the preadvance sediment, whereas those of the previous continuum involve variations in the thickness of the glacial sediment of the last advance, so the stratigraphy must be known. In addition, compression ridges are much larger, have shorter radius of curvature, and are much rarer than washboard moraine.

Degree-of-postglacial-modification continuum. The amount of postglacial erosion on a hummock can be determined by measuring the volume of postglacial sediment filling intervening depressions. Where all of the products of erosion are washed into the depressions and about 20% of the glacial landscape is covered with postglacial sediment, the average erosion is one fifth the average thickness of the sediment in the depressions. The proportion of the original hummock that has been eroded away can be used to quantify the postglacial-modification continuum: none (0 to 0.1), slight (0.1 to 0.3), partial (0.3 to 0.9), complete (0.9 to 1.0). The first category is absent in North Dakota. The last category is no longer a glacial landform and so is outside the scope of this legend.

All of the above seven continuums are defined genetically (even though they are recognized descriptively). The following two continuums are defined descriptively and have unknown genetic significance. They are included here because they represent conspicuous elements of the glacial landscape.

Shape-of-hummocks continuum. Hummocks range from simple mounds to circular disintegration ridges. This continuum can be quantified using the ratio of the depth of the hole in the circular ridge to the height of the hummock. A ratio of 0 would represent a simple hummock with no hole and a ratio of 1 would represent a circular ridge that is as high as its hole is deep. However, we will here divide this continuum into two categories at the point where the ring becomes conspicuous on air photos: simple hummocks and circular ridges.

Shape-and-size-of-shear-mark continuum. The size of shear marks might be quantified using width, and the shape can be quantified using the length-to-width ratio. Instead we will indicate the size and shape of the ridges using a line symbol.

The legend. How can these nine continuums, or the categories we have broken them into, be combined into a legend? The obliteration of the preadvance element could be indicated using colors varying from dull for no obliteration to bright for complete obliteration, and obliteration of the subglacial element could be indicated using colors varying from warm for no obliteration to cool for complete obliteration (Fig. 11). No subdivisions of complete obliteration of the subglacial element (blue) is needed because the degree of obliteration of preglacial relief by subglacial shearing can not be easily recognized if the subglacial element has been completely obliterated. Similarly, differentiation of partial and complete obliteration of the preadvance element is difficult if there has been partial obliteration of the subglacial relief (green). The thickness and fluidity continuums can be combined, because they coincide, under the conditions outlined above. They can be indicated by the darkness of the colors given in Figure 11. Because thickness and fluidity are irrelevant if there is no obliteration of subglacial relief, this applies only to gray, green, and blue. Assuming that the thickness category "none" (fluidity category "very highly fluid"; slope angle category "level") applies only to the obliteration-of-subglacial morphology category "none," only three grades of darkness are needed: light (thin, highly fluid, and undulating), medium (medium thick, medium fluid, and rolling), and dark (thick, slightly fluid, and hilly). In the nonuniformity and compression continuums, the "isotropic" category can be ignored, and the transverse lineations can be indicated with a line pattern. In the postglacial-modification continuum, the "slight" category can be ignored, and "partial" can be indicated with a dashed-line symbol. In the shape-of-hummock continuum, the simple hummocks can be ignored, and the circular ridges can be indicated using a dot pattern. In the shape-and-size-of-shear-mark continuum, the shape and size can be indicated by a crossed-line pattern. So, it can be seen that the nine continuums can be combined to give a fairly simple legend (Fig. 12).

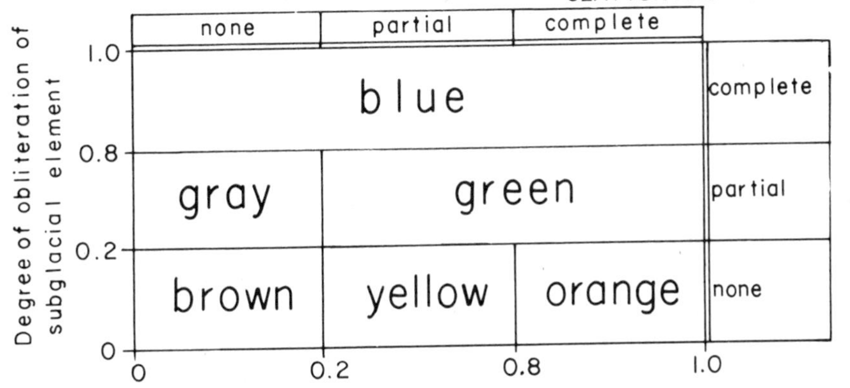

Figure 11.

Map colors for the combined continuums indicating the degree of obliteration of subglacial and preadvance morphologic elements.

brown	Preglacial morphology
yellow	Preadvance morphology, subglacially modified
orange	Subglacial morphology.
medium gray	Superglacial morphology (medium thick) with partially buried preadvance morphology.
light gray	Superglacial morphology (thin) with partially buried preadvance morphology.
medium green	Superglacial morphology (medium thick) with partially buried subglacial morphology.
light green	Superglacial morphology (thin) with partially buried subglacial morphology.
dark blue	Superglacial morphology (thick).
medium blue	Superglacial morphology (medium thick).
light blue	Superglacial morphology (thin).
	Subglacial or superglacial morphology partially modified by postglacial erosion.
	Longitudinal subglacial shear ridges; azimuth, length, and width of ridges are indicated.
	Transverse compressional ridges.
	Washboard moraine.
	Circular disintegration ridges.

Figure 12.

A glacial geomorphology model: the legend for a medium-scale map.

CONCLUSIONS

We have perhaps put emphasis in somewhat different places than many geomorphologists would. In considering glacial erosion, we have placed emphasis on the temperature of the bed and on the influence of groundwater pressure caused by the ice loading on the shear strength of the subglacial sediment. In considering glacial deposition, we have emphasized the importance of superglacial sediment and the unimportance of subglacial sediment. We have emphasized the need to know the glacial stratigraphy before the glacial geomorphology can be understood. We have emphasized the need for mapping all glacial features of an area rather than considering a single landform as an isolated element. We have emphasized that glacial terminology and map legends should reflect the idea that individual glacial landforms must be explained in terms of several process continuums; each landform does not have a completely independent interpretation.

These views are certainly biased by our midcontinent point of view, but we consider our glacial process-form model to have broader application. We would like to have these models criticized for their applicability by geologists working in other areas.

ACKNOWLEDGEMENTS

We thank C.L. Matsch, G.W. White, David M. Mickelson, Walter L. Moore, Samuel S. Harrison, William B. Bickley, A.F. Jacob, John A. Brophy, and Ernest H. Muller for their help in making this more readable. We especially thank John A. Cherry for his contributions to the section on groundwater.

REFERENCES

Albertson, M. L., Barton, J. R., and Simons, D. B. 1960. *Fluid mechanics for engineers.* Prentice-Hall, Englewood Cliffs, N. J., 561 p.

Arndt, B. M. In press. Geology and groundwater resources of Cavalier and Pembina Counties, North Dakota--Geology: North Dakota Geol. Survey, Bull. 62, pt. 1.

Bickley, W. B. Jr. 1972a. Stratigraphy and history of the Sakakawea Sequence, south-central North Dakota: Univ. of North Dakota, Unpub. Ph.D. dissertation, 142 p.

Bickley, W. B. Jr. 1972b. Quaternary history of south-central North Dakota: Geol. Soc. Amer., Abstracts with Programs, v. 4, no. 5, p. 309.

Bluemle, J. P. 1970. Anomalous hills and associated depressions in Central North Dakota: Geol. Soc. Amer., Abstracts with Programs, v. 2, p. 325-326.

Boulton, G. S. 1971. Till genesis and fabric in Svalbard, Spitsbergen, in Goldthwait, R. P. (ed), *Till, a symposium:* The Ohio State Univ. Press, p. 41-72.

Boulton, G. S. 1972. The role of thermal regime in glacial sedimentation: Institute of British Geographers, Special Publication No. 4, p. 1-18.

Clayton, Lee. 1966. Karst topography on stagnant glaciers: Jour. Glaciology, v. 5, p. 107-112.
Clayton, Lee. 1966. Notes on Pleistocene stratigraphy of North Dakota: North Dakota Geol. Survey, R. I. 44, 25 p.
Clayton, Lee. 1972. Glacial geology: a continuum model: Geol. Soc. America, Abstracts with Programs, v. 4, p. 312.
Deal, D. E. 1972. Geology of Rolette County, North Dakota: North Dakota Geol. Survey, Bull. 58, 89 p.
Gravenor, C. P. 1953. The origin of drumlins: Amer. Jour. Sci., v. 251, p. 674-681.
Gravenor, C. P., and Kupsch, W. O. 1959. Ice disintegration features in western Canada: Jour. Geol., v. 67, p. 48-64.
Harris, K. L. 1973. Pleistocene stratigraphy of the Red Lake Falls area, Minnesota: Univ. of North Dakota, unpub. M. Sc. Thesis, 117 p.
Harris, K. L., Moran, S. R., and Clayton, Lee. 1974. Late Quaternary stratigraphic nomenclature in the Red River Valley, North Dakota and Minnesota: North Dakota Geol. Survey, Misc. Series 52, 47 p.
Harrison, P. W. 1957. A clay-till fabric: its character and origin: Jour. Geol., v. 65, p. 275-308.
Harrison, P. W. 1958. Marginal zones of vanished glaciers reconstructed from the preconsolidation-pressure values of overriden silts: Jour. Geol., v. 66, p. 72-95.
Hubbert, M. K., and Rubey, W. W. 1959. Role of fluid pressure in mechanics of overthrust faulting: Geol. Soc. Amer. Bull., v. 70, p. 115-206.
Moran, S. R. 1969. Geology of the Hudson Bay area, Saskatchewan: Univ. of Illinois, unpublished Ph.D. dissertation, 194 p.
Moran, S. R. 1971. Glaciotectonic structures in drift, in Goldthwait, R. P. (ed.), *Till, a symposium:* The Ohio State Univ. Press, p. 127-148.
Moran, S. R. and Clayton, Lee. 1972. Lake Agassiz and the history of the Des Moines Lobe: Geol. Soc. Amer., Abstracts with Programs, v. 4, p. 602-603.
Parizek, R. R. 1969. Glacial ice-contact rings and ridges: Geol. Soc. Amer., Special Paper 123, p. 49-102.
Salomon, N. L. In press, Glacial stratigraphy of northeastern North Dakota Acad. Sci., Proc.
Terzaghi, K. 1950. Mechanism of landslides, in Paige, S. (Chairman), *Application of geology to Engineering practice--Berkey Volume:* Geol. Soc. Amer., p. 83-123.
Ulmer, J. H., and Sackreiter, D. K. 1973. Late Cenozoic stratigraphy of the Lake Sakakawea bluffs north and west of Riverdale, North Dakota: North Dakota Geol. Survey, R.I. 51.
Weertman, J. 1961. Mechanism for the formation of inner moraines found near the edge of cold ice caps and ice sheets: Jour. Glaciology, v.3, p. 965-978.
Welty, J. R., Wicks, C. E., and Wilson, R. E. 1969. *Fundamentals of momentum, heat and mass transfer:* John Wiley and Sons, New York, 697 p.
White, G. W. 1971. Thickness of Wisconsinan tills in Grand River and Killback Lobes, northeastern Ohio and northwestern Pennsylvania, in Goldthwait, R. P. (ed.), *Till, a symposium:* The Ohio State Univ. Press, p. 149-163.
Wright, M. E. 1971. Retreat of the Laurentide ice sheet from 14,000 to 9,000 years ago: Quaternary Research, v. 1, no. 3, p. 316-330.

CHAPTER 4

THE MORPHOLOGIC SEQUENCE CONCEPT AND DEGLACIATION OF SOUTHERN NEW ENGLAND

By Carl Koteff

ABSTRACT

The morphologic sequence concept as evolved from Jahns' original definition (1941, 1953) now classifies eight basic chronologic groups of forms composed mostly of meltwater deposits and referred to as *sequences.* Sequences are distinguished according to differences in depositional environment, such as fluvial, lacustrine, and marine, and by the presence or absence of an ice-contact head of outwash. The construction of profiles to show depositional gradients enhances the mapping and interpretation of sequences.

Because of the scarcity of morainal features in southern New England, the distribution of morphologic sequences provides the best means of recognizing retreatal positions of the last ice sheet. The distribution and physical features of sequences strongly support the view that ice recession was characterized by stagnation-zone retreat, influenced greatly by topography. The width of the stagnant zones, crudely measured by the length of eskers or ice-channel fillings, was at least 1.5 mi (2.4 km). The major source of sediments that comprise the morphologic sequences is suggested to have been a shear zone at the live ice-stagnant ice interface. Debris from this zone, transported by meltwater from the surface of the live ice, was deposited as outwash in, but chiefly beyond the main edge of the stagnant front. Stagnant-ice blocks appear to have contributed only minor sediments to the mass of meltwater deposits in southern New England.

INTRODUCTION

Surficial geologists in New England during the last 60 yrs have used different methods with varying degrees of success to show the distribution of stratified drift. Since about 1940, members of the U. S. Geological Survey have applied the sequence concept that was first used by Jahns (1941, 1953) in areas of Massachusetts. The concept has since been extended to Rhode Island, Connecticut, and New Hampshire as part of the cooperative mapping programs the Survey has had with these states. Jahns was the first to demonstrate a new concept for tracing the progressive retreat of the edge of the last ice sheet by careful mapping of landforms composed of meltwater deposits of sand, gravel, silt, and clay. These landforms, such as eskers or ice-channel fillings, kames, kame terraces, kame plains, etc., are grouped into units called sequences. A sequence refers to a downstream progression of landforms from a head of outwash and thus a sequence has no time

Approved by the Director, U.S. Geological Survey. Prepared in cooperation with the State of New Hampshire, Department of Resource and Economic Development and the Commonwealth of Massachusetts, Department of Public Works.

connotation in itself except in relation to other sequences. The term may have been an unfortunate choice because the word sequence also connotes time within the unit. However, the word sequence is so well established in the literature that it seems unwise now to substitute another term to describe the concept. Instead, it is proposed here that such a group of landforms be called a "morphologic sequence".

Whether deglaciation in New England was characterized by regional stagnation of the ice sheet or by progressive retreat of the ice margin has long been debated (Antevs, 1922; Brown, 1933; Flint, 1930; Goldthwait, 1938; Lougee, 1940). As so often happens in scientific debates, the introduction of new products can shift thought heavily in one direction or even in a different direction. The introduction in the late 1930's of accurate large scale 7 ½-minute topographic maps (scale 1:31,680, contour interval 10 ft) replacing the more generalized smaller scale 15-minute maps (scale 1:62,500, contour interval 20 ft) of the early 1900's enabled Jahns to examine the glacial deposits in much more detail than previously possible and to develop what is now called the morphologic sequence concept. His findings support the theory of progressive retreat of the ice margin rather than that of regional stagnation. One can only speculate on what the additional detail shown by 3 ¾-minute topographic maps with a five-foot contour interval would do to current thought. Morphologic sequences have also been outlined on recent 15-minute topographic maps, but it seems clear that the modern concept was a logical derivative of the larger scale mapping.

Since Jahns' original work, the morphologic sequence concept has been enlarged and modified as new areas in New England have been mapped and as additional depositional environments have been recognized. New ideas have also evolved regarding the glacial regime at the edge of the ice sheet and the chief source of sediment laid down by melt water.

DEFINITION OF SEQUENCE

The original definition of a sequence by Jahns dealt chiefly with sand and gravel deposited by meltwater streams in a fluvial or subaerial environment in front of and in contact with a retreating ice sheet. Each sequence was composed of the landforms most commonly found in ice-contact meltwater deposits, such as eskers or ice-channel fillings, kames, kame terraces, kame plains, and other features associated with ice-contact deposits. Proglacial outwash plains were either not included or their position left ambiguous in the original definition because Jahns felt that a sequence was restricted to the stagnant zone of ice bordering the live ice. The approximate width of the stagnant zone, therefore, was indicated by the length of a sequence measured parallel to the direction of ice recession. The meltwater streams that deposited each sequence in a single valley were graded to a constant base level, generally a bedrock threshold. With the opening of successively lower outlets during ice retreat, new sequences were formed.

Since Jahns' original work in Massachusetts, his analysis has been found to be applicable to a variety of morphologic sequences. The concept now also includes deposits laid down in lacustrine or subaqueous environments , as well as proglacial deposits not in contact with the retreating ice sheet but in places associated with detached stagnant ice blocks. Along the coast of Maine, Borns (1968) has shown that moraines and associated meltwater deposits were laid down in a marine environment, thus adding sea level as a factor in controllingdeposition of sequences. It has also been demonstrated in many places that several sequences can share the same base- level control, and it is not always necessary for a later and lower outlet to be uncovered in order for a new sequence to be deposited in a particular valley or drainage system.

The common elements that all types of morphologic sequences share include (1) composition of the deposits (2) morphologic features of the deposits (3) specific base-level control (4) internal contemporaniety, and (5) similarity of the distribution of textures.

1. Deposits that compose a sequence are almost entirely products of meltwater deposition laid down during retreat of the ice sheet in any given area. The materials are chiefly sand, gravel, silt, and clay deposited as fluvial, lacustrine, or marine sediments. Till is not ordinarily included as part of a sequence except where it is part of an end moraine (which in New England is most often closely related in time to water-laid deposists) or where it occurs as flowtill. In theory there should be some component of till laid down directly by live ice beneath the active ice sheet that is equivalent in time to a given meltwater deposit laid down beyond the stagnant-ice margin. This cannot be demonstrated in the field, however, and therefore it is not practical to include most till as part of a morphologic sequence.

2. The morphologic features in a sequence are those commonly found in meltwater deposits, and as a general rule range from more collapsed forms such as eskers or ice-channel fillings and kames at the head of outwash, to kame terraces farther downstream, and to generally less collapsed forms such as outwash plains at the lower end of deposition. These forms are more typically fluvial deposits. Lacustrine deposits in a sequence include kame deltas, outwash deltas, and lake-bottom deposits. Sequences can comprise only one landform or can include a combination of them. For example, somelacustrinesequences are composed of only a single kame delta or of a delta with an apparent feeding esker or ice-channel filling. Some fluvial sequences have only kames and kame terraces; others may include outwash plains at their lower end. Some sequences consist only of ice-channel fillings. In fact, sequences that exhibit the full range of possible morphologic features are much less common than those composed of only a few morphologic forms.

3. All morphologic sequences are graded to and were controlled by a specific base level that can generally be identified in the field. The base-level controls for some deposits in coastal areas are now under the sea and a few inland sequences appear to have had stagnant ice for a base- level control, and these control points must be inferred. Jahns' original definition suggested only bedrock thresholds as controls, and although this is a very common and clearly the most durable type,

other types of thresholds are just as common and certainly just as important, if not more so. These other types of thresholds include standing water bodies (for the most part glacial lakes), till ridges, previously deposited sequences of sand and gravel, end moraines, and more rarely stagnant blocks of ice. There has been no evidence found so far in southern New England to indicate that live ice acted as a base-level control for sequences.

4. All landforms that make up a morphologic sequence are considered to result from contemporaneous deposition, laid down at nearly the same time by meltwater. Although the sand, gravel, silt, and clay of each sequence were laid down as aggrading layers, with the bottom layer the oldest, it is the surface and youngest layer that is most visible and thus best defines the sequence in terms of gradient and retreatal ice position. The time span of any given sequence is unknown, but tens of years or less seems reasonable. Obviously some sequences were active for a much longer time than others. For example, one valley may contain three or four sequences, but a neighboring parallel valley may have only one sequence that represents the same amount of ice retreat. Whatever the time interval, one entire sequence appears to represent the smallest practical time interval that can be used to measure ice retreat in any given valley or drainage system. Therefore, no individual part of a sequence is considered to be older or younger than any other part.

5. The distribution of textures of the deposits in a sequence generally shows decreasing coarseness downstream from the source or head of outwash of the unit, although some sequences are too short to exhibit a clear textural gradation. Even in highly collapsed or kettled deposits, there is generally no change in the textural pattern, indicating that the stagnant ice blocks were not a significant source of material for the meltwater deposits.

TYPES OF MORPHOLOGIC SEQUENCES

Morphologic sequences can be divided into eight basic types (Table 1). Three of these (FC, FNC, LC) are distinguished on the basis of meltwater deposition primarily in a fluvial or a lacustrine environment and whether or not an ice-contact head of outwash exists. Three types (FLC, FLNC, LFC) represent a combination of both fluvial and lacustrine environments. Two less common types (EMO, M) are essentially similar to any of the others and are deposits associated with an end moraine or are deposits laid down mostly in a marine environment. Further subdivision of this classification is possible by combining various parts of the eight basic types but is not done here for the sake of clarity.

Fluvial ice-contact sequence (type FC)

This sequence (Fig.1) can be considered the prototype as it fits nearly all elements of the definition first outlined by Jahns(1941, 1953) The only changes have been the addition of different kinds of base-level controls and the addition of an outwash plain or valley train as morphologic features. The possible morphologic

Table 1. Types of morphologic sequences

FC	Fluvial ice-contact sequence
FNC	Fluvial non-ice-contact sequence
LC	Lacustrine ice-contact sequence
FLC	Fluvial-lacustrine ice-contact sequence
FLNC	Fluvial-lacustrine non-ice-contact sequence
LFC	Lacustrine-fluvial ice-contact sequence
EMO	End moraine and associated outwash
M	Glaciomarine deposits

features now recognized are eskers or ice-channel fillings (generally indistinguishable from each other in the field), kames, kame terraces, outwash plains, and valley trains. The head of outwash or upper end of deposition was generally in contact with the edge of the stagnant zone; eskers and ice-channel fillings are possibly the only features of this type of sequence laid down totally within the stagnant zone. No head of outwash has been recognized that appeared to have been in contact with live ice, i.e., no shove structures or evidence of overriding ice have been noted in the water-laid deposits. Base-level controls for the gradients of this type of sequence are thresholds in bedrock, till less commonly sand and gravel, and more rarely stagnant ice blocks. Examples of fluvial ice-contact sequences (type FC) have been mapped in one form or another by Goldsmith (1962),Jahns(1953), Pessl(1970), Robinson(1961),and many others.

Fluvial non-ice-contact sequence (type FNC).

This type of sequence (Figure 2) is the same as a fluvial ice-contact sequence (type FC) except its head of deposition began downstream from the edge of the stagnant zone. This situation occurred when the ice margin extended along a divide or across ground whose slope away from the ice was too steep to permit accumulation of water-laid deposits at the ice edge. Deposition began where the surface became gentle enough to allow meltwater to drop its sediment load. Some sequences have a gradient on their upper surface of as much as 30 ft/mi (5.7 m/km), and some erosional meltwater channels in which there has been no deposition slope as low as 40 ft/mi (7.6 m/km). Therefore, it appears that the maximum slope on which depostion of meltwater sediment can take place is less than 40 ft/mi (7.6 m/km). The morphologic features possible in fluvial non-ice-contact sequences (type FNC) are the same as those in fluvial ice-contact sequences (type FC), except that eskers are absent. Ice-channel fillings are rare, but still possible where enough detached stagnant blocks of ice remained in the depositional area. Base-level controls for this sequence are the same as those for sequence type FC. Examples of fluvial non-ice-contact sequences have been mapped by Feininger (1965), Pessl (1966), Richmond (1953), and Schafer (1961, 1968a).

Lacustrine ice-contact sequence (type LC).

This sequence is composed of deposits laid down by meltwater directly into a lake in contact with the stagnant edge of ice (Fig. 3). In most examples the meltwater streams emerged into the lake next to the valley wall. Less commonly the head of deposition was at the ice edge farther out into the lake. The landforms are mostly kame deltas (with or without feeding eskers or ice-channel fillings) and lake bottoms. Even though topset beds of a delta are fluvial, they directly overlie subaqueous foreset beds and thus the entire sequence is considered to be part of a lacustrineenvironment. The lake that acted as a base level for the topset beds was itself controlled by a spillway in bedrock, till, sand and gravel, or stagnant ice. Bedrock thresholds are by far the most common type for this sequence because other types are more easily eroded and would therefore generally not have lasted long enough to permit deposition of extensive deltas.

Many lake outlets have been mapped showing till in the bottom of the spillway, but it is suspected that thinly veneered bedrock was the actual threshold in these places. One notable exception is the apparent spillway for the glacial lake that occupied Cape Cod Bay. There may have been more than one spillway for this lake, all of which are underlain by sand and gravel and perhaps strengthened by buried stagnant ice blocks. Examples of lacustrine ice-contact sequences (type LC) have been mapped by Koteff (1964a, 1966), Koteff and Volckmann (1974), and Pessl (1970).

Fluvial-lacustrine ice-contact sequence (type FLC).

This type of sequence (Fig. 4) combines forms built contemporaneously in both fluvial and lacustrine environments. The upper end of the sequence is in a valley tributary to a lake basin and consists of fluvial deposits in the form of eskers or ice-channel fillings, kames, and kame terraces in contact with the stagnant-ice margin. The lower end of the sequence is in the lake basin and consists of deltas and lake bottoms. The lake sediments may or may not have been in contact with the lobate ice edge that dammed the glacial lake. The deltas in most examples of this type of sequence are outwash deltas rather than kame deltas and do not show any appreciable sediment that was supplied directly from ice into the lake. Base-level controls for the glacial lakes were the same as for lacustrine ice-contact sequences (type LC). Examples of this type of sequence have been mapped by Hartshorn and Koteff (1967), Koteff and Volckmann (1974), and Koteff (1973).

Fluvial-lacustrine non-ice-contact sequence (type FLNC).

This type of sequence (Fig. 5) is similar to type FLC, except that the upper end of deposition was not in contact with the stagnant-ice edge. Eskers or ice-channel fillings therefore are not found in this type. Kames and kame terraces generally are less common and less extensive in this sequence. Typically the sequence is composed of a pitted (kettled) outwash plain that grades into an outwash delta. Base-level controls are thresholds for glacial lakes as in fluvial-lacustrine ice-contact sequences (type FLC). Examples of this type of sequence have been mapped by Koteff (1970), and Randall and Pessl (1968).

Lacustrine-fluvial ice-contact sequence (type LFC).

This type is similar to lacustrine ice-contact sequences (type LC) in that it started out primarily as a kame delta with sediments laid down into a glacial lake in contact with the stagnant-ice edge (Fig. 6). However, in this situation the valley containing the lake sloped up-ice preglacially and was dammed completely by the ice edge, the previous divide at the head of the valley served as a threshold and spillway for the lake. With time the entire basin was filled by the prograding delta, and the upper surface of the sequence became entirely fluvial, graded to the threshold that originally acted as the glacial lake spillway. In this type of sequence the enclosing valley has to be small enough to allow complete filling with meltwater deposits before the ice edge retreated far enough for sedimentation to be directed elsewhere. The most typical morphologic feature of this sequence at the surface is a valley train, perhaps kettled, with a sharp ice-contact head. Examples of this type of sequence have been mapped by Koteff (1970), La Sala (1961), and Schafer (1965).

End moraine and associated outwash (type EMO).

Because of the sparseness of end moraines in southern New England, this type of sequence is less common. The most prominent feature is the morainal ridge itself, generally having an associated outwash plain which is with or without an ice-contact head. The outwash plain is slightly younger than the moraine, but because together they define an ice position, they are combined here into a separate type of sequence. If all the outwash plain is in front of the moraine, then clearly the meltwater deposits can not be considered ice contact. However, in places it appears the moraine either grades laterally along the ice margin into the outwash, or the outwash slightly later overwhelmed part of the moraine and consequently the head of outwash was in contact with the ice edge. The ice edge in the latter situation then can be determined to have changed from a live to a stagnant condition. Examples of this type of sequence have been mapped by Chute (1959), Koteff (1964b), and Schafer (1961, 1965).

Glaciomarine deposits (type M).

This sequence is distinguished solely on the basis of deposition in or near a marine environment. Examples of this type inNew England arerestricted to coastal areas from northeastern Massachusetts to Maine. All the characteristics of previously discussed types can be found in type M except that the sea is the base-level control. Good examples of this sequence have been described by Borns (1968) in coastal Maine.

PROFILES

Meltwater deposits that make up most morphologic sequences have nearly all the characteristics of aggrading streams associated with ice so it is possible to represent the surface of any sequence by a profile to show the depositional gradient of the

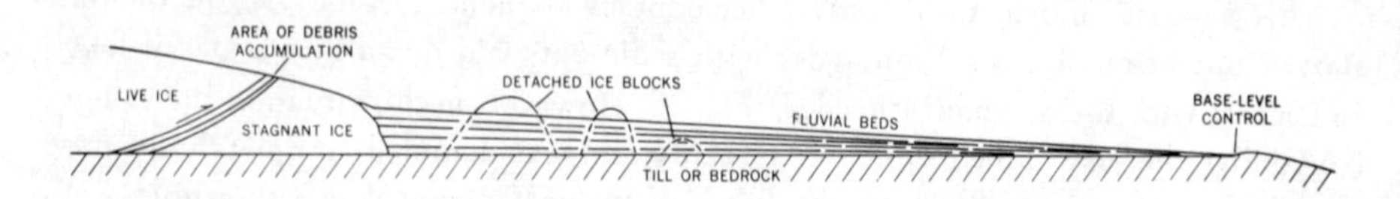

Figure 1. Diagrammatic profile of a fluvial ice-contact sequence (type FC). Detached ice blocks represent sites of future ice-contact slopes.

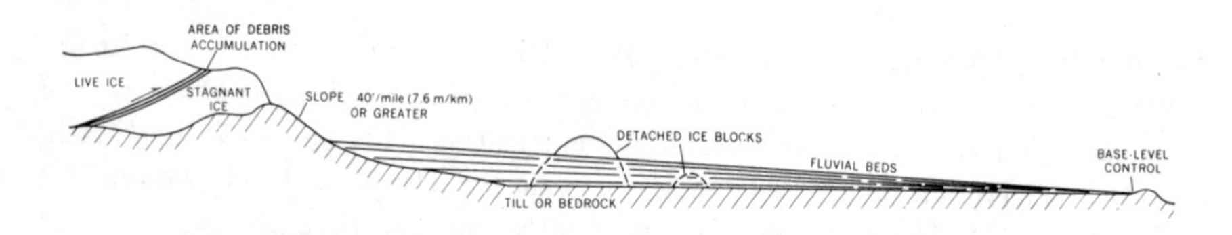

Figure 2. Diagrammatic profile of a fluvial non-ice-contact sequence (type FNC).

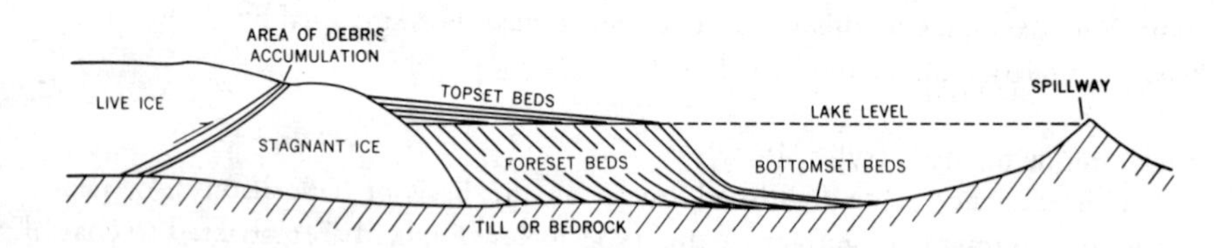

Figure 3. Diagrammatic profile of a lacustrine ice-contract sequence (type LC).

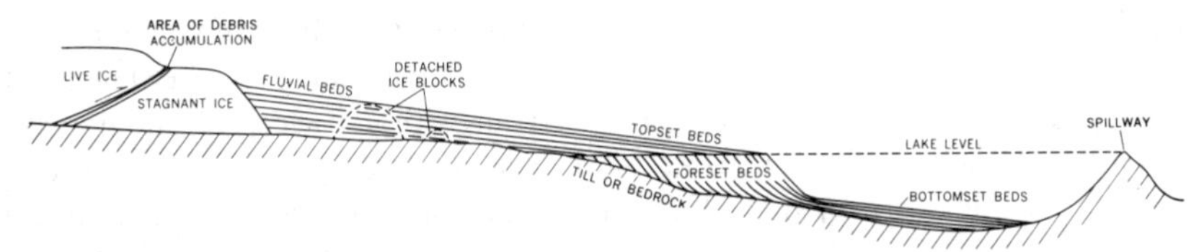

Figure 4. Diagrammatic profile of a fluvial-lacustrine ice-contact sequence (type FLC).

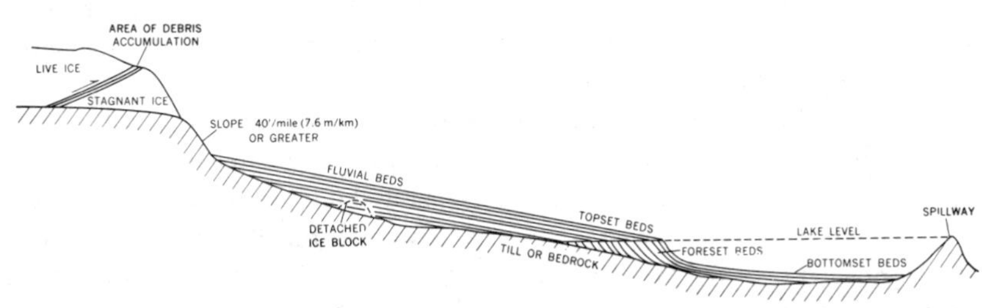

Figure 5. Diagrammatic profile of a fluvial-lacustrine non-ice-contact sequence (type FLNC).

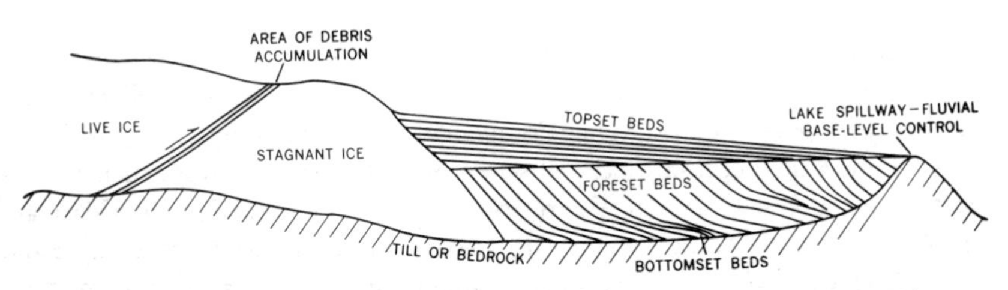

Figure 6. Diagrammatic profile of a lacustrine-fluvial ice-contact sequence (type LFC).

unit. In places where the deposits are highly collapsed from subsequent melting of stagnant ice blocks this becomes more difficult, and a line drawn directly down the surface slope is too irregular to show a meaningful profile. In order to better reconstruct the pre-collapse surface, map points plotted along a center line parallel to the slope and projected at right angles from within a width of about 2000 ft (610 m) each side of the centerline have been used successfully. This technique smooths out the irregularities of collapsed surfaces and allows a more accurate depiction of the pre-collapse surface. Figure 7 is a profile from the Pepperell quadrangle, Mass-N.H. (Koteff and Volckmann, 1974). It shows relationships of four sequences in an area near Dunstable, Mass., where Jahns (written communication, 1967) first began to formulate the sequence concept. It is suggested that a vertical exaggeration of x20 in areas of moderate relief such as New England is needed to demonstrate clearly the differences between several sequences that may be present in the same area. Most profiles of sequences are closer to straight-line tangents than curved segments, although a few show an increasing upward curve near the head of outwash.

Profiles can help in distinguishing more than one sequence in a valley having a complex history of ice retreat. Landforms composed of meltwater deposits that project above a particular gradient can be separated out as probably members of an older sequence; a lowering of base-level control for deposition can also be determined more easily with the help of profiles. Perhaps the most dramatic characteristic shown by profiles is the "shingle" effect of succeedingly younger sequences in a drainage basin having the same base-level control. The surface of each succeeding sequence becomes slightly higher up-valley, which is illustrated in Figure 7 by sequences 3 and 4 that probably shared the same base-level control. If the sequences were long enough, the downstream parts of the profiles would tend to merge.

IMPLICATIONS OF THE MORPHOLOGIC SEQUENCE CONCEPT

The distribution of morphologic sequences and the physical character of the deposits have important implications concerning ice front positions and ice regime during retreat. The use of morphologic sequences to determine retreatal positions is especially important because of the absence of end moraines over much of New England. End moraines are largely restricted to coastal areas of southern and eastern New England. These include the massive end moraines from Long Island, N.Y., to Cape Cod, (Fig. 8). Other end moraines are found only a short distance from the coast in Connecticut, Rhode Island, southeastern Massachusetts, and Maine. Ice-readvance localities inland, which are scarce, include the Fresh Pond moraine near Boston (which could almost be considered coastal), the Middletwon readvance in the Connecticut Valley, a probable readvance near Mt. Tom in Mass., a minor readvance near Manchester, N.H., and possibly some others elsewhere in southern and central New England. The readvances appear to have only slightly overrridden earlier meltwater deposits, suggesting they were only minor positive pulses during a period of general ice recession. Therefore, it is the heads of outwash

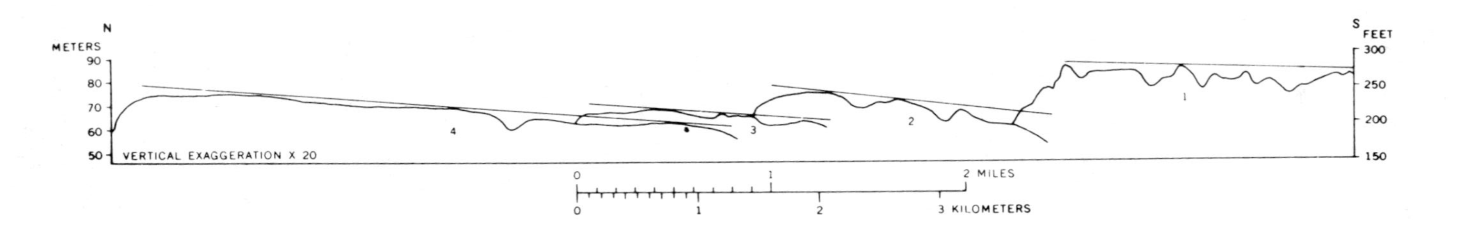

Figure 7. Profile of four sequences (type FC) in the Pepperell quadrangle, Mass. N.H. Sequence 1 is the oldest. Tangents indicate an idealized gradient. Base-level controls are to the south and east of the quadrangle. Modified from Koteff and Volckmann (1974).

of the morphologic sequences that are the most important indicators for ice retreatal positions in southern and central New England and perhaps in northern New England as well.

Two major conclusions about the morphologic sequence concept in view of the relative paucity of end moraines are (1) the retreat of the last ice sheet was systematic, and (2) the edge of the live ice sheet was bordered by a zone of stagnant ice which acted as a buffer zone between the live ice and meltwater deposits. The term "stagnation-zone retreat" has been used for many years to describe this process of ice wastage.

The rate of ice retreat is somewhat problematical because too few reliable indicators such as radiocarbon dates are available. Schafer (1968b) has calculated rates of retreat in New England at about 400 ft (122m) per year in the southern part and about 1000 ft (305m) per year in the northern part. Nearly all of New England probably was deglaciated in less than 3000 radiocarbon years from the initial time of retreat from Martha's Vineyard and Nantucket. This is based on radiocarbon dates of 15,300 $\pm$ 800 B.P. (Kaye, 1964) for retreat from Martha's Vineyard and about 12,700 B.P. for the time the ice front stood on the south side of the St. Lawrence Valley. Most of the few other radiocarbon dates associated with deglaciation in New England and surrounding areas are progressively younger northward, which strongly suggests systematic retreat of the terminus of the last ice sheet.

The systematic nature of ice retreat is also demonstrated by the distribution of morphologic sequences, particularly where several successively younger sequences whose profiles show a "shingled" pattern (see Fig. 7) are found in the same drainage system. Jahns' original concept was formulated in an area where successively younger sequences were controlled by successively lower base levels. Shingled sequences have been found in many areas since, however, that had the same base-level control, including areas that were formerly glacial lakes. One example of this is the area near Concord, Mass. that was the site of glacial Lake Sudbury, first described by J. W. Goldthwait (1905). Later work on larger scale maps (Koteff, 1963, 1964a) shows a series of three deltas progressively younger northward that were laid down in Lake Sudbury; each delta was controlled by essentially the same spillway (Fig. 9). Other deltas, controlled by another spillway to the northeast, were formed in glacial Lake Concord. Each of these deltas has an ice-contact head and shows coarse-texture materials grading to finer texture downstream. In profile the Lake Sudbury deltas exhibit very well the shingled arrangement. These lacustrine ice-contact sequences (type LC) were deposited directly in the lakes in contact with a stagnant-ice edge. The combination of the physical characteristics of collapsed ice-contact heads of outwash having no evidence of overriding ice, downstream change from coarse to fine textures, and shared base-level control, along with a shingled or en echelon profile for several of the deltas, strongly supports the concept of systematic retreat of the ice sheet here. This situation has been demonstrated for both fluvial and lacustrine sequences in a large number of other areas in southern New England.

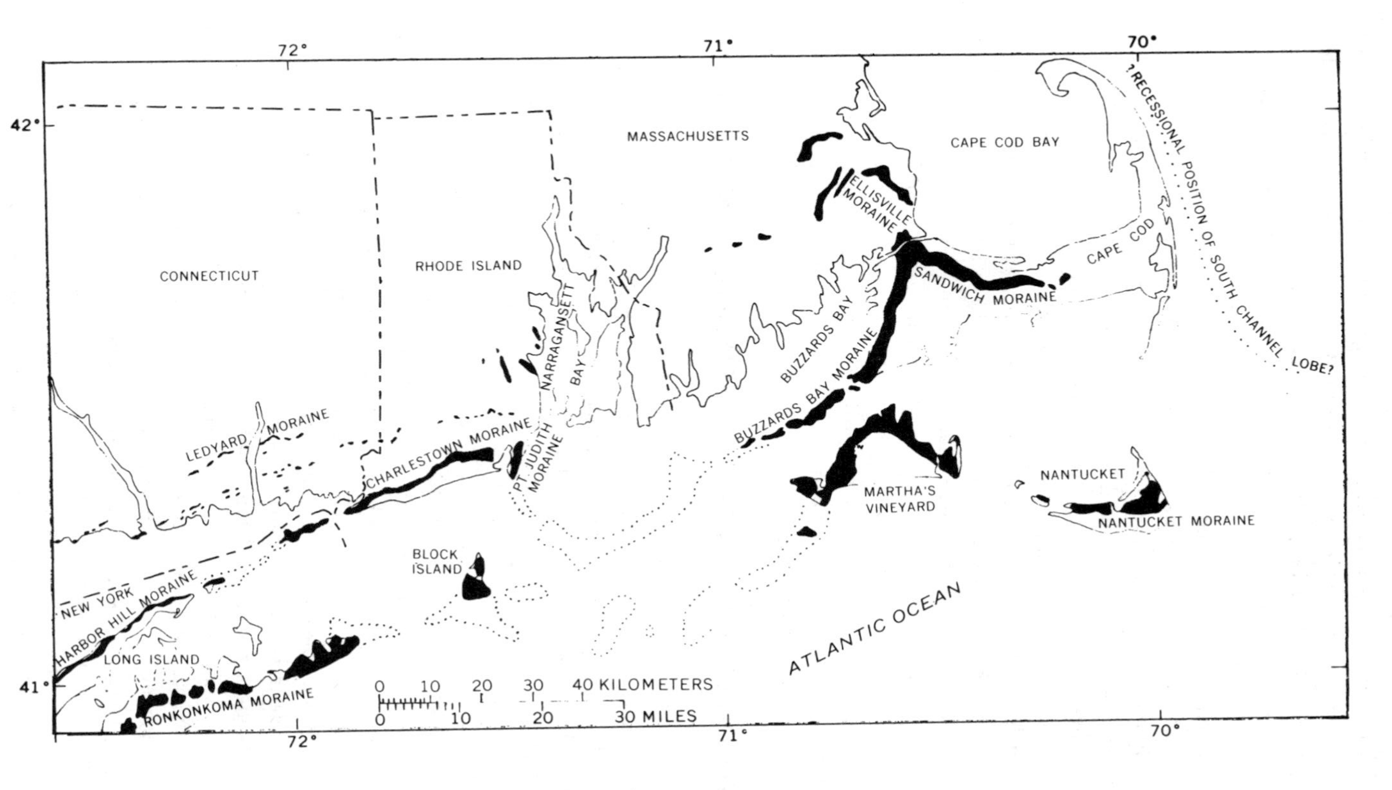

Figure 8. Maps showing distribution of moraine systems from Long Island, N.Y. to Cape Cod, Mass. Adopted from Schafer and Hartshorn (I965).

The nature of ice retreat between positions indicated by morphologic sequences such as shown in Figure 9 is not clear. The only evidence available is the deposits themselves and, therefore, one individual sequence is considered to be the smallest practical time unit that can be used to measure ice retreat at the present scale of mapping.

The disappearance of the last sheet by stagnation-zone retreat is also strongly suggested by the physical characteristics of the morphologic sequences. With remarkably few exceptions, the heads of outwash of hundreds of sequences mapped in southern New England were deposited in contact with stagnant ice or were deposited downstream away from the ice edge; evidence for overriding ice is known only for those few readvance localities previously mentioned. Commonly an ice-contact head of outwash has an esker or ice-channel filling extending upstream. The esker tube or ice-walled channel probably acted as a feeder for the major part of the morphologic sequence. These features also show no evidence of having been overridden by live ice and thus give a crude measure of the width of the stagnant zone. Several such ice-channel fillings are as much as 1.5 mi (2.4 km) long. This view of the width of the stagnant zone differs from Jahns' original view in that he defined the stagnant zone as being measured by the length of any given sequence away from the stagnant ice edge.

TOPOGRAPHIC CONTROL OF SEQUENCE DISTRIBUTION

The distribution of morphologic sequences in southern New England were controlled more by both large-scale and small-scale topographic features rather than by climatic fluctuations of the retreating ice sheet. Climatic fluctuations were important in controlling sequence distribution only near the maximum extent of the last ice sheet, in the areas outlined by the morainic systems and associated outwash that extend from Long Island, N.Y., to outer Cape Cod, Mass. (Fig. 8). It seems clear, however, that topography exerted a large influence in these areas as well because the major ice lobes that formed the moraines were themselves related to broad topographic depressions. Although moraines can be physically traced along a more or less continuous line, the morainal segments from each lobe overlap eastward and also are progressively younger eastward by perhaps a total of at least 2000 years from Long Island to Cape Cod. This strongly suggests that each ice lobe remained relatively more vigorous toward the east and was controlled by successively more pronounced topographic basins in that direction. The few other minor ice readvances previously mentioned appear to be the only other evidence for climatic fluctuations of the ice sheet during deglaciation of New England.

Several mapped areas of both low relief and high relief illustrate how the local topography influenced the distribution of moraines and morphologic sequences. On the west side of the Narragansett Basin, an area of broad low to moderate relief, Schafer (1961) (Fig. 10) has shown the very close relation of successive morainal positions to meltwater deposits as the Buzzards Bay lobe (Fig. 8) retreated back

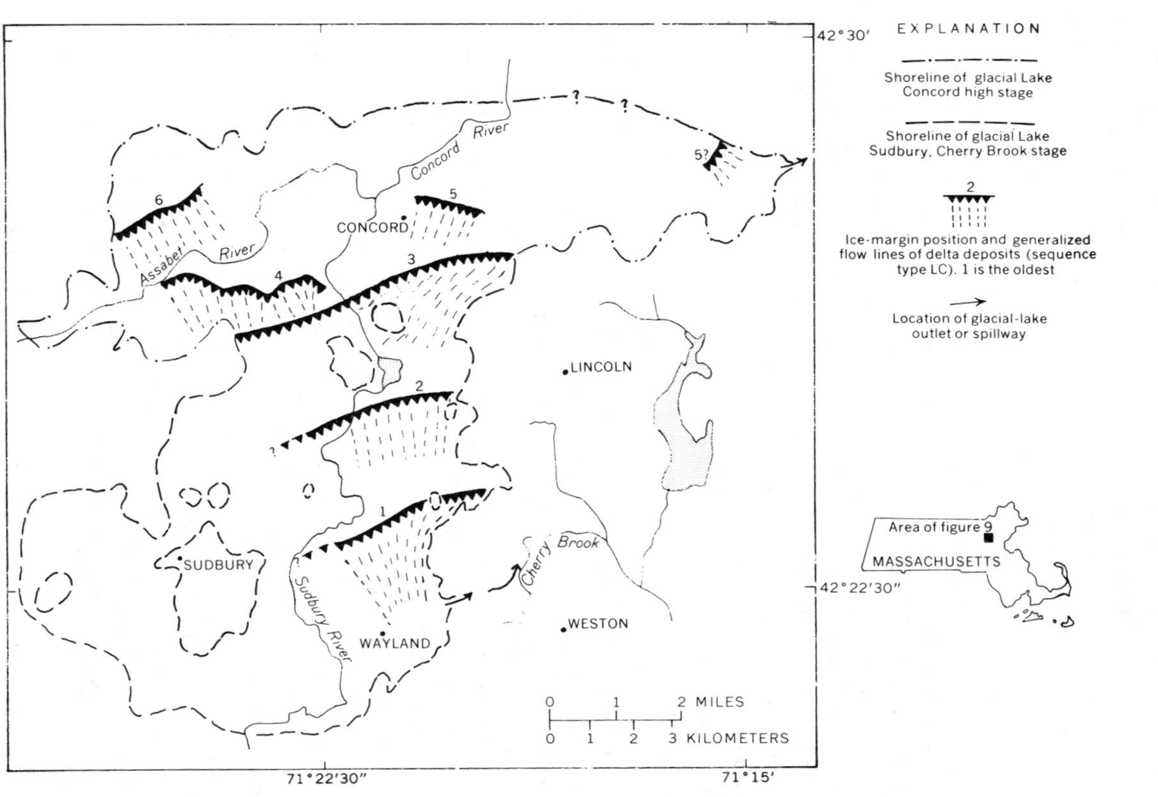

Figure 9. Distribution of several deltas (sequence type LC) and related ice-margin position in glacial Lakes Sudbury and Concord in the Concord, Mass. area. Ice-margin positions based on heads of outwash of the deltas. Extension of position 3 indicates final northward ice dam of glacial Lake Sudbury. Modified from Koteff (1963).

from the Point Judith moraine. The west side of the basin trends north-south, and the mapped units outline a northwest trend of the lobate ice edge as the ice retreated northeastward into the basin. The upland immediately adjacent was then free of ice. The distribution of moraines, constructed in part by live ice, and meltwater deposits in this area is transitional between coastal areas, where ice built prominent moraines, and the interior, where distribution of morphologic sequences shows stagnation-zone retreat was dominant.

In the Connecticut Valley near Springfield, Mass. (Fig. 11), Hartshorn and Koteff (1967) have shown the relation of meltwater deposits to topography where ice recession was characterized by stagnation-zone retreat and recessional moraines have not been recognized. The valley at this locality is over 15 mi (24 km) wide and the local preglacial relief along the east wall is about 900 ft (275 m). The ice lobe extended down valley about 5 mi (8 km) at two different positions as outlined by stagnant-zone meltwater deposits and a slightly later fluvial-lacustrine ice-contact sequence (type FLC). As measured by the stagnant-zone deposits, the width of the zone here was at least 1 mi (1.6 km) separating live ice to the north from glacial Lake Hitchcock to the south.

The Merrimack River valley in southern New Hampshire is essentially a smaller version of the Connecticut Valley both topographically and glacially. The valley north of Nashua, N.H. is about 5 mi (8 k) wide and the local relief is about 400 ft (122 m). Ice lobation down valley during deglaciation of the Nashua area was about 1.5 mi (2.4 km) at several stagnant-ice positions as shown by the distribution of delta deposits in the valley and fluvial meltwater deposits on the upland (Koteff, 1973) (Fig. 12).

In an area of smaller relief in Townsend, Mass., near the New Hampshire border, several morphologic sequences are found to have been controlled by meltwater channels eroded in till (Fig. 13). Together the sequences and channels define successive retreatal positions of the stagnant- ice margin. Local topographic control of sequence distribution is shown by the uncovering of successively lower base-level controls. Examples such as found in the Townsend area abound throughout southern New England, although the amount of development of the channels in this area appears to be rather exceptional.

Although the distribution of sequences is the best indicator of retreatal ice positions in New England, regional correlation should be done with care. The example given previously of neighboring parallel drainage systems, one with several sequences and the other with only one sequence representing the same amount of ice retreat, illustrates the necessity of careful detailed mapping across divides.

SOURCE OF THE MELT WATER DEPOSITS (The Dirt Machine)

Perhaps the most stimulating result of treating meltwater deposits as mappable morphologic sequences is the view that nearby live ice was the major source of sediment for each sequence. The mechanism suggested is that during

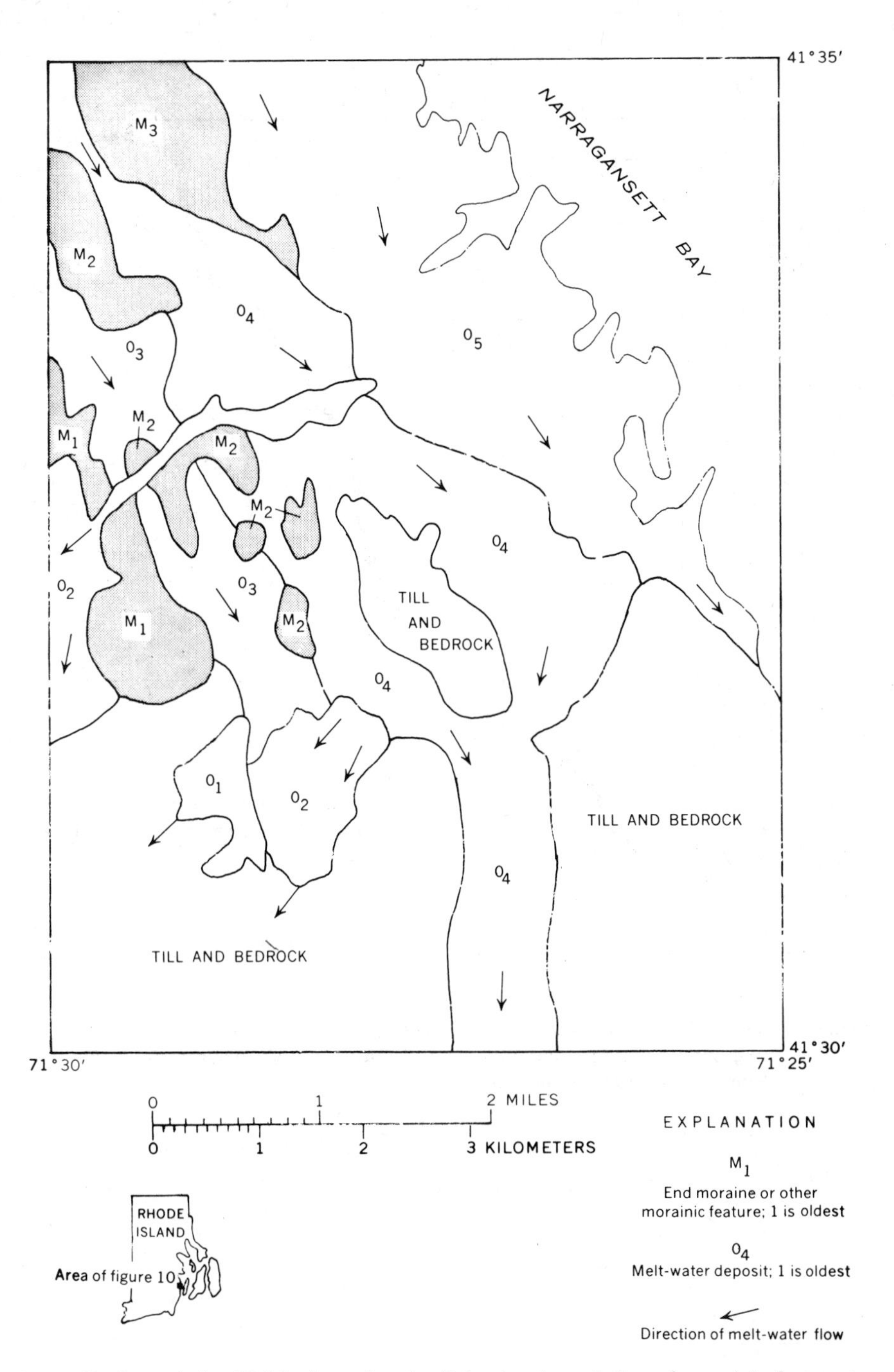

Figure 10. Part of the Wickford quadrangle, R.I., showing relation of morainic feature to meltwater deposits. Modified from Schafer (1961). For regional relations see also Figure 8.

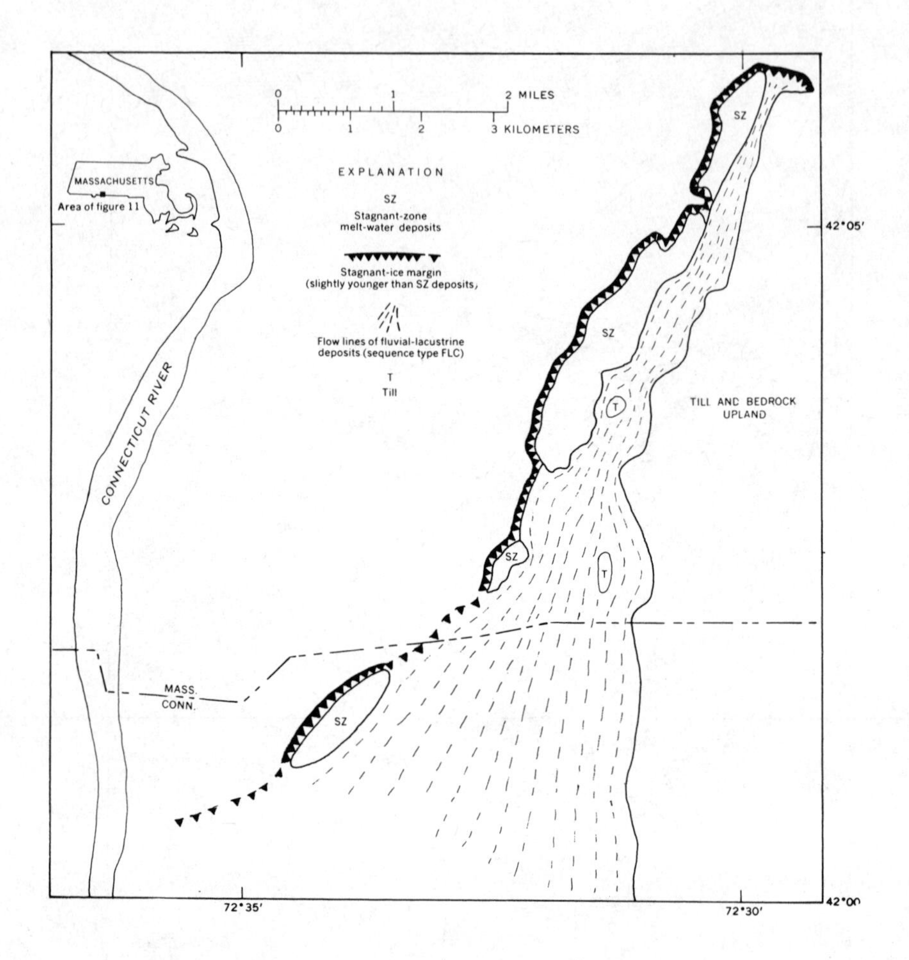

Figure 11. Part of the Springfield South quadrangle, Mass.–Conn. showing relation of stagnant-zone deposits and melt water deposits to an ice margin position. Modified from Hartshorn and Koteff (1967).

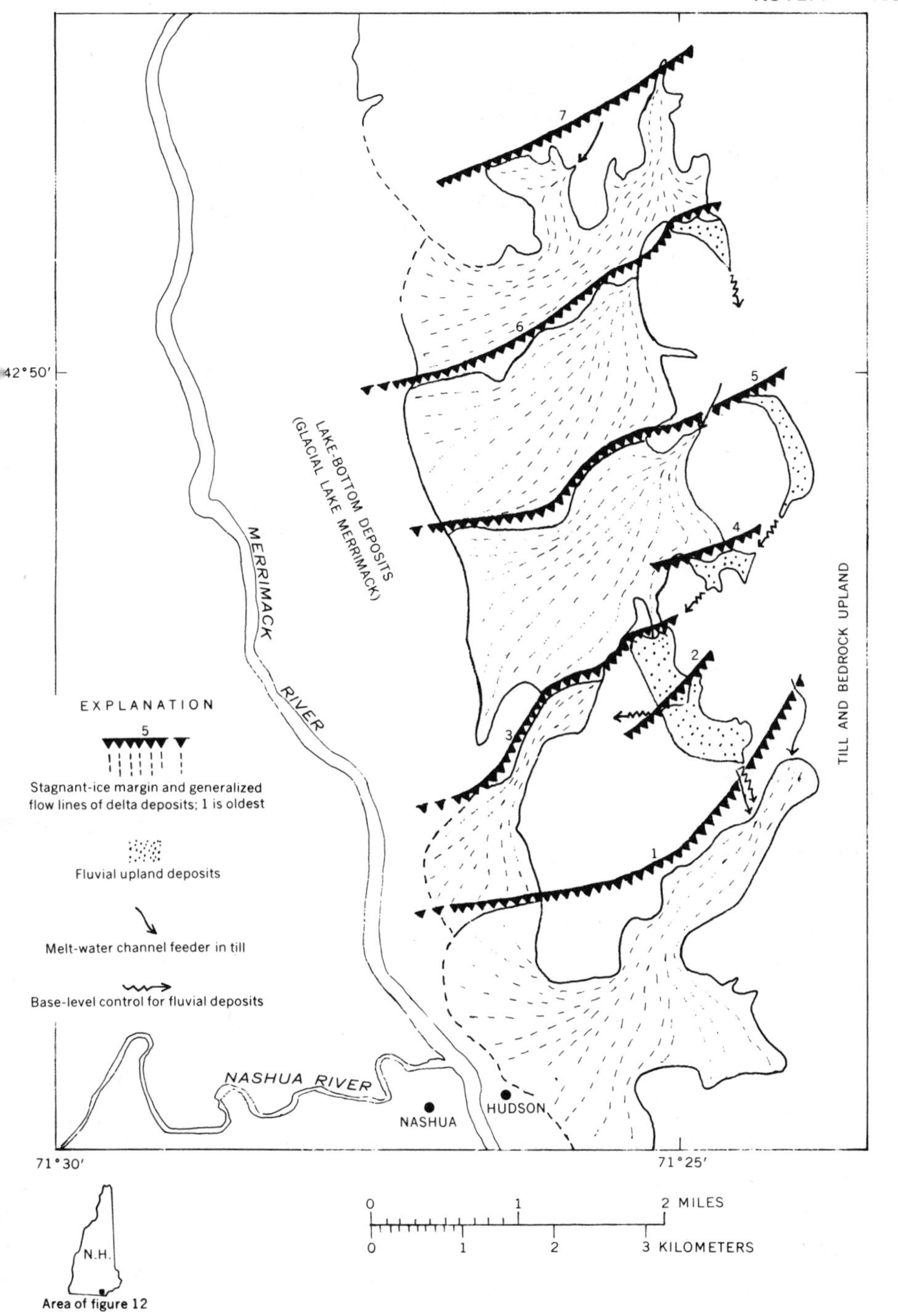

Figure 12. Part of the Nashua North quadrangle,N.H., showing relation of successove stagnant-ice margin positions to meltwater deposits. Modified from Koteff (1973).

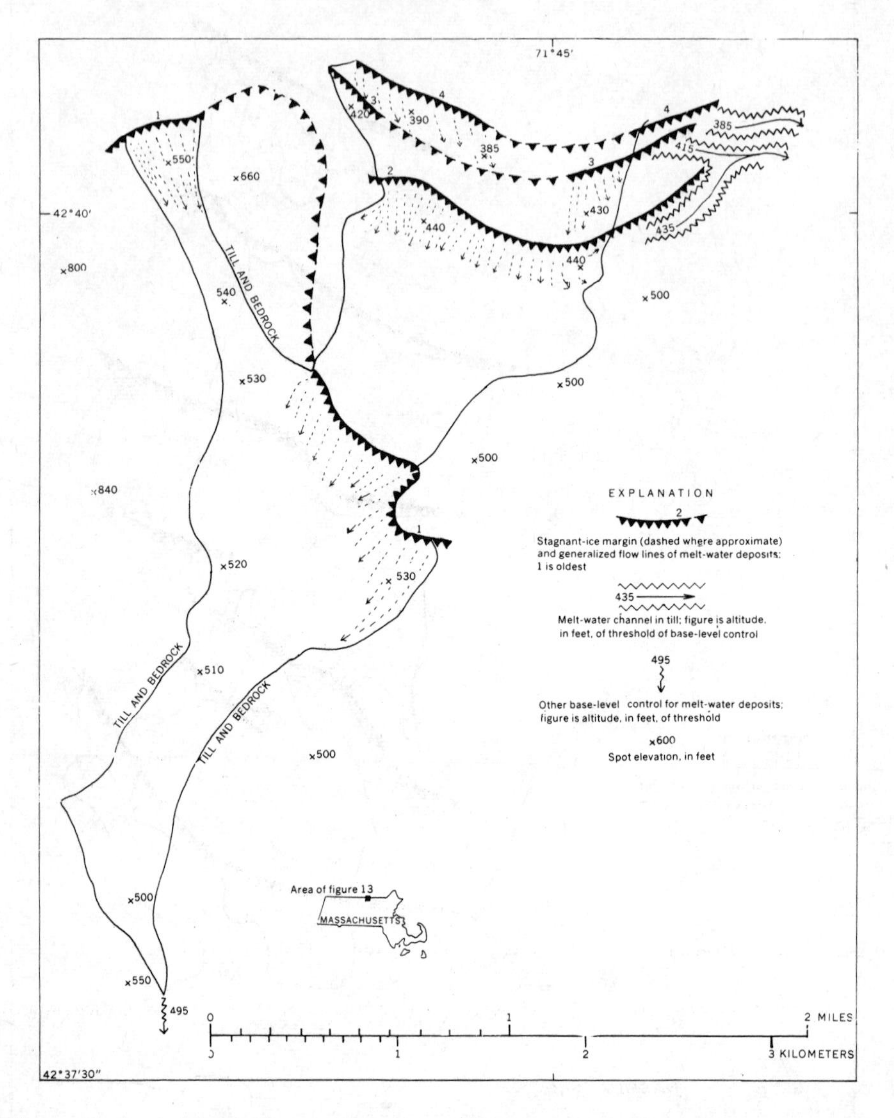

Figure 13. Part of the Ashby and Townsend quadrangles, Mass.-N.H., showing relation of stagnant-ice margin positions to melt water deposits and base-level controls.

stagnation-zone retreat material actively eroded by the ice sheet was brought up along a shear zone near the live ice--stagnant ice interface (see Fig.1). Meltwater, principally from the surface of the live ice, picked up material as it passed through the area of debris accumulation near the interface, and deposited it as outwash both within, but chiefly beyond the stagnant zone. The still moving ice is thus viewed as a conveyor belt, or "dirt machine", constantly replenishing the supply of material to be reworked and eventually laid down as meltwater deposits.

There appears to be little quantitative data, but a few studies, summarized by R.P. Goldthwait (1971), indicate that much of the material abraded by a continental glacier is carried forward near the bottom 10 ft or few meters of the ice sheet and that the surface is relatively free or clean of debris. The heads of outwash for many morphologic sequences on the other hand are commonly 150 ft (50 m) thick or more, which suggests that the abraded material was sheared up to at least that height at the edge of the live ice, where it became available to meltwater streams.

In southeastern Massachusetts and Cape Cod, large volumes of meltwater deposits are closely associated with the presence of nearby live ice. The retreatal ice positions are marked by end moraines, presumably deposited in large part from live ice, and by massive outwash plains. This area is one of generally low relief that has been nearly overwhelmed by meltwater deposits. There are very few higher hills of till and no bedrock outcrops that could have supplied material to the outwash. Thus ice erosion is the only logical source for the sand and gravel deposits. The orientation of the few drumlins and ice-molded features that do project above the moraines and outwash plains indicates a very close relationship between time of formation of the drumlins and retreatal positions of the ice (Fig. 14). Only till derived from the last ice sheet is known here. As the Buzzards Bay lobe retreated northwestward, the trend of the ice-lobe margin swung from nearly north to northeast. The change in ice-margin trend is marked not only by various meltwater deposits and morainic features, but also in several places by drumlin orientations. This suggests that at least the final molding of the drumlins was accomplished near the edge of the retreating ice. Perhaps the drumlins were constructed during the relatively short time that the nearby edge of the live ice occupied an adjoining position. If so, this further shows that the vicinity of the edge of the live ice was generally a zone of debris accumulation; this is consistent with the "dirt machine" concept. Elsewhere in New England drumlin orientations appear to be as closely related to local topography as is the outwash, but in the absence of morainal positions the relation between drumlins and ice edge is not as clear.

The "dirt machine" concept does not explain the source of all meltwater deposits, as other processes obviously were responsible for some deposition, but these appear to have been only minor when compared to the bulk of deposition. For example, flowtill found on or in meltwater deposits probably was derived from stagnant ice blocks. Also, in very few places material from stagnant ice blocks appears to have been deposited as water-laid sand and gravel. The deposits that occur in the vast majority of kettles and in ice-contact slopes of morphologic sequences, however, show no discernible differences in texture pattern from that expected in a typical ice-contact head to base-level control situation. If stagnant ice blocks were a significant source of debris, the pattern of coarse to fine textural distribution found in nearly all sequencs would have been interrupted in these places. This has been found to be so only in isolated examples. The textural pattern and position of melt-water deposits in southern and central New England appears to be most readily explained by the morphologic sequence concept.

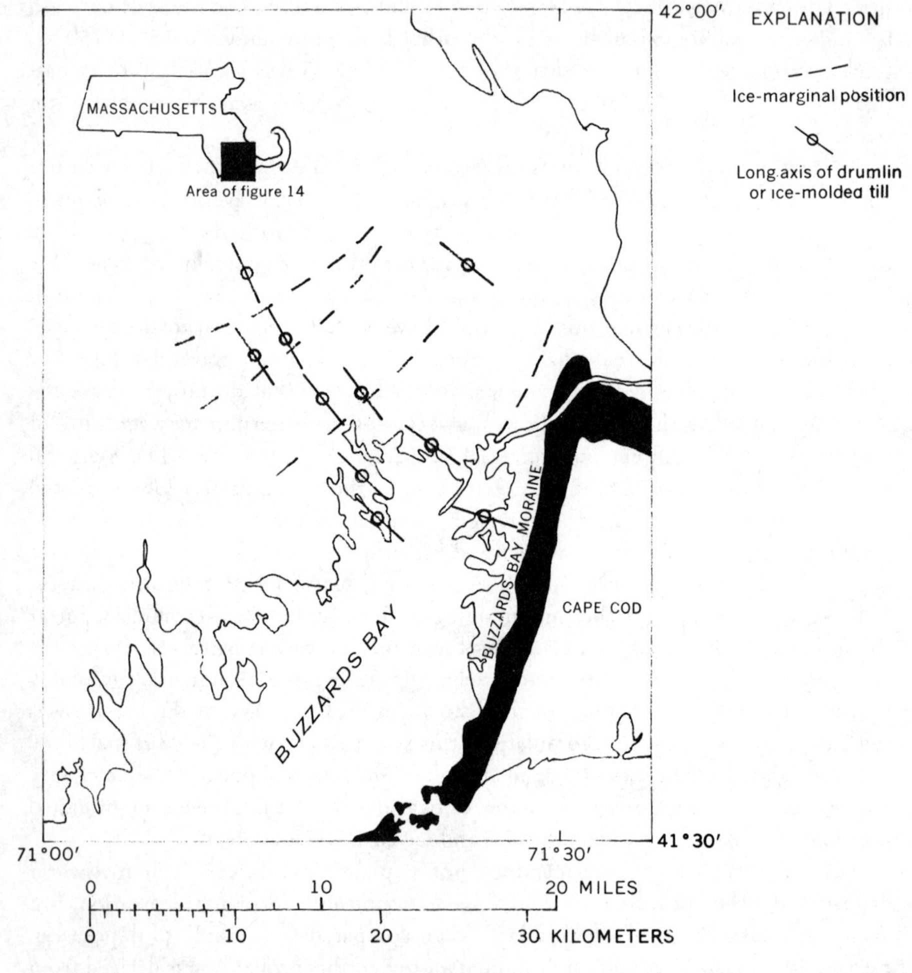

Figure 14. Relation of drumlin orientation to changes in trend of the receding ice margin in southeastern Massachusetts.

CONCLUSION

This paper is an attempt to collate the work of numerous people since Jahns originally proposed the sequence concept. The evolution of ideas over the last 20 to 30 years is the result of geologic mapping of many quadrangles, and it is too difficult to ascribe where proper credit is due for the major ideas presented here. However, John P. Schafer and Fred Pessl, Jr. have contributed greatly through discussions and friendly arguments. Thanks are due to Charles S. Denny and Jack B. Epstein for very helpful critical reviews of the manuscript. The errors, or course, are solely due the author. In summary, this paper shows that:

1. The morphologic sequence concept has been modified and enlarged since Jahns' original definition. While all morphologic sequences show some common characteristics, distinctive types of sequences can be classified according to differences in depositional environment and the presence or absence of an ice-contact head of outwash.

2. Mapping and interpretation of morphologic sequences are greatly aided by construction of profiles showing the pre-collapse depositional gradient of each sequence.

3. The concept provides a means of recognizing retreatal ice positions in areas where recessional moraines are absent.

4. Stagnation-zone retreat, influenced largely by topography, is the dominant process of ice wastage suggested by the morphologic sequence concept.

5. The source of the bulk of meltwater deposits that compose morphologic sequences was probably a shear zone near the live ice- stagnant ice interface. Isolated blocks of stagnant ice contributed little except morphology to the meltwater depositional history in southern New England.

REFERENCES

Antevs, Ernst. 1922. The recession of the last ice sheet in New England: Amer. Geogr. Soc. Research ser. no. 11, 120 p.

Borns, H. W., Jr. 1968. An end-moraine complex in southeastern Maine [abs.]: Geol. Soc. Amer. Spec. Paper 101, p. 249-250.

Brown, T. C. 1933. The waning of the last ice sheet in central Massachusetts: Jour. Geol., v. 41, n. 2, p.144-158.

Chute, N. E. 1959. Glacial geology of the Mystic Lakes-Fresh Pond area, Mass: U. S. Geol. Survey Bull. 1061-F, p. 187-216.

Feininger, Tomas. 1965. Surficial geologic map of the Voluntown quadrangle, Conn.—R.I.: U.S. Geol. Survey Geol. Quad. Map GQ-469.

Flint, R.F. 1930. The glacial geology of Connecticut: Conn. State Geol. Nat. Hist. Survey Bull. 47, 294 p.

Goldsmith, Richard. 1962. Surficial geology of the Montville quadrangle, Conn: U. S. Geol. Survey Geol. Quad. Map GQ-148.

Goldthwait, J. W. 1905. The sandplains of glacial Lake Sudbury: Harvard College Mus. Comp. Zoology Bull. v. 42, (geological series 6), p. 263-301.

———————1938. The uncovering of New Hampshire by the last ice sheet: Amer. Jour. Sci., 5th ser., v. 36, n. 215, p. 345-372.

Goldthwait, R. P., ed. 1971. *Till, a symposium:* Ohio State Univ. Press, 402 p.

Hartshorn, J. H. and Koteff, Carl. 1967. Geologic map of the Springfield South quadrangle, Hampden County, Mass, and Hartford and Tolland Counties, Conn: U. S. Geol. Survey Geol. Quad. Map GQ-678.

Jahns, R. H. 1941. Outwash chronology in northeastern Massachusetts [abs.] Geol. Soc. Amer. Bull., v. 52, n. 12, pt. 2, p. 1910.

———1953. Surficial geology of the Ayer quadrangle, Mass: U. S. Geol. Survey Geol. Quad. Map GQ-21.

Kaye, C. A. 1964. Outline of Pleistocene geology of Martha's Vineyard, Mass: U. S. Geol. Survey Prof. Paper 501-C, p. C134-C139.

Koteff, Carl, 1963. Glacial lakes near Concord, Mass: Art. 96 *in* U. S. Geol. Survey Prof. Paper 475-C, p. C142-C144.

———1964a. Surficial geology of the Concord quadrangle, Mass: U. S. Geol. Survey Geol. Quad. Map GQ-331.

———1964b. Geology of the Assawompset Pond quadrangle, Mass: U. S. Geol. Survey Geol. Quad. Map GQ-265.

———1966. Surficial geologic map of the Clinton quadrangle, Worcester County, Mass: U. S. Geol. Survey Geol. Quad. Map GQ-567.

———1970. Surficial geologic map of the Milford quadrangle, Hillsborough County, N. H.: U. S. Geol. Survey Geol. Quad. Map GQ-881.

———1973. Surficial geologic map of the Nashua North quadrangle, Hillsborough and Rockingham Counties, N. H.: U. S. Geol. Survey open file map.

Koteff, Carl and Volckmann, R. P. 1974. Surficial geologic map of the Pepperell quadrangle, Middlesex County, Mass, and Hillsborough County, N. H.: U. S. Geol. Survey Geol. Quad. Map GQ-1118.

La Sala, A. M., Jr. 1961. Surficial geology of the Southington quadrangle, Conn: U. S. Geol. Survey Geol. Quad. Map GQ-146.

Lougee, R. J. 1940. Deglaciation of New England: Jour. Geomorph., v. 3, n. 3, p. 189-217.

Pessl, Fred, Jr. 1966. Surficial geologic map of the Fitchville quadrangle , New London County, Conn: U. S. Geol. Survey Geol. Quad. Map GQ-485.

———1970. Surficial geologic map of the Woodbury quadrangle, Litchfield and New Haven Counties, Conn: U. S. Geol. Survey Geol. Quad. Map GQ-896.

Randall, A. D., and Pessl, Fred Jr. 1968. Surficial geologic map of the Danielson quadrangle, Windham County, Conn: U. S. Geol. Survey Geol. Quad. Map-660.

Richmond, G.M. 1953. Surficial geology of the Georgiaville quadrangle, R.I.: U.S. Geol. Survey Geol. Quad. map GQ-22.

Robinson, C. S. 1961. Surficial geology of the North Scituate quadrangle, R. I.: U. S. Geol. Survey Geol. Quad. Map GQ-143.

Schafer, J. P. 1961. Surficial geology of the Wickford quadrangle, R. I.: U. S. Geol. Survey Geol. Quad. Map GQ-136.

———1965. Surficial geologic map of the Watch Hill quadrangle, R. I. - Conn.: U. S. Geol. Survey Geol. Quad. Map GQ-410.

———1968a. Surficial geologic map of the Ashway quadrangle, Conn. - R. I.: U. S. Geol. Survey Geol. Quad. Map GQ-712.

———1968b. Retreat of the last ice sheet in New England [abs.] : Geol. Soc. Amer. Spec. Paper 115, p. 291.

Schafer, J. P., and Hartshorn, J. H. 1965. The Quaternary of New England: p. 113-128 in Wright, H. E., Jr., and Frey, D. G., eds., *The Quaternary of the United States,* Princeton Univ. Press, 922 p.

PART 2

GLACIAL LANDFORMS

No other geomorphic process produces such a wide range of landforms as those caused by glaciation. One reason for the great diversity is that glaciers have fewer constraints than most other agents. For example the work of streams is restricted to valleys and in a downslope direction as are gravity movements; groundwater sculpture is largely confined to carbonate terrane; waves and currents are significant only in the narrow band of shoreline, and; work of wind is generally subordinate to other geomorphic forces. Glaciers can produce major topographic changes in the widely contrasting regions of mountains, plains and plateaus, and in coastal areas. The variety of landforms occurs in regards to scale, type, shape, erosional forms, depositional features, in rock and in unconsolidated materials. Although some landforms such as drumlins (see Muller) can apparently be formed by a combination of erosion and deposition, many others are clearly the product of only a single mode of development. The largest and most spectacular of the erosion forms are ciques and U-shaped valleys (and fiords). Walcott cirque in the Antarctic is 10 mi wide with a 10,000 ft headwall, and Himalayan cirques are 1,000's of ft in relief. Many U- shaped valleys as in Yosemite, are more than 3,000 ft deep whereas erosion in several fiords exceeds 4,000 ft. The residual alpine erosion forms yield an impressive array of features including materhorns, aretes, hanging valleys, rock steps, and truncated spurs. Roches moutonnees provide some of the smaller-scaled landforms created by glacial erosion. There is a gradation from this form to those with a reversal of asymmetry as they pass through the various stages of crag and tail, rocdrumlins, drumlins, and till shadow hills. Such features provide a blend of the erosion and depositional processes. Depositional landforms also contain a large size range. Moraines may be measured in 100's of ft in relief and others may be followed continuously for 10's of mi. Even stratified deposits may create forms of large continuity. Eskers and kame terraces may be miles in length, and outwash plains (as on Long Island) may be 100's of sq mi in size or valley trains may be 10's of mi in length.

Chapter 5 by King analyzes the morphometry and harmony of glacial landforms. She provides a rationale and indicates the importance of such measurements because quantitative data can be statistically tested and aid in pointing the way for the development of meaningful relationships. Such insight can lead to greater understanding for methods of formation and origin of the feature.

Chapter 6 by Goldthwait covers a topic that is seldom treated. He synthesizes the work of many researchers in an unusually rewarding approach that quantifies the rate of landform development. It is significant that many features represent only the very late stages of glaciation.

Chapter 7 by Muller reviews one of the most controversial topics in glacial geomorphology. Drumlins have stimulated the imagination of many and the hotly contested debate between erosion and deposition theories of formation has raged for decades. Their polygenetic origin places them in a special landform category, that hopefully is fairly limited.

Chapter 8 by Coates reassesses several traditional ideas of the Glaciated Appalachian Plateau and concludes there are many misconceptions about the region. Recent literature is summarized and new data is added concerning the landforms for this large area.

CHAPTER 5

MORPHOMETRY IN GLACIAL GEOMORPHOLOGY

Cuchlaine A.M. King

INTRODUCTION

Morphometry consists of the measurement of form either from maps or in the field. It provides quantitative data that can then be analyzed to establish significant aspects of the morpology of the features investigated. As geomorphology is concerned with the understanding of the shape of features, morphometry provides essential material on which to base a valid analysis. From morphometric measurements it is possible to obtain the mean dimensions of specific features and to obtain a quantitative range within which they exist. The features in one area can be compared and contrasted with those in another area, providing often valuable clues as the the cause of differences in dimensions. In this contribution some examples of morphometric analysis on glacial geomorphology will be examined to illustrate the value of this method of study.

A great deal of work has been done in fluvial geomorphic morphometry, based on the early work in this field by Horton. This work has concentrated on the characteristics of drainage patterns, based on the ordering of streams and an analysis of the relationships between stream orders and other variables. Much work has also been done on morphometry in relation to the dimensions of stream channels and discharge, as well as other variables. There has been very much less work along these lines in relation to glacially modified landscapes. One of the reasons for this difference may be that rivers tend to become adjusted by negative feedback to conditions in which they are in balance with the processes operating to control their morphometric characteristics. Thus where conditions are comparable the fluvial morphometry will be similar, and under the controls of negative feedback, a state of dynamic equilibrium can be set up, leading to consistent morphometric relationships. In the instance of glacial landforms positive feedback seems to be an important aspect in the processes operating (King, 1970). The result of this dominance of positive feedback is that irregularities tend to be exaggerated to a considerable extent so that there is less uniformity among some glacial features. For this reason morphometric laws are not so readily established, and each valley or area must be considered as a unique situation. Recently, however, some attempts have been made to establish relationships somewhat similar to those established in fluvial morphometry, for example by V.M. Haynes (1972) in her work in west Greenland.

In a discussion of morphometry in glacial geomorphology it is necessary to consider both a variety of techniques of measurement, and also a variety of different forms, including erosional and depositional features. The paper is subdivided on this basis: first some topological relationships will be considered, in which the network characteristics are explored and the exact distances and orientations are not the values analyzed; secondly geometrical properties will be examined. In the analysis of geometric properties, the results can be applied in one, two, or three dimensions. One dimensional values are of relatively limited use, so two and three dimensional situations will be considered. The two dimensional situations are of two types: (1) the analysis of profiles, with one vertical and one horizontal dimension, including both longitudinal and cross profiles. (2) reference to two horizontal dimensions, in which the emphasis is on outline form and orientation patterns. The three dimensional situation applies to the volumes of features, which can be either erosional, such as cirques, or depositional, as drumlins. Examples of these different dimensions of morphometric analysis will be considered.

Morphometric analysis can be applied to both erosional and depositional forms of glaciated terrane. The erosional features that have been analyzed most intensively so far are cirque basins, including the shape, orientation and elevation; cirques can be readily measured from good topographic maps. One problem is that these features usually occur in mountainous areas where there may be an absence of large scale accurate topographic maps. There are, however, enough examples to enable interesting relationships to be established, which provide useful evidence of the processes forming the features and the controls on which their dimensions depend. Glacially eroded valleys also provide froms that can be measured and analyzed by morphometric methods. The fitting of geometric curves to these features can provide useful evidence concerning their formation, and the processes that modify them. They are large units on the landscape, and as such numbers of examples must necessarily be fairly small in any one region. This aspect is related to the problem of sampling, which is often a significant aspect of morphometric analysis. Where a few large features are involved it is often possible to measure all the individuals, for example all the cirques in one mountain group, or all the longitudinal profiles of glaciated valleys in the same area. Cross profile analysis does, however, involve a sampling problem, as there are an infinite number of possible samples along the valley length that could be measured.

Smaller-sized features can also be measured, often in the field, and these usually must be measured on the basis of a sampling plan. An example is a set of measurements on the shape of glacially eroded stones, or the orientation pattern of glacial striae. In both instances it is necessary to select a random sample of features to measure, so that the results of a statistical analysis can be applied to the entire population of features.

Depositional features resulting from glacial action are many and varied. Again some are fairly large, while others are small, also both map data and field measurements may be involved. An example of the analysis of larger features would refer to the gradient of large outwash plains, which could be measured in their entirety, i.e. all plains could be included, but sampling of each one would be required. At the medium scale there are features such as drumlins and moraines. Drumlins in particular have been studied by mophometric techniques because they usually occur in large numbers in distinct drumlin fields so that sampling is necessary, but there are sufficient numbers to allow a significant statistical relationship to be established.

On the smaller scale in studying glacial deposition the analysis of till fabrics provides a method that has been extensively used and studied and it has provided much valuable data. This is essentially a field method, but statistical analysis is required to obtain maximum information from the data, and to produce valid results applicable to the whole deposit rather than only to the very small number of individual stones that can be measured. The sampling problem becomes really vital in such studies, as indicated by the work that has been done on the variability of till fabrics over short distances and in connection with variation in the size and shape of the stones measured. Till fabric analysis is a large topic in itself and will not be further discussed in this chapter, which concentrates on the type of morphometric analysis that can be obtained mainly from measurements made on detailed topographic maps.

TOPOLOGICAL RELATIONSHIPS

Features forming networks are suitable for topological analysis. An illustration of the application of topological relationships to esker patterns in Baffin Island has been discussed by Doornkamp and King (1971 p. 316-320). This study shows that it is possible to differentiate two distinct types of esker ridge patterns in terms of simple graph theory, using the relationships between nodes, arcs, and regions. The mu (μ) index, given by $a - n + p$, where a is the number of arcs, n is the number of nodes, and p is the number of subgraphs. The value of mu is higher for a complex pattern, in which ridges bifurcate and rejoin, so that this index can be used to separate complex patterned eskers from a simpler linear type of feature.

Topological analysis has been successfully applied to fluvial systems in connection with network analysis of stream patterns, whereby relationships between streams of different orders have been established. The situation with respect to glacially modified networks is complicated by the fact glaciers frequently form diffluent and transfluent channels,diverging from the original valley. The pattern of glacial valleys is thus more complex than that of fluvial valleys, where only minor divergence occurs, for example in alluvial fans, deltas, and braided streams. Little work has been attempted as yet to analyze the pattern of glacially modified troughs in areas of complex glacier flow. The mapping of glacial troughs, for example by K.M. Clayton (1965) in the Finger Lakes region of New York, yields a pattern that shows many linkages and bifurcations. The glacial erosion accompanying the formation of this complex pattern has had a profound effect on the relief of the area and on its present fluvial morphometry. This area is one near the margin of a large continental ice sheet, where the ice was impinging on and eroding into an upland region in its path.The major troughs are, therefore, intrusive, and where the ice was concentrated major through valleys have been eroded.

In other areas, such as the English Lake District, the area formed a center from which ice dispersed in all directions. Transfluent glaciers flowing from one center produce a pattern markedly different from the normal stream pattern in which small tributaries unite to form one major stream. In such a transfluent system, the ice streams spread out from a small central area, producing major bifurcations and imposing a radial pattern on the major valleys of the area.

Detailed studies of the topological relationships between the different elements of complex glacial trough patterns could well reveal useful information concerning directions of ice-flow, snow lines, and the relationships between neighboring glacial centers. In the normal fluvial situation the bifurcation ratio gives the relationship between streams of two orders. N_1/N_2 and N_2/N_3 for example; the result is always greater than unity. In the case of diffluent glacial systems the ratio could be less than unity, where many glaciers flow out from one major trough. The study would need to be based on past glaciers rather than on present streams occupying the glaciated valleys. The patterns visible on air photos of parts of Antarctica and Greenland illustrate the type of pattern that could be analyzed. It would be necessary to determine the direction of ice-flow, which is not always immediately apparent.

GEOMETRIC RELATIONSHIPS

Two dimensional–profiles

When profiles are considered there is one horizontal dimension and one vertical dimension. In this aspect of morphometry the actual survey profiles can be used or profiles can be constructed from good topographical maps with small contour intervals, such as the 1:24,000 U.S.G.S. maps or the 1:25,000 British Ordnance Survey maps, the former usually have contour intervals of 10, 20, or 40 feet depending on the total relief, while the latter all have a 25 feet contour interval, thus providing adequate control, particularly in steep mountain areas.

One aspect of morphometry that can be usefully applied to glacial landforms is the fitting of specific geometrical forms to the natural profiles. Following the early work of Svensson (1959) in this field, various attempts have been made to fit curves of specific geometric form to the cross profiles of glaciated valleys.

Cross profiles of glaciated valleys

Svensson showed that for a valley in Sweden the cross profile closely approximated a parabola, a curve given by $y = ax^b$, where y is the vertical distance from the valley bottom, and x is the horizontal distance from the mid-point of the valley. a and b are constants for any one curve, b, the exponent should be 2 for a true parabola, which is defined by $y = ax^2$. The equation in this form can be fitted simply by trial and error, but a more useful result is obtained if the equation is used in its logarithmic form, and the curve fitted by means of least squares regression. The equation is used in the form $\log y = \log a + b.\ \log x$. The value of b is now obtained from the regression equation, and a can also be ascertained in this way.

The value of r, the correlation coefficient, provides a measure of the goodness of fit of the parabolic curve to the actual hill-side profile. The value of b, the exponent in the equation, is the slope of the regression line through the points, and provides a measure of the closeness of the fit to a true parabola; it seems to indicate a degree of glacial adjustment of the cross profile. The closer b approaches the true parabola value of 2, the more the profile has been eroded to fit the form of the glacier moving down the valley. Examples of the fit of a number of glaciated valleys provide the following results: Bishopdale in Yorkshire, modified by a diffluent glacier, produced a relationship $\log y = -2.924 + 1.6785 \log x$. Grisedale in the Lake District gave an equation $\log y = -3.5384 + 1.953 \log x$, while the Nantlle valley in Snowdonia, Wales gave $\log y = -2.9235 + 1.7483$, the r values for the three profiles respectively were 0.9931, 0.9899 and 0.99545, showing a good fit in all three areas. The results suggest that Grisedale provides a good example of a glacially modified valley cross profile that has been adjusted to fit the parabolic form. This valley is situated near the center of the Lake District ice mass and was actively eroded. Both the other two valleys were cut by diffluent glaciers, which may not have carried as much ice, and which would also have required more erosion, owing to the much smaller preglacial streams occupying these valleys compared with Grisedale. Graf (1970) has also obtained similar results from an analysis of this type in the Beartooth Mountains, obtaining values of b between 1.5 and 2.0, the value increasing with the degree of glacial erosion, until almost a true parabola fit the glaciated valley sides.

It is also possible to fit other mathematical curves to glaciated valley slopes, as exemplified in the analysis of the slopes of Albion and Kiowa in the Front Range of the Colorado Rocky Mountains (Doornkamp and King, 1971 p. 281-285). These slopes were surveyed in the field so that more accurate data were available for analysis. The testing of a number of different types of curves showed that the closest fit for a slope, extending from 3770 m to 3900 m, which was that of Albion, was provided by the equation $y^2 = a + b^2$, with an r value of 0.9994. A good fit was also provided by the equation $\log y = a + bx$, with $r = 0.9972$. Both these curves have the important feature in common with the parabola that the scale is contracted vertically upwards, indicating a steepening of the slope upwards from the valley floor. The second equation applies to the long slope of Kiowa, which extends from 3770 m to 4050 m. These rather more complex curves allow a wider range of gemetrical forms to be examined and related to the surveyed slope measurements.

Longitudinal profiles of glaciated valleys

The most significant characteristic of the long profiles of glaciated valleys is their irregular nature compared with those of fluvial valleys. It has been shown that in many instances the longitudinal profiles of river valleys, where they have not been affected by rejuvenation, fit a logarithmic equation, given by $y = a - k \log(p-x)$, where y is the height, x is the distance from the mouth, and $p-x$ is the distance from the source. a and k are constants that must be found for each stream.

It is rare for a simple logarithmic equation to fit the profile of a glaciated valley, because of the irregularities that develop by positive feedback loops between the variables that are involved in the processes of glacial erosion at the bed of the glacier. Any preglacial variations in slope, caused by knick points, variations in discharge due to tributary junctions, or to variations in rock resistance are enlarged by glacial erosion. Glaciers have the ability to enhance these irregularities and to convert them into rock basins with the intervening bars or riegels. As suggested by I.S. Evans (1969) a form of pressure release could help to account for the great depth to which glaciers can erode the rock basins. Joints parallel to the surface develop due to unloading, when overlying rock is removed. Horizontal stresses are greater than vertical loads in the earth's crust. Thus the walls of a deep glacial troughs are pushed laterally widening the valley. Compression is relieved along the walls by jointing, which further enhances the widening process. Meanwhile the floor of the trough is under increasing compression. The excessive compressions causes uparching of the rocks in the deeper parts of the trough, which again produces weaknesses in the form of joints that the glacier can exploit in its erosive capacity. As deepening procedes the stresses become greater, more than counter-balancing the pressure due to increasing ice weight, thereby causing the process to continue by positive feedback. Thus erosion increases with ice thickness when the ice is confined to a channel. Most of the major glacial troughs must have been cut when the ice was at its thickest. Many of those still occupied by glaciers are not being deepened so effectively at present. Even now, however, the large quantities of glacial flour in meltwater streams testifies to the effectiveness of glacial scouring.

A morphometric study of the relationship between the depth of glacial erosion, as recorded in the level of the glaciated valley floor at specific distances from the head of the valley, and the size of the glacier occupying the valley, as recorded in the area of the catchment area at its head, revealed a close correspondence. The measurements were made in the Front Range of the Colorado Rocky Mountains where the glacial valleys are fed from cirque basins to the east of the continental divide. Similar relationships have been recorded by V.M. Haynes (1972) in a discussion of the relationship between the cross sectional area of glacial troughs in west Greenland draining from the Sukkertoppen ice cap. She showed that the cross sectional area of the trough was closely related to the longitudinal cross sectional area, giving the relationship $\log y = 0.2893 + 1.2827 \log x$, where y is the cross sectional area and x is the longitudinal section area. The value of the correlation coefficient was 0.92 for 19 sets of observations. There was also a fair correlation of $r=0.71$ between the area of the catchment basins for the glacier and the cross sectional area of the troughs, implying also a relationship between the longitudinal sectional area. In this instance the glaciers were of the Icelandic type so that the troughs were of the outlet type and not the Alpine type, as in the Colorado Rockies. Another important variable in the Greenland troughs was the continentality, which would determine the volume of ice to be discharged, and hence the rate of flow. When this variable was included in the analysis the value of r increased to 0.93. Thus it is unlikely that purely morphometric variables, such as valley elevation and catchment area will explain all the variability in the relationship between them; glaciologic factors may also be important. In any one small area, however, where conditions of glaciation are comparable, it is possible to demonstrate significant relationships between the morphometric variables related to the processes of glacial erosion, whereby the longitudinal profile of the glaciated valley is shaped.

Cirque elevations

Cirques form distinct units that are amenable to a variety of morphometric analyses. Those that have been studied include; (1) elevation, either of the lip or summit of the head wall, (2) the orientation, which is a directional value, involving two horizontal dimensions, (3) cirque area, also expressed in two horizontal dimensions, and (4) cirque profiles, using a vertical and horizontal dimension. The first and last are relevant to this section, the middle two are dealt with in the next.

Cirque elevations are closely related to the elevations of the mountains on which they occur, but within some mountain groups there are systematic variations in the elevations with respect to orientation or position, while in others there is no systematic pattern. The contrast between the cirques of the Snowdonian mountain group and those of the Lake District in England illustrate some of the variables on which these relationships depend. In Snowdonia in north Wales, as pointed out originally by B. Seddon (1957), and treated in further detail by Doornkamp and King (1971, p. 285-293) and more recently by D.J. Unwin (1973), the cirque altitudes form a systematic pattern. There are four mountain groups in the area, and Unwin has shown that the lowest cirque elevation and mean elevations vary from southwest to northeast as follows:

	No.	Lowest Height m	Mean Height m	Coefficient of variation m
Hebog group	19	238	366.6	12.3
Snowdon	25	270	473.3	31.3
Glyderau	21	315	538.7	23.4
Carneddau	19	436	640.8	16.1

Grand mean: 496.6 m

Crest altitudes vary in a similar way, with a mean for all 81 sites of 795.1 m. Unwin has fitted trend surfaces through the data to show that there is a systematic eastward slope of the elevations at 13.3 m/km. This is a steep slope compared to that recorded in other mountain groups, such as the 4.9 m/km recorded in the 175 corries analyzed in Scotland by G. Robinson et al. (1971). A still steeper gradient was found by J.T. Andrews et al. (1970) of 17 m/km in east Baffin Island. In the southwest Lake District, on the other hand, different mountain groups showed little systematic variation in altitude with respect to orientation. An analysis of variance gave insignificant F values of 0.17 for between aspect values and 0.58 for between groups values of elevation.Thus the individual mountains produce different results, some being systematic while others are not. The reason for the difference in elevations in the north Welsh and southwest Lake District hills is due to the arrangement of the hills with respect to the incidence of precipitation-bearing winds. In the Welsh hills the cirques are lowest in the southwestern groups, which are the wettest. In this hill group the individual mountain masses are arranged in a southwest to northeast sequence, each providing a measure of rain shadow for that further northeast, and thus reducing the precipitation. In the Lake District Hills the groups are arranged more on a northwest to southeast pattern, so that each group received about the same precipitation, while the elevations of the groups also did not favor the development of rain shadows so effectively. In Scotland the mountains cover a wider area, but the same pattern applies, with cirque altitudes increasing to the northeast where conditions are dryer. Thus an important variable in the cirque altitude is its position in the mountains with respect to the incidence of snow-bearing winds, which were probably the same as those which now bring the maximum precipitation.

Cirque profiles

The degree to which glaciers modify the hillside hollows to form cirques depends on a number of variables, and there is a complete range of features from minor nivation hollows, which are shallow scallops of the hillside occupied by snow patches for a relatively long period during spring and summer, to fully developed cirques, which have a deep ice-eroded rock basin, often containing a cirque lake held in by the rock rim. The variables that determine which stage of development is reached include: (1) the length of time of glacial action, (2) the relationship between snow accumulation and ablation, which in turn determines the thickness of the snow cover and the length it can survive, and (3) the temperature relationships, which determine the type of glacial action. True cirques can form only where true glaciers exist. a minimum thickness of snow is needed to form ice and initiate movement. Ice must be 25 to 30 m before flow becomes effective, and flow is very slow even when the ice is 50 m thick, as evidenced by the observations in the tunnel dug through Vesl-Skautbreen in the Jotunheim, Norway, and described by McCall ((1960) in Lewis (ed.)). This small basin, however, has the form of a typical cirque with an almost circular long profile in which a small lake is dammed. It closely resembles the longitudinal profile of Blea Water in the Lake District, which is also a typical cirque basin.

A morphometric measure of the degree of cirque modification is the relationship between the maximum gradient of the headwall at the back of the cirque basin and the slope of the cirque floor, which in a well-developed cirque has a negative slope, producing the closed basin. Cirques and cirque-like features can be identified by the gentler basin floor slope, which lies between the steeper backwall and the steeper valley slope immediately below the cirque lip. These gradient differences can usefully be expressed as the ratio of the cotangent of the floor slope to that of the headwall slope, a value that will always exceed unity. It is better to express this value as the difference between the two slopes in degrees, as the reversal of slope can then be included more readily. A value of this type has been suggested by I.S.

Evans (1969 p. 373), who calls this aspect of cirque geometry the closure in profile. Well developed cirques can have values between 45° and 90° depending on the verticality of the backwall and the degree of reverse slope, which is often difficult to obtain from a map if contours are not given in the cirque lake. Theoretically the value could amount to well over 90°. Evans also uses the value of the closure in plan to denote the horizontal form of the cirque. The value is determined by the range of azimuth of the longest contour. Values can exceed 270° for complex cirques deeply cut into the valley head; such features usually consist of a number of tributary cirques forming one major valley head.

In another study the variation in cirque gradient ratios provided one of the most useful means of distinguishing different cirque types and degrees of development in various mountain groups in the United States, ranging from the poorly developed cirque-like forms of the Catskill Mountains of New York and Pennsylvania to the well developed cirques of parts of the Rocky Mountains in Montana and Utah.

Table 1 summarizes some of the results of this study. The variable most diagnostic of degree of cirque development is the flatest/steepest ratio, which is the ratio of the maximum horizontal distance covered by 200 ft vertically divided by the minimum horizontal distance covered by 200 ft vertically. The former represents the cirque floor and the latter the back wall. Values range from 28.9 in Utah cirques to 4.7 for those in the mountains of New England. Some t-test and analysis of variance tests are given in the table, showing that the degree of cirque development can be statistically differentiated in the different mountain areas on this basis, the F value for all areas being 27.8. Even for the two most similar values, those recorded in the Catskill Mountains and in the mountains of New England, the t value of 2.40 is signficant at 98%. The two-way analysis of variance results show that the differences in the flatest/steepest ratios cannot be explained by variations in overall gradient, or cirque area, although there is a tendancy for larger cirques to show a higher flatest/steepest ratio.

Curve fitting can also be applied to cirque longitudinal profiles. It has been shown that a well developed cirque approximates a circular arc, the radius of which provides a measure of the cirque capacity. The circular arc becomes perfected as the rotational flow of the glacier gradually modifies the cirque floor to fit the slip lines of glacier flow. Vesl-Skautbreen and Blea Water both have a similar circular profile with a radius of about 240 m; these are small cirques but they are well developed and adjusted to the glacier cutting them so that they represent an equilibrium form of geometric type.

Two-dimensional–areal and orientational–horizontal dimensions

Two horizontal dimensions involve the projected ground shape of features, their outlines, as well as the orientational characteristics. The shape involves the size also, including the area covered by specific forms. Spacing problems can also be considered in this category from the point of view of morphometric analysis. Again

Table 1
CIRQUE MORPHOMETRY

Areas measured: New England, Utah, Montana, Catskills (New York)

Variables measured:
- Location – °, ′
- Area – $miles^2$
- Perimeter – miles
- Highest elevation – feet
- Lowest elevation – feet
- Relief – feet
- Length – feet
- Width – feet
- Length/Width ratio
- Length/Relief ratio – overall gradient
- Flattest Gradient – measured over 200 ft vertically
- Steepest Gradient – measured over 200 ft vertically
- Flattest/Steepest Gradient ratio
- Orientation – eight directions, N, NE, E, SE, S, SW, W, NW.

Some results.
Means and standard deviation.

Means

Regions	Area	Overall gradient	Flatest/Steepest ratio	length/width ratio	Relief	Length
New England	0.73	4.32	4.70	1.50	1500.7	6113.3
Montana	0.47	2.49	11.20	1.19	1632.8	4253.0
Utah	1.08	4.60	28.94	1.64	1559.3	7090.3
Catskills			5.62			
All regions		3.81				

Standard deviations

Regions	Area	Overall gradient	Flatest/Steepest ratio	length/width ratio
New England	0.374	1.3321	1.936	0.4535
Montana	0.324	0.7992	9.659	0.4352
Utah	1.322	1.5184	31.58	0.5208
All		1.5588	2.137	0.5087

t-tests

Area	New England–Montana	t = 2.872 99%
	Catskills–Montana	t = 0.484 NS
Length/Width ratio	New England–Utah	t = 1.13 NS
	New England–Montana	t = 2.667 99%
Flatest/steepest ratio	Catskills–New England	t = 2.40 98% 138 d.f.
Analysis of variance. All areas	Flatest/steepest ratio.	F = 27.8058 99% F 3, 195
Analysis of variance. All areas	Areas	F = 20.1169 99% F 3, 195

Number in sample Catskills 110, New England 30, Montana 30, Utah 29

Analysis of variance two-way.
Areas and overall gradient in terms of Flatest/steepest ratio

	Flat 4	Medium	Steep 2.6
New England	4.8	4.6	3.5
Montana	24.0	14.0	8.8
Utah	27.6	27.4	49.0

F between gradients 0.233 NS
F between areas 7.25 95%

Cirque area and overall gradient in terms of Flatest/steepest ratio

	Flat	Medium	Steep
	Flat	Medium	Steep
Large 1	31.9	15.5	38.0
0.5–1.0	9.4	16.9	8.5
Small 0.5	6.4	11.5	11.2

F between gradients 0.229 NS
F between area 4.328 NS (F 2, 4, 95% = 6.94)

the smaller features, such as cirques and drumlins provide a sufficient number of examples to enable meaningful relationships to be established between the morphometric variables and the formative processes. Specific examples that could be discussed include the area of cirques glacier in relation to characteristics of the valleys into which they discharged ice. An example of the type of result that such a study can reveal has already been commented upon in the discussion of relationships between longitudinal profiles of glaciated valleys and their catchment areas. This matter will, therefore, not be considered further. Two other relationships will, however, be mentioned, one dealing with the orientation pattern of cirques, the other considers the outline and orientation of drumlins.

Cirque orientations

Cirque orientation patterns are related to their altitudinal patterns which have already been mentioned. Cirques are most likely to form in those areas where snow can accumulate most readily and remain unmelted for longest. In some areas these sites are likely to coincide. In the northern hemisphere mid- to high-latitude areas where cirques are normally found in the largest numbers, the most favorable orientation for the preservation of snow is on the northeast side of a hill mass, because this direction receives the minimum effective insolation. The afternoon and evening sun is more effective in melting snow than the morning sun, which shines into the east-facing cirques more effectively. In these areas also, particularly those on the western side of the land masses, the northeastern aspect is in the lee of the precipitation-bearing westerly winds. Thus the northeastern aspect is the most favorable for both criteria and cirques would be expected to be best developed and most numerous with this aspect. This is indeed found to be true.

An analysis of the orientation pattern of cirques in the most marginal zones of their development, for example in the Dingle Peninsula of southwest Ireland, showed that of 34 cirques all but one faced to the northeast. This area with its mild oceanic climate is marginal for cirque formation so that only the most favorable situations were able to accumulate enough snow to form cirque glaciers during the last major ice advance, from which period these cirques data. An analysis of the orientation pattern gave the following results: vector length 21.65, strength 63.68, vector mean $026.69^{\circ} \pm 19.08^{\circ}$ at 95% confidence, standard deviation 54.65°.

In the Welsh hills and the Lake District a wider range of orientations is present, but in the Lake District, as already mentioned, there is no relationship between aspect and the elevation of cirque lips. There is, however, a well established relationship in Snowdonia, with the majority of cirques facing northeast. Unwin gives the mean orientation for 84 cirques in north Wales as 47.5°, with only three cirques in the southwest quadrant. There is also a relationship in this area between cirque altitude and aspect, but it is not highly significant, the F value being only 3.92. The means values for the northeast-facing cirques is 1566 ft, for the southeast-facing ones 1833 ft, and the northwest-facing 1933 ft.

An analysis of the frequency of cirques facing in different directions in the Lake District provides more significant results. Five different groups of hills and four different quadrants were used to provide the following data:
Frequencies of cirques in the 5 groups were 8, 28, 11, 15, and 11, while the frequencies in the four quadrants were northeast 38, southeast 13, northwest 18, and southwest 4, giving a total of 73 cirques in all. A two-way analysis of variance of the values showed that there was no significant difference between the numbers in the different hill groups, but that the F value of 4.94 between the numbers facing in different directions was significant at the 97.5% level for 3, and 12 degrees of freedom. There is thus a significant difference in the numbers of cirques facing in the different directions, but no significant difference in their altitude. This result implies that cirques will only form in those situations where the aspect is favorable, both from the point of view of receiving snow, that is in the lee of the hill crest where snow can accumulate, and in areas where melting is minimized. Thus most of the snow received can survive to add to the last year's snow, forming firn and later ice, until finally a cirque glacier is created and a cirque is eroded. Cirques probably only form effectively where the snowline is sufficiently below the hill crest for snow to accumulate in the favored areas. Differences in altitude of the cirques are only significant over considerable distances, as across Scotland, or where one hill group forms a rain-shadow to those in its lee, as in Snowdonia, in north Wales.

In the Catskill Mountains the orientation frequency of cirque-like forms at the head of the valleys showed a significant predominance for those facing south, unlike the results obtained in other mountain groups, including New England and the Rocky Mountains as shown in Table 2. This result suggests that the features in the Catskills were not true cirques. They were probably formed by nivation processes, which would operate best on south-facing slopes where thaw-freeze activities would take place most effectively.

All mountain groups represented in Table 2 show strong preferred orientations, the values of x^2 being largest for the cirques in Montana. The most favorable quadrant is the northeast one in New England and Montana, while Utah shows a predominance of the southeasterly quadrant.

Table 2
Orientation frequency of cirques in four mountain areas of U.S.A.

	Catskills	New England	Montana	Utah	New England Montana, Utah
N	21)	4)	2)	1)	7)
	) 38	) 11	) 16	) 7	) 34
NE	17)	7)	14)	6)	27)
E	8)	2)	5)	11)	16)
	) 14	) 9	) 7	) 15	) 29
SE	6)	7)	2)	4)	13)
S	28)	5)	1)	3)	9)
	) 40	) 5	) 7	) 6	) 18
SW	12)	0)	6)	3)	9)
W	13)	4)	0)	2)	6)
	) 18	) 5	) 0	) 3	) 8
NW	5)	1)	0)	1)	2)
Totals	110	30	30	31	89
Chi^2	31.964	12.80	40.93	19.84	37.292
Significance	99.9%	90%	99.9%	99%	99.9%

Thus orientation morphometry may in some circumstances provide a diagnostic clue concerning the origin of cirque-like features, according to the requirements for the optimum operation of different processes.

Drumlin outlines and spacing

Drumlins normally occur in fields in which very large numbers of individuals occur. They, therefore, provide useful data for morphometric analysis. The analysis can concentrate either on the form of individual drumlins or on the distribution of drumlins within a drumlin field, taking each drumlin as a point and analyzing the pattern of points. An example of each type of analysis will be outlined to indicate some of the possibilities of this type of morphometry.

The degree of regularity of drumlin spacing can be assessed by counting the number of drumlins in suitable sized square areas. The frequencies can then be tested by means of the chi square test to ascertain the probability of regular or random spacing. Three areas were analyzed in this way, two in the Eden valley drumlin field of northwest England and one in the large drumlin field of New York State around Newark. One of the Eden valley areas was in the vicinity of Appleby where the drumlin frequency ranged between 4 and 34 per 2.25 mi^2. The pattern was very diverse and far from regular. This is an area where the ice flow was variable; there were basal ice divides, major diffluent ice streams, and zones of concentration and dispersal. Flow directions were also varied.

The other Eden valley area was around Wigton, where the ice from the Lake District was joined by Scottish ice, the combined ice stream flowing westwards, vigorously and consistently. The resulting drumlin pattern was much more regular, and the drumlins were less numerous, but much more uniform in distribution. The mean number per 2.25 mi^2 was 15.8 for the Appleby area, 9.3 for the Wigton area, and 7.0 for the Newark, New York area.

The New York area also showed a very regular distribution pattern. The ice in this area was far from its source and moving fairly uniformly as one major ice sheet towards the Allegheny Escarpment. The relative spacing of drumlins over a considerable area, thus appears to reflect the constancy and uniformity of the ice sheet that shaped them. The differences between the three areas in drumlin frequency can be shown to be statistically significant by means of the t-test. The t value was 7.48 and highly signifcant at the 99.9% level.

Another method by which drumlin spacing can be compared in different areas is to record the variations of orientation within different drumlin fields. This criterion was the only one which could be used to differentiate between the drumlin field of New York State and that in Wisconsin. Other criteria such as drumlin length, width, elongation ratio, and parallel and perpendicular spacing were not significantly different in the two drumlin fields.

The shape of individual drumlins can be measured from maps or air photos with reasonable accuracy, and an analysis of their outline form can provide valuable evidence of the formative processes. Drumlins can be treated individually by fitting curves to their outline. A variety of curves have been used, such as ellipses, rose curves and lemniscate curves (See Muller, this volume). There is an advantage in using the two latter forms as these reflect the streamlined outline of the drumlin, which is one of its most significant features; their streamlined forms show that the drumlins were shaped by flowing ice, which adjusted the shape to conform most suitably to the flowing medium. Tests on a number of drumlins of the Eden valley drumlin field have shown that the narrower rose curve, given by the equation $r = a \cos k\theta$, provides a better fit than the broader lemniscate curve, given by $r = a^2 \cos k\theta$. r is the radius vector, a is the long axis length, θ is the angle of the radius vector r, and k is a dimensionless value derived from $k = a^2 \pi / 4A$, where A is the drumlin area.

The value of k is closely related to the elongation of the drumlin, given by L/W, where L is the length and W is the width. This measure is much easier to obtain by map measurement, and can be used to assess variations in drumlin elongation in different drumlin fields and in different parts of the same drumlin field. The elongation increases as the ice forming the drumlin flows more rapidly and consistently in direction. The relationship has been demonstrated for various parts of the Eden valley drumlin field. The maximum elongation was recorded in the Wigton area, where the drumlins were formed by vigorous uniform ice flow. In areas of variable flow and ice divergence the elongation ratio fell to low values, for example near Carlisle, where ice diverged to flow both east and west (Embleton and King, 1968 p. 325).

Three dimensional–volumes

The effects of ice on the landscape over which it is passing are related to the volume of ice flow as well as to the speed of flow (see also Boulton), both factors being involved in calculating the discharge, which is the important variable. Thus a study of the volume of previously occupied glacial valleys, cirques and diffluent channels can provide useful evidence of the effectiveness of glacial erosion. The relationship between source volume and volume of glacial erosion could usefully be explored if these values could be accurately obtained. It is more difficult, however, to obtain former estimates of ice volumes than of areas covered, thus introducing further uncertainties into the analysis. Few attempts have yet been made to assess such relationships. The volume of cirques, drumlins, moraines and similar features could usefully be assessed in order to relate them to the methods and capacity of ice erosion and deposition.

CONCLUSIONS–ANALYTICAL RELATIONSHIPS BETWEEN MORPHOMETRY AND PROCESS

The aim of morphometry is to establish relationships between the measured variables of the features under consideration and the processes that created them. Some examples of relationships that can be established by this means have been discussed. Morphometry provides a quantitative method of assessing significant differences in the character of glacial landforms in different situation, thus throwing light on the local controls that determine the development of the features under different conditions. The measures of cirque altitutde and orientation provides examples of this sort of analysis, whereby the significance of environmental conditions can be assessed.These include changeable variables such as climatic controls, as well as the basic environmental setting, such as rock type and jointing characteristics. Both types of control can exert an effect on the resultant morphometry.

Another advantage of morphometric methods is that they provide quantitative data that can be analyzed by a variety of statistical techniques, to provide significance levels whereby the likelihood of the relationships tested can be assessed. There is, however, a considerable problem in some types of study in the difficulty of measuring a suitable sample of the variables being considered. In some instances the features under consideration are so few that a total population can be analyzed, in others, such as drumlins, sampling becomes very important. A truly random sample must be achieved if the results are to be statistically valid for the complete population under consideration.

Morphometric data can provide information on the range of forms associated with features in varying stages of development, both from the point of view of time and of process. There is, for example, a range from incipient glacial cirques to degraded forms, as well as from true glacial cirques, through nivation cirques to nivation hollows.

As maps and air photos of glaciated areas become increasingly available it should be possible to develop glacial morphometry to provide many valuable data in studying the processes that created the forms of both glacial erosion and glacial deposition.

REFERENCES

Andrews, J.T., Barry, R.G. and Drapier, L. 1970. an inventory of the present and past glacierization of Home Bay and Okoa Bay, east Baffin island, N.W.T., Canada, some climatic and paleoclimatic considerations: J. Glaciol. 9:337-362.

Andrews, J.T. and Dugdale, R.E. 1971. Quaternary history of north Cumberland Peninsula, Baffin Island, N.W.T. Part V. Factors affecting corrie glaciation in Okoa Bay: Quaternary Research 1, 552-561.

Andrews, J.T. 1972. Glacier power, mass balances, velocities and erosion potential: Zeit. fur Geomorph. N.F. Supplement Bd. 13, 1-17.

Clayton, K.M. 1965. Glacial erosion in the Finger Lakes Region (New York State, U.S.A.) Zeit. fur Geomorph. N.F. 9 (1) 50-62.

Doornkamp, J.C. and King, C.A.M. 1971. *Numerical analysis in Geomorphology*: Arnold, London.

Embleton, C. and King, C.A.M. 1968. *Glacial and periglacial geomorphology:* Arnold, London, 608 p.

Evans, I.S. 1969. The geomorphology and morphometry of glacial and nival areas: In *Water, Earth and Man.* R.J. Chorley, ed. 369-380, Methuen, London.

Graf, W.L. 1970. The geomorphology of the glacial valley cross section: Arct. and Alp. Res. 2 303–312.

Graf, W.L. 1971. Quaternary analysis of Pinedale landforms, Beartooth Mountains, Montana and Wyoming. Arct. and Alp. Res. 3 (3) 253-261.

Haynes, V.M. 1972. The relationship between the drainage areas and sizes of outlet troughs of the Sukkertoppen Ice Cap, West Greenland. Geog. Ann. A 54 (2) 66-75.

King, C.A.M. 1970. Feedback relationships in geomorphology: Geog. Ann. A 52 (3-4) 147-159.

Lewis, W.V., ed. 1960. Norwegian cirque glaciers: R.G.S. Res. Ser. 4 104 p.

Miller J.W. 1972. Variations in New York drumlins: Ann. Assoc. Amer. Geogs. 62 (3) 417-23.

Robinson, G., Peterson, J.A. and Anderson, P.M. 1971. Trend surface analysis of corrie altitude in Scotland: Scot. Geog. Mag. 87 (2) 142-146.

Seddon, B. 1957. Late glacial cwm glaciation in Wales: J. Glaciol. 3 (22) 94-99.

Svensson, H. 1959. Is the cross section of a glacial valley a parabola? J. Glaciol. 3 362-63.

Unwin, D.J. 1973. The distribution and orientation of corries in northern Snowdonia, Wales: Trans. Inst. Brit. Geog. 58 85-98.

CHAPTER 6

RATES OF FORMATION OF GLACIAL FEATURES IN GLACIER BAY, ALASKA

Richard P. Goldthwait

ABSTRACT

Since 1958 a continuing glaciologic and stratigraphic program has disclosed a time perspective for the formation of seven types of glacial deposits. The broad basins of ice wastage in Muir Inlet and outer Lituya Bay are so similar to hills and valleys in northern America that analogies to the last Pleistocene ice sheet are good.

Changes in fabric direction of basal till, corresponding to known changes in ice flow, indicate that basal till accumulated at rates of 0.5 to 2.5 cm/yr. Taking a 1-5 m thickness of most Neoglacial till layers, the deposition occurred only during the last two centuries of thinning moving ice.

Swarms of nesting drumlins occur where Neoglacial ice rose abruptly out of deep valleys. The elongate inverted spoon-shaped cores formed in 2000 yrs from horizontally bedded Hypsithermal outwash gravels. Till was added in the last three centuries of thinning Neoglacial ice.

Ablation till ended up as thin washed sands, mudflows, or a profusion of scattered angular stones and boulders. This till was spread as a sheet in the direction of the retreating thin ice edge at rates of 20-150 m per ablation season.

Minor moraines were observed to form in one to five summers where juicy till squeezed up into crevasses or into shear zones. They melt out at about one per year. The major end moraine rampart at Lituya Bay contains carbon-dated materials that disclose several ice reoccupations separated from each other by hundreds of years.

Marginal hillside channels were made along the ice edge and left high and dry within a year or two. Main channels were cut in a month, and lower thresholds to these channels were made that same season. The vertical till walls collapse to a V-shape annually.

The larger eskers formed in tubes in 1-5 yrs. The slightly active ice closed in on deeper sections of esker prematurely embalming them. Across a basin a graded ice surface canyon complex gathered sorted materials in a decade. After two more decades these were still dividing into an even greater ridge complex as the buried ice core collapsed in the basin. In the lee of each bedrock hill, groups of small eskers, kames, and terraces formed for 10-15 yrs in and under a mass of lingering stagnant ice.

The only dated outwashes accumulated in Hypsithermal time. They encroached slowly on interstadial forests up the hillsides for at least 4800 yrs. Carbon-14 dates give rates of thickening of 1-5 m per century. Modern valley trains are 30 to 90 percent complete before they appear out from under thinning ice.

INTRODUCTION

How long does it take for a glacial deposit to form? Multiple drift layers and carbon-14 dates give the impression that since the ice of "a glaciation" covers an area for thousands of years, the deposits which record it represent all these millenia. Indeed they do not. They are the tail-end consequence of centuries of glacial erosion, or at most, the well-separated short episodes of glacial deposition which lead to till-coated moraines, thrust stacking, re-deposition, and all kinds of two-phase events.

One ideal place to document the duration of deposition is Glacier Bay National Monument, Alaska, because a large Neoglacial ice mass expanded southward at least 100 km and then retreated over that same terrain. Wood for carbon-14 dating is abundant. Confining high ridges and mountains channelled the major inflow of ice from high mountain sources to the west (4000 m) and north (2000 m). The ice spread out in a lop-sided triangle 30 km on a side. Within this area the hilly topography has the same local relief (300 m) as does much of New York, New England, or Puget Sound. Since temperate ice comprises most of these peripheral Pleistocene ice sheet areas, many of the features formed by the ice are similar and topographically identical (Fig. 1).

During 16 summer research parties from Ohio State University have searched Muir Inlet and Lituya Bay areas to establish stratigraphic chronologies and to find evidence of how the glacial features were made. These efforts have been described in several Institute of Polar Studies reports (numbers 3,9,20,25,40) and several little known dissertations. This paper will bring together scattered facts and figures by these many authors bearing upon the time it takes to form some glacial deposits. Some material (e.g. drumlins) is mostly unpublished.

BASAL TILL

In 1970 Mickelson (1971) hit upon a neat and effective way to gage the rate of till accumulation on gentle hillslopes under Burroughs Glacier. In a series of maps and air photographs from 1892 onward it was evident from contours, crevasses, and moraine banding exactly which way the ice flowed. In the lee (southeast) of a major nunatak, Bruce Hills, which appeared in 1910 to 1940, and in the lee of lesser nunataks (A,B,C,D and E in Fig. 1) appearing from 1950 to 1958, it was apparent that the thinning ice had to change course around rather than over each hill. This was confirmed (1959-1960) by measurements of stakes in the ice surface (Taylor, 1962), and by a change in provenance of pebble lithologies in the till (Mickelson, 1973). Pebble fabrics were made at seven critical localities where change had occurred (Fig. 2). These were done in depth units 5 cm to 40 cm thick, and involved pebbles that were more than 1.7 times as long as wide. The strong fabric concentration indicated the turning of basal ice motion.

When each year of established ice motion is plotted against median depth of the corresponding fabric, the rates of deposition (slope of line) come out from 0.5 cm/yr to 2.5 cm/yr (examples, Fig. 3). The average rate, 1.5 cm/yr, might be explained by basal melting; Mickelson calculates that a 25 percent ice-till mixture at the sole could yield 1.5 to 3 cm/yr. The usual depths of deposited Glacier Bay till by several detailed studies vary from 1-5 m. The deposition began on the order of one to five centuries ago. Possibly deposition was interrupted; most likely it began in depressions as found under Casement Glacier tunnel (Peterson, 1969) or shown in theory by Nobles and Weertman (1971).

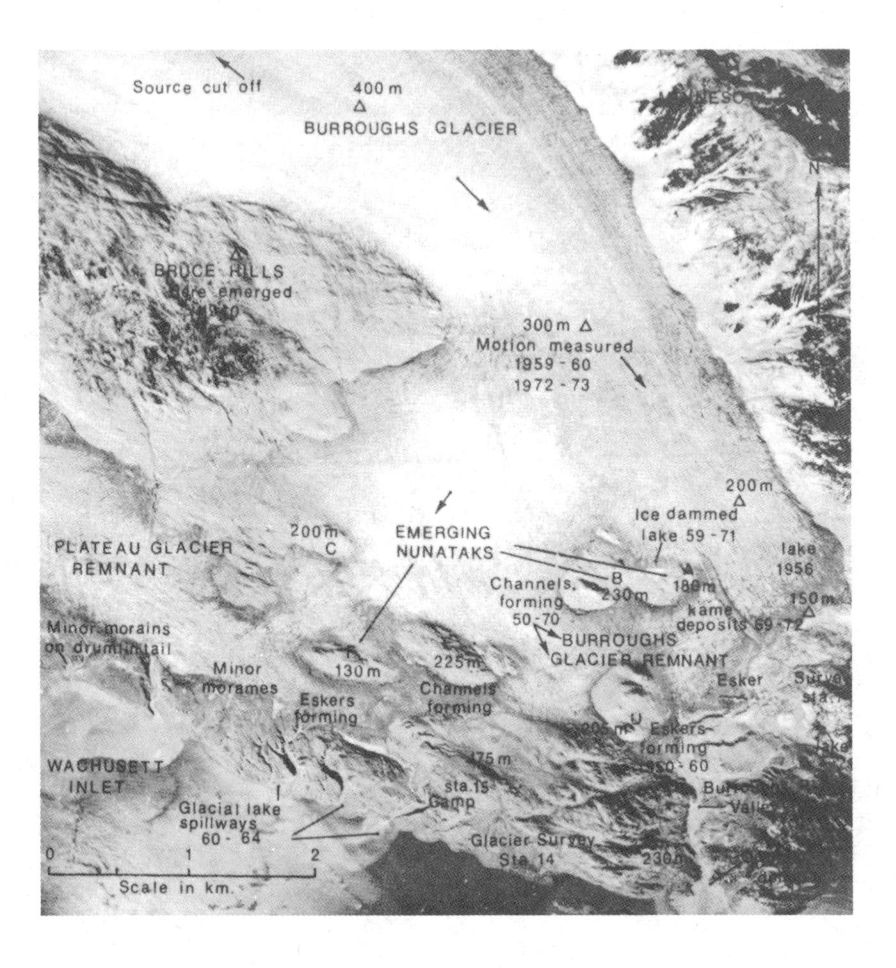

Figure 1. Vertical air photo, August 1963, of emerging nunataks at southeast end of Burroughs and Plateau Glaciers, Glacier Bay National Monument, Alaska. Scale approximately 1:70,000. Approximate elevations in meters. Many of the specific study areas are indicated.

DRUMLINS

As in other drumlinized areas, the drumlins near Muir Inlet occur in four clusters of 10 to 30 each (restricted areas). They form the most regular and striking feature of lowland ground moraine. They were uncovered by ice between 1958 and 1973 north of Wachusett; between 1930 and 1960 northeast of The Nunatak; between 1920 and 1950 southeast from Wachusett, (Fig. 4); and between 1875? and 1900 around the northwest edge of Mt. Wright. In each case the long axis exactly parallels the slope (and thus movement) of ice on H.F. Reid's contour map in 1892. At each group the basal ice was rising out of a deeply scoured main valley (now fiord). If Neoglacial maximum trimline of two to three centuries ago is properly mapped, the deepest ice over these groups was about 1000 m, 1000m, 850 m, and 700 m respectively.

In northeast Glacier Bay the general stages of drumlin origin are reasonably clear. Most of the drumlins, exposed internally by deep late-glacial gullying or slides, have a core of bedded sandy gravel and some lacustrine units. Some do encase or butt up to bedrock knobs. Furthermore, stumps are in place between the drift beds (see Flint, 1971, Fig. 15-2) and this wood is dated from 7050 to 2500 cyr B.P. Surely this is Hypsithermal glacial outwash which once had a graded

Figure 2. Glacier Bay till (Neoglacial) 1 m thick overriding interstadial stumps in place, 2520 cyr old. Note the forest litter and paleosol on Wisconsin age till below. This is just northwest of Nunatak A with Burroughs Glacier showing above. One till fabric-age correlation made here.

surface some 100 m above present tide. From it Neoglacial ice fashioned symmetrically spaced drumlin shapes between 2500 and 200 yrs ago. That it took no more than 20 centuries is likely. Each drumlin is smoothed like the inverted bowl of a spoon but it is covered by a meter or more of till. Pebble fabrics are only moderately strong and parallel the drumlin axis. In one very high drumlin (Fig. 5) the till thickens toward the crest to round off an otherwise flat (graded?) outwash top. Even the truncated beds of the lower side have a coating of basal till. This was duplicated in at least 25 sections in all groups; so, there was surely a second and later depositional phase. Based on the till deposition rates calculated in nearby areas, deposition took only a century or two of thinning ice.

This was not the very last shaping. These drumlins are all in areas of late-melting "remnant" ice on the gentle side or divide of a broad valley. The ice was stagnant, but air photos and maps (original 1948 Juneau D-sheet) showed ice

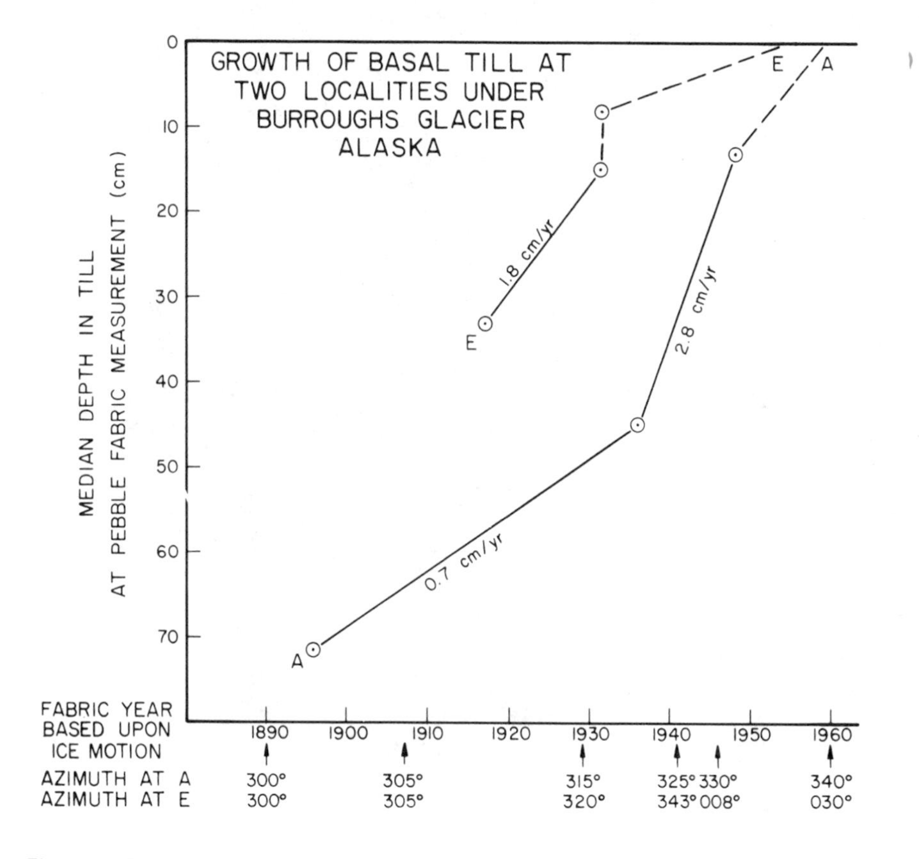

Figure 3. Graph of the till fabric-age correlation against the median depth of 50 pebbles measured. Slope of the line represents the rate of deposition of basal till providing motion is perpendicular to ice surface contours and strong orientation does indicate basal motion.

Figure 4. Group of 30 elongated drumlins south of Wachusett Inlet (left) in a broad saddle to Hunter Cove (right) on Muir Inlet (in distance). Stagnant ice lingered over the drumlins to the left until 1950.

Figure 5. Deep gully section in a 50 m-high drumlin just south of Wachusett Inlet. Note the carapace of till (darker) down the slope, right; the bedded sandy gravels (Hypsithermal age) below; and the horizontal top of those gravels. The top, left, is rounded off by thicker till. It shows a 2 m boulder for scale.

surface sloping into the inlet. North of Wachusett Inlet Taylor (1962) measured motions up to a meter a year on just such "dead ice". This induced a last slight cross motion exhibited on the till surface now by till "tails" (crag-and-tail form) behind every big boulder, and like ribs onto drumlin sides. This is judged to be a final insignificant modification of the juicy surface till.

ABLATION TILL

Where the ice is melting rapidly back (100 m/yr) and down (2 to 10 m/yr), a thin veneer of debris, usually a centimeter or two thick and rarely up to 50 cm (noted on Fig. 6) accumulates. Price (1964), Mickelson (1971), and others report that this till has a more sandy texture than basal till nearby; fine silt and clay are winnowed out by melt seepage. Half as many pebbles bear striae, and angularity is 0.1 to 0.2 greater on a scale of 1.0. Fresh broken surfaces mark 23.7 percent of the pebbles instead of just 15 percent. Pebble fabrics are all weaker than in basal till, although the see-saw shaking effect of let-down from thin ice seems to leave long pebble axes nearly horizontal. These data favor intense frost-splitting during and

Figure 6. Remnant ice of Burroughs and Plateau Glaciers looking west up Wachusett Inlet. Nunataks were hilltops appearing through the ice from 1950 (D) to 1958 (F). One indicator rock (marble) is shown. An oblique aerial photograph in September, 1969, by Austin Post.

just following emergence onto the ice surface. Accumulation on dirtiest basal ice often produced more than a five centimeter cover of till. Therefore more silt-clay is retained and mudslides occur every sunny day. These gather even thicker masses with locally oriented pebbles at the land-base of every ice depression. In all these cases actual emplacement takes place within a few minutes or hours, but it proceeds *seriatim* across the basal till or rocky land.

Bands of specific lithologies are dispersed through the basal ice down the line of ice motion. The higher the rock knob or source the deeper the dispersal up into lower ice; these appear as medial moraines "from nowhere" as the ice thins. Seen 20–40 yrs ago from The Nunatak, or 5–15 yrs ago from lettered Nunataks above Burroughs Glacier, these looked smooth on ice (Fig. 1). Today these have mostly come to rest on drumlins and hillslopes; in spite of their rise and fall they are remarkably straight and narrow. The outstanding white marble band across the sides of Nunataks E and D and down Bob Valley (Fig. 6) is composed of scattered sharp pebbles and boulders with occasional sand pockets.

The rate of irregular retreat of the dirty ice margin controls the rate at which this ablation till is emplaced. A sloping ice wedge on land, which thins 1–8 m a year, produces horizontal (map) retreats of 20–150 m/yr. On steep hillsides retreat was slowest (Klotz Hills, 1892 to 1910; Curtis Hills, 1906 to 1940; Bruce Hills, 1907 to 1970; The Nunatak, 1920 to 1940; lower Nunataks A through F, 1950 to 1975?). Fastest have been the detached "glacier remnants" in broad high valley situations (Adams Remnant, 1929 to 1940; Casement foreland, 1910 to 1950; McBride Remnant, 1930 to 1965; Muir Remnant, 1940 to 1980?; and Burroughs Remnant, Figs. 1 and 6, 1950 to 1980?). Each one is placing its lithologic bands, and within two to four decades each will become covered by mosses, and engulfed in *Dryas* or alder vegetation.

MORAINES

The major moraine in Glacier Bay Monument is a monolithic sharp lateral ridge, 100–400 m high, called "Solomon's Railroad". It surrounds Lituya Bay where the bay extends beyond the mountains (Fig. 7). It's Pleistocene analogy would be in the Rocky, Cascade, or Sierra Nevada Mountains. The main original ridge must be Wisconsin in age, for it enclosed stumps in stratigraphic section which are over 9000 carbon-14 years old (Goldthwait and others, 1963). One lower older moraine, with cavernous boulders and trimmed by higher interglacial sealevel, lies outside of it (Derksen, 1974). Complex alternating layers of till and outwash, exposed at every landslide, suggest multiple occupancy by expanded ice. Most significant are logs buried 9 m and 12 m below the north crest (345 m altitude) which carbon date 1090± 100 and 2600± 160 year old. The massive Wisconsin moraine held back Neoglacial ice nearly 8000 yrs later. Reconnaissance in 1974 (Post and Goldthwait) show low ridges and till grooving formed by later ice which

pushed tongues through low places in the Wisconsin moraine loop. Dead ic
remains under deep spruce forest against the northern bulge (Post and Streve
press). The total construction span may be as much as twenty-four milleni
actual deposition from ice may be just a few scattered centuries.

Minor morianes abound near Glacier Bay shores, especially up Muir Inlet
8). They are 0.5–2.5 m high and may be traced for 10–1500 m. They co
groups and are subparallel with some Y-junctions. Most are parallel to the rec
ice edge and they are separated by just about the distance of annual retreat
these reasons my first presentation (Ohio Academy of Science, 1963) accounted for them by the annual forward ice push every winter. However, two careful studies (Haselton, 1966, at McBride Remnant; and Mickelson, 1971, at Plateau Remnant) have demonstrated that most of these small moraines are fashioned in a crevasse, or a shear zone within one kilometer of an ice cliff. Air photos show that these fall exactly where deep tension cracks once existed (Mickelson, 1971). Apparently these are released by ice retreat at about one per year, but they may build up in the crevasse for 1–5 yrs earlier.

Squeeze-up of basal till accounts for nearly all the material. Ridges have grain-size-distribution almost identical to the basal till beneath. Pebble fabric here is weakly parallel to moraine slopes. No significant ablation till is detected on the clean ice above or on the ridge except in the form of scattered angular stones and boulders. As temperate ice disappears the till released most places is "knee-deep" and soupy; it can be seen bulging up into some crevasse floors (Fig. 9).

Figure 7. Moraine around south side of Lituya Bay, left, cutting across interglacial and postglacial marine benches (4) stretching to the right. A tongue of Neoglacial ice poked through the Wisconsin-age ridge at the left. Oblique air photo by Bradford Washburn, 1934.

Figure 8. Minor moraines on top of drumlin and rock-based ridge between Goose Cove and Anchorage Cove on the east side of Muir Inlet, 1958. These first appeared about 1929-1930 when ice retreat equalled ridge separation. They form undrained squares with till grooves, at first.

Figure 9. Basal till squeezed up from below to form a ridge in the middle of a crevasse. The pressure of ice right and left near the terminal cliff of McBride Glacier appears to produce this extrusion.

Figure 10. Box-cut channel in till below a dirty ice margin of Plateau Remnant Glacier in September, 1958. Now this is high on the slope above Camp Creek and Lake, just northwest of Survey Station 15.

CHANNELS

On many middle and lower hillslopes in Glacier Bay box-cut channels form in till just under the dirty ice margin (Fig. 10). These occur mostly around bulging convex ridges; elsewhere meltwater runs on ice and marginal terraces. Each year a new threshold entrance to old channel,or a whole new channel, was cut lower down on the slope as the ice surface lowered (Fig. 11). Some streams following rapidly lowering ice (Nunatak B, Northwest side, 5 m/yr) annually cut as many as three or four separate channels. Note that the streams around Station 15 (Fig. 12) produced ten thresholds with little sandy platforms and cut four large gullies in eight years. Only a careful appraisal of channel arrangement would allow an estimate of rates of ice wastage.

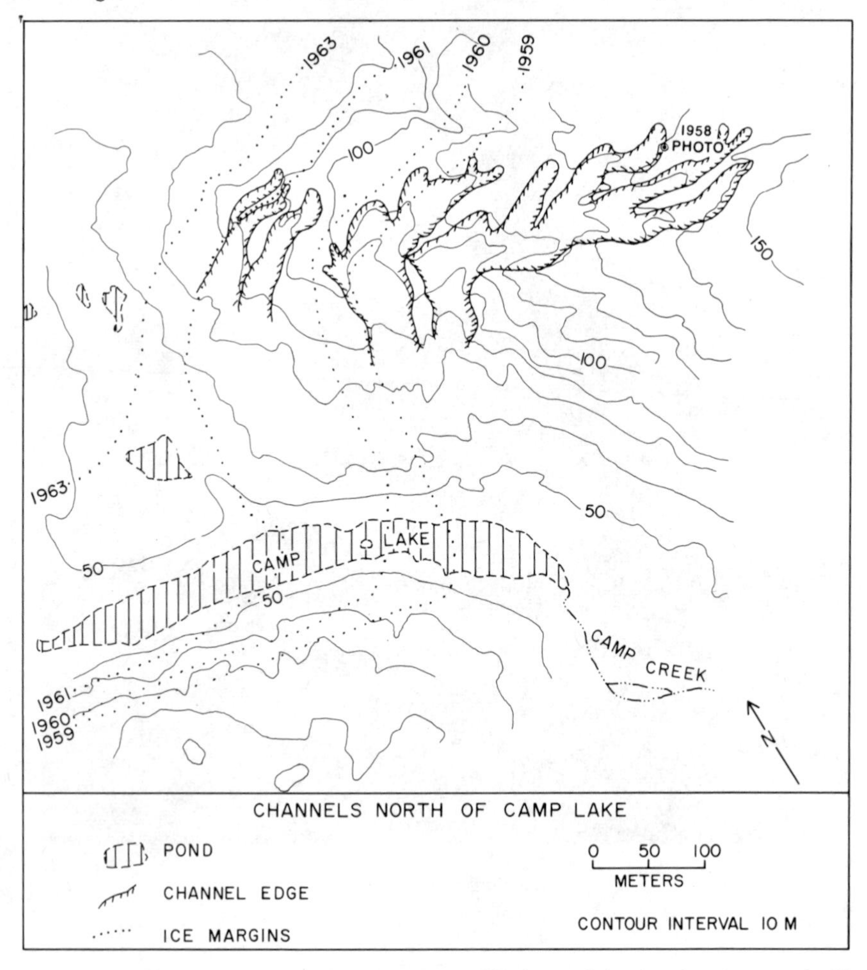

Figure 11. Map of channels north of Camp Lake and Wachusett Inlet showing photographed ice margins and 10-m contours. D.M. Mickelson, 1971.

Channel size is more a function of mean maximum discharge than of time. All cutting is caused by floods during one to four months of summer; discharge fluctuates wildly (10 x) on a diurnal cycle. Recent studies of the hydrology of Burroughs Glacier (Larson, 1972, and in preparation) explore these fluctuations and variations of water table in ice showing how routing in the ice makes a world of difference. John Creek (Fig. 6) for example started in 1965 on high slopes to the northeast then dropped channel by channel into the valley floor eroding it 3–5 m wide and 0.5 m deep with 15 cfs on sunny days. In 1971 the flow shut off suddenly and completely as the glacier ice ramp above dropped below the level of a col threshold; then water flowed subglacially southwest to Bob Creek. Large rivers like Burroughs River, which exceeds 200 cfs on many summer days, strip the cover 4–6 m wide and cut 1–2 m into bedrock in a year.

Hillside lateral channels are left high and dry when stagnant ice melts across the valley floor. The freshly exposed till is still saturated with 30 to 60 percent water; at the least jarring or oversteeping it liquifies and flows away in noisy mudflows. Within a year after abandonment channel walls slump to a V-shape, and wider reaches get filled with flow diamicton. In a few years the V-shape stabilizes.

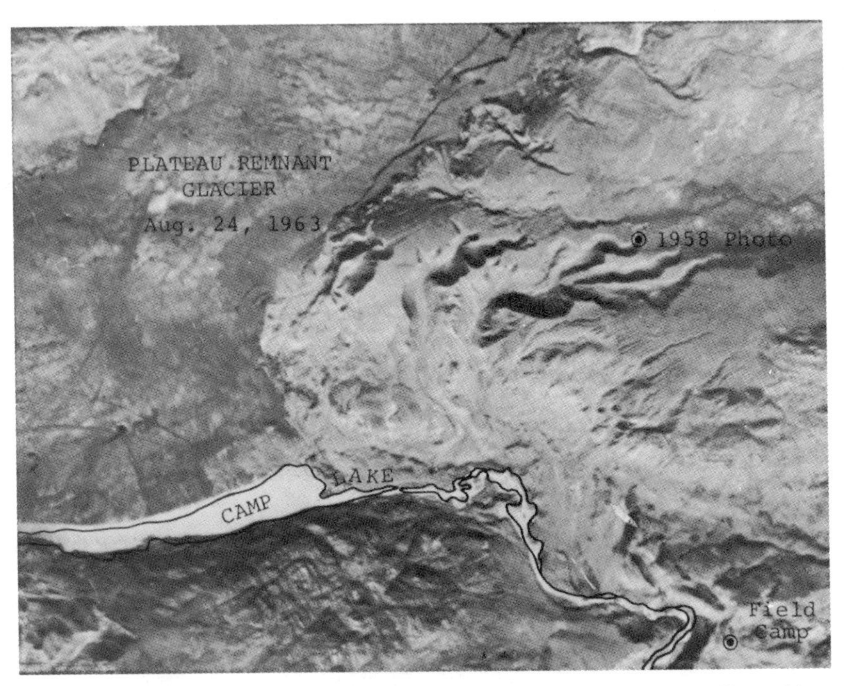

Figure 12. Enlargement of vertical air photo showing the area of channels in Figure 11 as of August, 1963. Photo by Austin Post.

ICE CONTACT DEPOSITS

Eskers have been studied in Muir Inlet area in three types of situations. The first is gradient systems at the bottom of the active ice along the north side of Casement Glacier. Tunnelling through basal ice toward the glacier center, Peterson (1969) encountered a gravel ridge tightly encased in glacier ice. This active moving ice had closed in by creep on an esker tunnel opened earlier by warm summer meltwaters. Other routes developed; at one exit for this system the steep active ice was receding and exposing a seven-meter-high ridge with a narrow smooth top (Fig. 13). Since water issued actively from the ice here for about 5 yrs the esker is deduced to have formed within that time.

A second situation involved a surface stream crossing stagnant ice which filled present Seal River Valley at the south edge of Casement Glacier. Price (1964) found this system depositing washed drift in ice canyons on 1948 airphotos. The complex graded streams entered the ice surface at nunataks near 500 m and exited at 180 m altitude into broad channels leading westward into Muir Inlet. In a decade gravels 3–6 m thick marked the several channels and protected that ice from melting. The ice core is still there (explored seismically in 1966 under flat- topped sections) but the whole system is sagging below the former outlet channels, and ridges are subdividing as cores sluff their gravel cover off. Seal River removed a large section in the middle of the esker, exposing the ice cores. Thus a supra-glacial esker is coming to rest in three decades as a complex of anastomosing ridges running both down and gently up grade.

The third situation carried esker development in stagnant ice masses one step further in the lee of each nunatak at Burrough Glacier Remnant (Figs. 1 and 14). The ice becomes honeycombed; first it carried a main sublateral drainage like Burroughs River as demonstrated by lithologic counts (Mickelson, 1971), and then appeared a host of leakages by ponding and a kame terrace on the high side of the ice just under the nunatak (Fig. 15). As new tubes opened up the old ones did not close here, and a fretwork of routings with kame terraces and subparallel ridges developed. This entire cycle occurred against Nunatak D from 1950 to 1965, near F from 1961 to 1969, and under A from 1958 to 1973. No individual ridge took more than a year or two to form. Near F, (Fig. 1), where water ponded and overflowed a ridge into deep channels, the energy level was so low that a gravelly esker grades downslope into sand, then upslope into silts.

Kame terraces formed only 3–6 m thick on top of ice on the south side of Adams Inlet in about 1930 (McKenzie, 1969). This may have been produced in an equally short 2–6 yrs, but the thermal regime, which is collapsing the 35 m thick ice buried beneath them is taking at least four decades to melt out. Most of the ice is going by back-melting and by developing ice caves and pits, because measured thermal conduction through the gravel top would take till 2100 A.D. Little of the original flat and covering vegetation will actually be left by 1980. Similar gravel layers, washed all over the northeast part of McBride Remnant (Haselton, 1966)

Figure 13. Esker 7 m high on left and ice tunnel melting back in the north side of lower Casement Glacier. The glacier ice above is covered with ablation moraine (till-to-be).

before 1940, had produced an entire kame field over flat terrain by 1966, but here nothing but moss and fireweed could get established on the shifting tops. The long part of kame formation is the loss of its buried ice.

OUTWASH

Unlike its counterparts thus far, outwash seems to grow rather slowly. During glaciation impressive amounts of erosion product go off in braided outwash. Immediately after retreat very active gullying of till introduces more load to outwash. In some senses outwash is the ultimate terrestrial resting place of coarse glacially eroded materials. Attempts to measure bed load give only short-term orders of magnitude.

One circumstance made the rate of outwash measureable over long terms in Glacier Bay. In Wachusett Inlet especially a ubiquitous sandy gravel is found under Glacier Bay (Neoglacial) till within 100 m of sea level. Because it is sandy or has lenses and current structures, and beds are nearly horizontal from bottom to top, it is interpreted as outwash, not as ice-contact deposit. The common logs and occasional forest duff layers always fall in the carbon-14 age bracket of 1700–7100 yrs. Although called "interstadial" they bracket Hypsithermal and Neoglacial time and clearly show presence of glaciers which retired upvalley a short distance.

Figure 16 shows each radiocarbon date in Wachusett Inlet plotted as a function of elevation. These fall on a remarkably straight line (Goldthwait, 1961). Other inlets have similar but not identical rates; likewise the highest gravels here

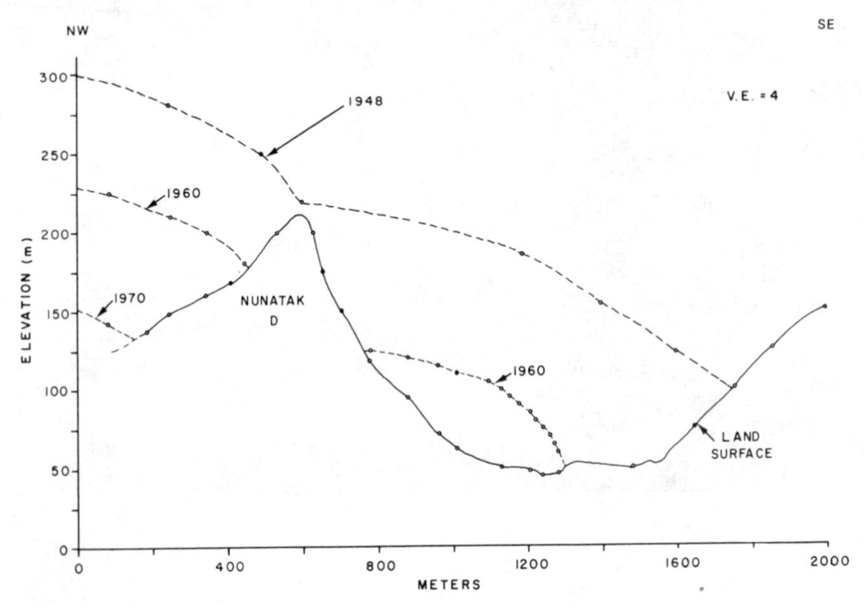

Figure 14. Profiles of the ice surface at the southeast edge of Plateau Remnant Glacier across Nunatak D at three photographed years. By D.M. Mickelson, 1971.

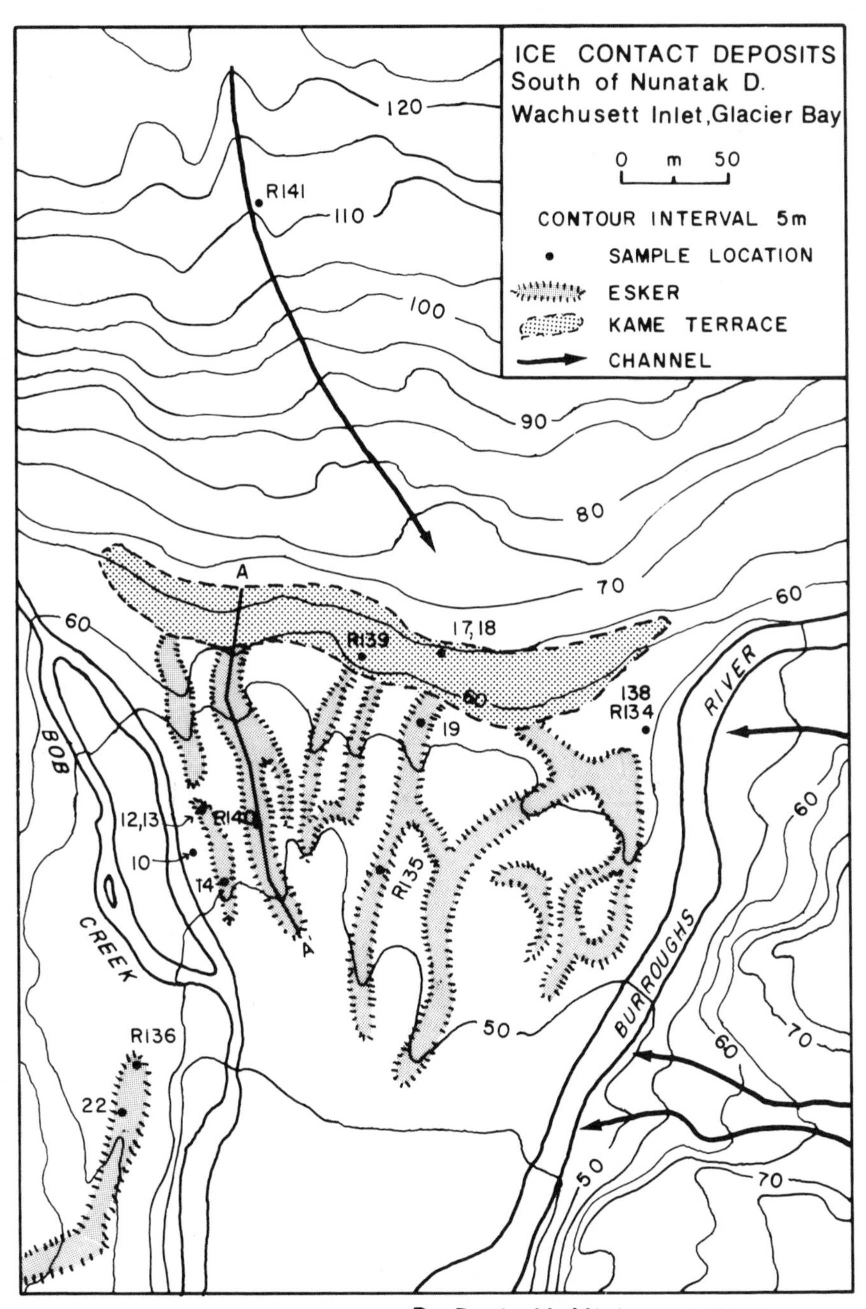
ICE CONTACT DEPOSITS
South of Nunatak D.
Wachusett Inlet, Glacier Bay
0 m 50
CONTOUR INTERVAL 5m
SAMPLE LOCATION
ESKER
KAME TERRACE
CHANNEL
120
110
100
90
80
70
60
50
R141
R139
17,18
138
R134
19
12,13
R140
10
14
R135
R136
22
A
A'
BOB
CREEK
BURROUGHS
RIVER

By David M. Mickelson 1971

were always near 100 m, but in some others, such as Adams-Endicott they were higher, or in some lower or inequal like Lituya Bay.Between wood-bearing layers at any one locality the rates of accumulation (including some thin lacustrine units) varied from 1–5 m per century. The average growth by extreme Wachusett dates from bottom to top is 1.4 m per century but the straight line growth most of the time was 1.6 m per century. The occasional insertion of layers of alder and young spruce (as at Camp Creek) confirm that this was a staccato building at any one place, and it is true today.

One interpretation has not been accepted by all, namely that most of these outwashes filled the valley wall-to-wall to give such altitudinal uniformity. This assumption is based upon uniformity of coarse material, presence of current structures rather than chaotic ice-contact dips, and absence of boulders and till masses. The matching dates on both sides of the valley argue that it did build up both sides uniformly, unlike the growth of a kame terrace. This implies removal of large quantities of gravel from the central valley during the drumlin-eroding Neoglacial time. Certainly there is room underwater where Neoglacial Glacier ended, and perhaps only some side valleys and inlets were filled so deep.

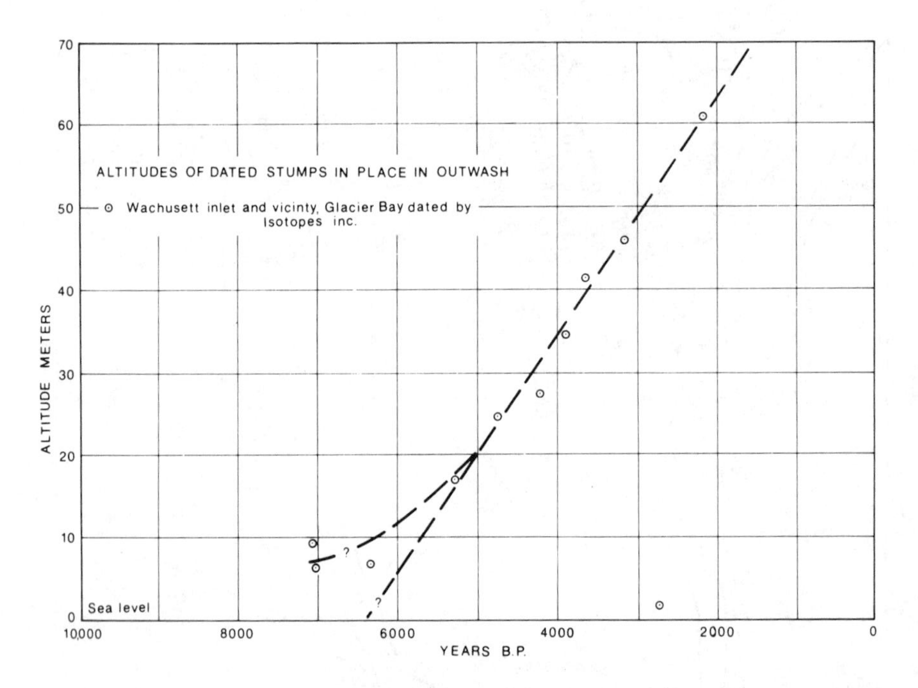

Figure 16. Graph showing altitudes of dated stumps in place in outwash (Hypsithermal-Neoglacial) of Wachusett Inlet, Glacier Bay. All dated by Isotopes, Inc. Slope of the dashed line represents rate of growth of outwash (1.6 m/century on straight-line phase).

Outwash trains and aprons in front of the present receding glaciers are already well built or half-built in the year the ice uncovers them. Always there is a broad stream, and frequently the ice surface sags and shows collapse fractures over these outlets. Needless to say outwash building is not all pro-glacial, out in the open. Much is begun as far back as a kilometer or two under such temperate ice.

SUMMARY

Glacial deposits and channels are notoriously rapid in their period of actual emplacement or occupation. All eskers, kames, and hillside channel witnessed were completed within one year or a decade. Of course the deposition of stony ablation moraine at any one place was almost diurnal but spreads over new terrain for as long as the ice retreats.

Repetition is commonplace and because of the cyclical expansion and contraction of nearly all glaciers the accumulated deposits become thick. Even the large lateral-loop moraines in and near mountains, tend to be stacked layers of repetitive tills sandwiched with marginal stream gravels, as at Lituya Bay. The topography is such as to impede slightly larger advances, thus compounding and regenerating the stack. Several short periods of deposition may actually span many thousands of years; most of the time record is is hiatuses and in paleosols or buried vegetation.

Two exceptions to the rapid speed of deposition are basal tills and outwashes. Till is widespread, but the till sheet of any one advance and retreat is usually thin and grew at one to five centimeters per year. Except for early deposition in depressions, or spot deposition later to be picked up and redeposited, basal till deposition occurs in the last few centuries of glaciation. Outwashes are restricted to gentle gradients. Due to the tendency of depositing streams to fill the channels with bars, to divide, and to flow over recently unoccupied parts of the valley plain, deposition is constantly shifting. Valley trains start long before the ice melts away and continue long after the ice uncovered the area, as long as the annual flood of meltwater feeds the river.

Thirdly we note that nearly all deposits, and all hillside channels, relate to the recessional phase of one glaciation. Basal till is deposited while the ice is thinning, ablation till is deposited just as the ice thins to zero. Drumlins were coated with till in the thinning-years of the glacier. Minor moraines squeeze up into crevasses or shears only in about the last five years of ice occupation when crevasses reach the bottom where the till is supersaturated. Hillside channels function only as long as ice is nearby. Chutes under the ice may function during and after the cutting of sublateral channels, but due to the apparent restriction in depth to which subglacial waters move easily this may be limited to the last 100 m of thinning. Even outwashes are basicly retreatal forms since those that are invaded by later expansion may be carved into drumlins as happened at Muir Inlet, beginning 2500 years ago.

ACKNOWLEDGEMENTS

William O. Field introduced me on a 1958 IGY project to the fine examples of glacial deposits forming in Glacier Bay. His invaluable collection of photographs, duplicated freely for many co-workers and students who have gone there since then, is deeply appreciated. To the seven graduate students, now MScs. or Ph.D's and teachers themselves, who did the hard work for two or three years apiece goes the real credit: S.J. Derksen, G.M. Haselton, G. Larsen, G.D. McKenzie, D.M. Mickelson, D.N. Peterson and L.D. Taylor. Co-workers from other universities contributed greatly too: George Burns of Ohio Wesleyan, I. McKellar of New Zealand, R.J. Price of University of Glasgow, N. Tenbrink of Grand Valley College and F.C. Ugolini of University of Washington. Some 25 other summer field assistants were good help and deserve thanks.

None of this work would be possible without the great logistic help and services rendered by National Park Service, Glacier Bay National Monument under L.J. Mitchell and then R. Howe, Superintendents. The crews of "M.V. Nunatak" and the many rangers made travel and communications pleasant and possible.

For 12 years this was generously supported by National Science Foundation (G–16023, G–24149, G–13842, GP–1058, GP–2537, GB–3364, GA–403, and G–12300). Of course, the Ohio State University through its Institute of Polar Studies and its Geology Department has injected funds and equipment and thousands of hours of labor, especially since 1970. The Atomic Energy Commission granted funds for Peterson's work and Sigma Xi made several grants when most needed.

REFERENCES CITED

Derksen, S.J. 1974. Raised marine terraces southeast of Lituya Bay, Alaska: Thesis, Ohio State University, 84 p.

Flint, R.F. 1971. *Glacial and Quaternary Geology:* John Wiley and Sons, New York, 892 p.

Goldthwait, R.P. 1961. Dating the Little Ice Age in Glacier Bay, Alaska: Internat. Geol. Congress XXI, Norden, Part XXVII, p. 37-46.

Goldthwait, R.P., McKellar, I. and Cronk, C. 1963. Fluctuations of the Crillon Glacier System, southeast Alaska: Internat. Assoc. of Science, Hydrology, Symposium at Obergurgl, Bull. VIII no. 1, p. 62-74.

Haselton, G.M. 1966. Glacial geology of Muir Inlet, southeast Alaska: Inst. of Polar Studies, Report no. 8, 64 p.

Larson, G.J. 1972. Internal drainage of stagnant ice, Burroughs Glacier, southeast Alaska: Thesis, Ohio State University, 95 p.

McKenzie, G.D. 1970. Glacial geology of Adams Inlet, southeastern Alaska: Inst. of Polar Studies, Report no. 25, 121 p.

McKenzie, G.D. and Goldthwait, R.P. 1971. Glacial history of the last eleven thousand years in Adams Inlet, southeastern Alaska: Geol. Soc. Amer. Bull. v. 82, p. 1767-1782.

Mickelson, D.M. 1971. Glacial Geology of the Burroughs Glacier area, southeastern Alaska: Inst. of Polar Studies, Report no. 40, 149 p.

——————1973. Nature and rate of basal till deposition in a stagnating ice mass, Burroughs Glacier Alaska: Arctic and Alpine Rsch. v. 5, no. 1, p. 17-27.

Mirsky, A. (ed.) and eight authors 1966. Soil development and ecological succession in a deglaciated area of Muir Inlet, southeast Alaska: Inst. of Polar Studies, Report no. 20, 167 p.

Nobles, L.H. and Weertman, J. 1971. Influence of irregularities of the bed of an ice sheet on deposition rate of till: in *Till, a Symposium* ed. R.P. Goldthwait and others, The Ohio State University Press, p. 117-126.

Peterson, D.N. 1969. Glaciological investigations on the Casement Glacier, Southeast Alaska: Ph.D Dissertation , Ohio State University, 183 p.

Price, R.J. 1964. Land forms produced by the wastage of the Casement Glacier, Southeast Alaska: Inst. of Polar Studies, Report no. 9, 41 p.

Taylor, L.D. 1962 Ice structures, Burroughs Glacier, southeast Alaska: Inst. of Polar Studies, Report no. 3, 106 p.

CHAPTER 7

ORIGINS OF DRUMLINS

Ernest H. Muller

ABSTRACT

Drumlins belong to that family of streamline glacial features that range from wholly erosional to wholly depositional. Whether formed by erosion or deposition, the tendency is to develop a form which imposes minimum resistance to glacier flow. Erosion commonly precedes deposition, but either may dominate in shaping of a particular drumlin field.

Drumlin form is affected by protrusions in the glacially overridden topography and by character and abundance of glacial load. The tendency of drumlins to occur in groups, their similarity of form, orientation and spacing within a group, and their progressive change in form and spacing in the direction of glacier flow, all point to dynamic controls in the relationship of glacier sole and subjacent deforming sediment as critical in drumlin formation.

Although many drumlins are rock- or drift-cored, others have no distinctive cores. Dilatance theory applies to all, for till is a typically dilatant system in that, under a deforming load, it must expand before continuous deformation can occur. Once dilatant, it continues to deform even under reduced load. An ice sheet thinning toward its terminus or wasting during the waning hemicycle imposes diminishing load until at some point it becomes inadequate to maintain subglacial sediment in a dilatant condition. The basal load collapses into a stable static state, too firmly packed to deform further without renewed dilatance. Till fabric analyses indicate depositional modification of such drift cores by accretion of successive shells building outward along slip lines of maximum shear stress.

The intimate but typically subordinate association of stratified drift with subglacial till in many drumlins accords with the essential role of basal meltwater in glacial flow theory. Enhanced plasticity and intensified shear adjacent to a basal obstruction account for a zone devoid of competing cores. In the absence of rock control, selective development of those randomly initiated cores which are favorably positioned relative to competing nuclei controls the form and spacing which characterize most drumlin fields.

INTRODUCTION

A drumlin is an elliptical hill composed of till and aligned parallel to the direction of glacier flow. This has long been an accepted definition of a drumlin, with composition, form and origin all implicit in the term. In the discussion which follows, however, it will be shown that drumlins are not ideally elliptical, that stratified drift as well as till is an integral component of drumlins, and that iceflow directions locally within a drumlin field rather than lying parallel to drumlin long axes ranges systematically through as much as 90°.

The state of knowledge of drumlins has been ably summarized at various stages by many investigators, among them, Ebers (1926), Gravenor (1953), Charlesworth (1957), Embleton and King (1968) and Flint (1971). Based largely on distribution and occurrence of drumlins as described in the published literature, the following relationships are recognized as relevant in development of a general theory of drumlin origins:

1. Drumlins and associated features are not uniformly present in glaciated landscapes. They are, in fact, the exception rather than the rule, and their existence denotes somewhat special circumstances during glaciation.
2. Drumlins exist in groups and are almost unknown as isolated individuals.
3. Some drumlins are shown to possess rock or drift cores. Many others, however, possess no nucleus distinguishable from the lodgment till of which they are primarily composed.
4. Some drumlin fields are located where glacial flowlines diverged radially in a glacier lobe.
5. Some drumlin fields are located in areas of reverse gradient where ice flowed up a gentle slope.
6. Some drumlin fields are notably located where previously deposited drift or nonresistant bedrock provided a source of abundant material in the subglacial load.
7. Some drumlin fields were developed as broad bands at a distance behind end moraines. These bands are broader than any ordinary moraine belt, and suggest instead a subglacial zone at some distance from the glacier margin in which conditions were suitable for drumlin development.
8. Within a given cluster, drumlins typically show marked similarity of form. Drumlins around the edge of the cluster are most apt to deviate from the form parameters of the group.
9. Within a given field, drumlin form tends to change consistently in the direction of iceflow.
10. Drumlins are characteristic of, but not entirely restricted to, areas of ice sheet (unconfined) as opposed to valley (confined) glaciation.
11. Drumlins are composed of lodgment till of diverse textures. Although silt and clay are always present in the till matrix, in western New York, at least, sparsely stony clay till is notably free of drumlins.
12. Because of their conformity to independently determinable glacier flowlines, the parallel long axes of drumlins are generally accepted as evidence that drumlins were produced beneath actively flowing ice.

13. Drumlin development involves something both of erosion and deposition. Any comprehensive view of drumlin origins must accordingly bring together these seemingly opposed processes.

With good reason, discussion has long ranged between investigators who support a depositional origin of drumlins as streamline features lodged beneath actively moving ice, and those who support an erosional hypothesis for their origin (Gravenor, 1953). Within a single drumlin field the evidence has commonly been contradictory.

In the New York drumlin field, for instance, Fairchild (1907a, 1907b, 1911, 1929) developed an "accretional" theory hypothesizing that drumlins develop by progressive lodgment on some commonly unrecognized core or nucleus. On the other hand, striking exposures of drumlin structure in shore bluffs of Lake Ontario stimulated contrasting views of drumlin origins (Slater, 1928, 1929). Some drumlins in New York are made of shale. Others include waterlaid sand and gravel. These demonstrably erosional drumlins occur in intimate association with others whose origins are not independently demonstrable.

Clayton (this volume) reports that field evidence shows erosion rather than deposition to be dominant in producing drumlins and other "longitudinal shear marks". Goldthwait likewise (this volume) describes drumlins in the Glacier Bay region, Alaska, shaped by erosion of outwash gravels containing forest beds which have been radiocarbon-dated as Hypsithermal, thus establishing both an erosional origin and a specific interval during which the features were shaped.

With such convincing evidence of the effectiveness of both erosional and depositional processes, one must assume that both play essential roles in shaping topographic forms which afforded minimum shear resistance to overriding ice.

A FAMILY OF STREAMLINE GLACIAL FEATURES

Where a process takes place too slowly for direct and continuous observation of resulting changes, it is a recognized and accepted though sometimes risky practice to infer genetic relationships from the assumed sequence of transitional or intermediate forms. Such, for instance, was the basis for the Davisian scheme of landscape development. The problem of drumlin origin may be approached similarly, but very cautiously. Here, too, it is impossible directly to observe the processes in action, less because of the time factor than because they take place deep beneath an ice sheet.

This kind of inductive procedure has been applied in two ways to the problem of drumlin origins. The first involves classification of drumlins and related glacial topographic features on the basis of inferred genetic relationships. The second involves interpretation of causes of sequential changes in form in the direction of iceflow.

Kupsch (1955) states that "roches moutonnées, crag and tail, rock drumlins, drumlins with till cores . . . form one series, all composed of modified nuclei with no, a little, or much glacial material lodged on their lee sides." This brief synthesis departs notably from the once conventional scheme of classifying erosional and depositional topographic features in opposition to each other.

Flint (1971), similarly, speaks of a family of streamline forms ranging from 100% bedrock to 100% glacial drift and molded through erosion, deposition or combination of the two processes. He identifies this continuum as ranging from stoss and lee topography at the bedrock end, through drumlins of bedrock with veneers of till, crag and tail, drumlins with bedrock cores, and drumlins with cores of stratified drift to drumlins of till only at the glacial drift end of the series.

Based on the relationships between form and composition on the one hand and assumed or apparent origin on the other hand, streamline glacial features may be classified as follows:

Table 1
Genetic Classification of Streamline Glacial Features (after Muller, 1963)

Erosion dominant	Erosion and deposition	Deposition after erosion	Deposition dominant
Crags and tails			
Crags with rock tails	Crags with drift tails	Cored drumlins	Uncored drumlins
Roches moutonnées		Rock-cored drumlins	
Rocdrumlins*		Drift-cored drumlins	

*Following Fairchild (1907a) the spelling of the Celtic roots is here retained in preference to the Anglicized and hybridized "rock drumlin".

Ideal roches moutonnées, characterized by abraded stoss end steeply quarried lee end, develop by the modification of knobs of massive, coarsely jointed rock beneath overriding ice. The same processes tend to produce drumlinoid form (rocdrumlins) in erosional modification of glaciated shale hills.

Consistent changes in drumlin morphology in the direction of ice flow have been noted in a number of drumlin fields. Chorley (1959) suggested that such gradual change in form may denote differences in flow resistance at the ice-rock interface. Muller (1963) attributed progressive lengthening and lowering of drumlin profiles southward in Chautauqua County, New York, to decreased abundance of material in the glacier load as it spread onto the Allegheny Plateau. Heidenreich (1964) found that drumlin width and length tend to increase together to a critical value, beyond which length may be greatly increased without corresponding change in width. Smalley and Unwin (1968) mention the transition in form from drumlins to irregular drift to end moraine to support their case for dilatancy of material in glacial transport as a fundamental factor in drumlin development.

Recently, Lundqvist (1969) has suggested a genetic relationship between drumlins and the so-called "ridged" or "rib" moraines which carries implications as to drumlin origins. Rogen moraine or "rib" moraine characterizes extensive areas in inland northern and central Sweden, and is similarly well developed in the divide area of central Labrador-Quebec (Henderson, 1959). "Rib" moraine consists of rather evenly spaced, gently arcuate, subparallel ridge segments oriented normal to ice flow. Though composed dominantly of lodgment till, and vaguely suggestive of the minor parallel ridges referred to as De Geer moraines, they are of much greater relief and more open spacing, being typically 10-20 m high and spaced 100-200m apart. Just as is true of drumlinized landscapes, ridge dimensions and spacing, though rather uniform within any given area, may be distinctly different in other areas.

Lundqvist (1969) recognizes a transition from laterally connected arcuate ridges of Rogen moraine to disjunct crescent ridges, to incomplete drumlinoids embayed at the down-ice end, to characteristic drumlins (Fig. 1). Such transitions are recognized as having developed in the direction of ice flow and as being reversed where the subjacent topography was concave, i.e., where the gradient steepened in the direction towards which the glacier flowed. It is reasonable to infer, in areas where such transition is recognized, that the drumlins were produced by initial placement and subsequent molding of till ridges beneath actively flowing ice. Conversely, the lack of features transitional

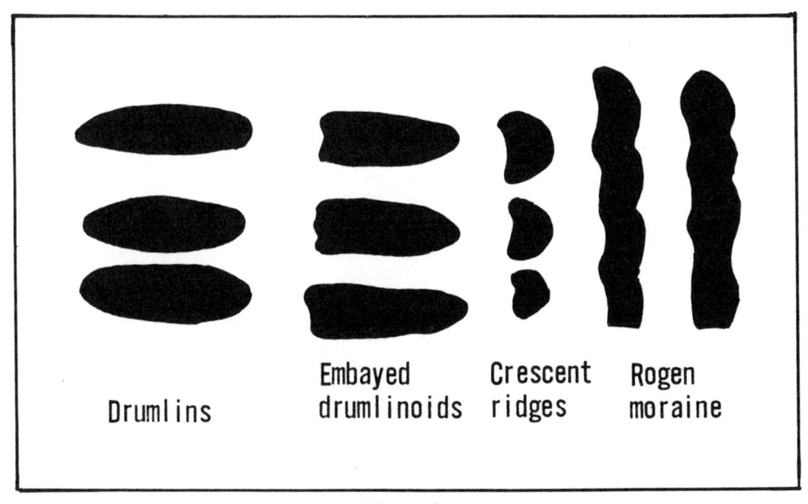

Figure 1

Transitional topographic shapes in the streamline glacial series from Rogen moraine to drumlins (after Lundqvist, 1969).

from "rib" moraine in most drumlin fields makes it questionable whether this is an essential sequence of development of all drumlins.

The possibility that interpretations of drumlin origins may be derived from variations in drumlin form leads naturally to statistical and theoretical modelling of drumlin morphometry.

DRUMLIN MORPHOMETRY

Similarity of drumlin form within a group, the tendency for regular change in prevailing form in the direction of ice flow, and the bearing of form on drumlin origins make drumlins rewarding objects for statistical analysis and theoretical modelling of form, orientation and spacing (King, this volume).

Length, width and height can be readily measured from a topographic map. The adequacy of data obtained in this manner relates primarily to the scale and contour interval employed in mapping. A convenient procedure is to define drumlins in terms of the configuration of the lowest enclosing contour line. This procedure necessarily undervalues all three parameters, but in comparisons of large groups of drumlins, the undervaluation of each parameter approaches a mean, though unknown value. Reasonably consistent values are obtainable, for instance, from New York maps at the 1:24,000 scale with 10 ft(approx.3m)contour interval, or for more prominent drumlins, even a 20 ft (approx. 6m) interval (Reed, et al., 1962; Miller, 1972). Comparisons between drumlin groups measured on maps with different horizontal scales, contour intervals or mapping procedures should be avoided where possible. Elongate drumlins, involving fluting and ridges (Table 2) are apt to be least adequately represented, both because of their low profile and because they are sometimes superposed on larger drumlins.

A classification of form applicable to all streamline topographic features and based on limiting ratios of length to width has been suggested (Muller, 1963) and is here recast in terms of width/length ratios now considered preferable because they range between the finite limits of 0 and 1.

Table 2

Morphometric Classification of Glacial Streamline Features (after Muller, 1963)

Class	Limiting W/L ratio	Name
1	more than .4	broad ovoid
2	.3 to .4	ovoid
3	.2 to .3	long ovoid
4	.12 to .2	ridge
5	less than .12	fluting

For ease of mathematical manipulation, drumlin form may be referred to best-fit geometric models such as the ellipsoid and lemniscate loop, each of which has its own advantages and disadvantages.

The ellipsoid (Reed et al., 1962) is described by an equation of the type:

$$\frac{x^2}{a} + \frac{y^2}{b} + \frac{z^2}{c} = 1$$

where x, y and z are the measured values of width, length and height, respectively. This model affords the advantages of dealing with three-dimensional space in terms of polar coordinates. The ellipsoid, however, has no direct relationship to the mode of formation of the feature described. The fit is without physical basis and only an approximation, departing from drumlin form particularly in the summit area.

The lemniscate loop (Chorley, 1959) is described by an equation of the type:

$$p = 1 \cos k\,\theta$$

where 1 = length, θ is the angle between arms of the loop, and k is a coefficient which expresses the elongation of the loop. Although this model involves two-dimensional space undefined by polar coordinates, it has the distinct advantage of possessing direct physical relationship to the origin of the feature described. A lemniscate loop is the form of an obstruction which generates least eddying or discontinuity of flowlines in a medium moving over it. Such, for instance, is the form of an airplane wing, or the drift built up by snow blown around a narrow obstacle aligned normal to wind direction. A second potential advantage of this model is the criterion which the dimensionless coefficient k provides for a rational classification of streamline glacial forms, analogous to that in Table 2.

In drumlin spacing analysis, the drumlin is assigned a point location, its center, usually taken as the midpoint on the long axis. Little difference will result, however, if the highest points on drumlins are consistently taken as centers, for adjacent drumlins usually are similarly shaped. Parallel spacing denotes the distances between centers of any two adjacent or in-line drumlins measured parallel to mean local long axis orientation, whereas perpendicular spacing is similarly measured normal to mean local long axis orientation (Reed et al., 1962). Vernon (1966) suggests a slightly different procedure for determining perpendicular and parallel spacing. Distance and azimuth angle are measured from each drumlin center to the two nearest drumlin centers in an upstream direction, one on either side of the extended long axis of the drumlin in question. These data are then converted to parallel and perpendicular components by use of a traverse table.

Spacing analyses have demonstrated a tendency toward modal concentrations both parallel to and normal to flow which differ widely from one drumlin field to another. Rather generally, as drumlin length increases within a given drumlin field, drumlin height and the perpendicular spacing between drumlins diminish. "Spacing between drumlins is more closely related to characteristics of moving ice than to bedrock lithology or topography." (Reed et al., 1962).

DRUMLIN CORES

Drumlins which can be shown to have been formed by localized lodgment of till on a subglacial obstruction are cored drumlins. Cores of rock and, in other cases, of unlithified drift are known in many drumlins and may be inferred to exist in many others. Such subglacial protrusions typically exhibit evidence of shaping and abrasion at the contact with overlying lodgment till.

Size and position of the core relative to the rest of the drumlin vary widely. At one extreme, drumlin-shaped hills composed entirely of rock without drift mantle are rocdrumlins. Because transitional forms exist between rocdrumlins and rock-cored drumlins, the terms have unfortunately sometimes been used interchangeably. The latter term is preferred for streamline hills such that rock is nowhere exposed in the natural profile, the inference being that localized till accumulation accounts for at least part of the topographic high.

The existence of drumlins composed solely of lodgment till without cores of other material poses a problem for those who adopt a depositional theory of drumlin origin. Perhaps in analogy to a braided stream pattern, Fairchild (1907a) suggested that drumlins result from shifting lines of temporarily diminished pressure in the basal ice, not controlled by external factors such as subglacial obstructions.

More often, the nucleus of accretion which is implicit in most depositional views of drumlin origin, has been attributed to tractional resistance in the basal load. Russell (1895) suggested that local concentration of debris in the basal ice decreases the rate of plastic deformation, inducing localized deposition of till. Localized freezing, or local escape of interstitial water in basal load in transport might similarly induce deposition of till as a nucleus for subsequent accretion.

Dilatance theory (Smalley and Unwin, 1968) affords a promising approach to the problem of accretion in drumlins without distinguishable cores. As a requisite for drumlin development, Smalley and Unwin hold that the glacier sole must be separated from the underlying stationary terrain by a layer of rock debris which is subject to continuous deformation. Such material, composed of a dispersion of large rock particles in a dense clay-water matrix is a dilatant system, for stress imposed on it in the stable or rest condition must first produce expansion to a less closely packed state before continuous deformation can take place. This relationship, as experimentally determined for a specific till sample, is illustrated in Figure 2. Increasing stress meets increasing resistance to deformation from 0 to A, at which point the loose packing necessary for deformation continues even with reduced stress in the range from A to B.

Beneath an ideal ice sheet, thickening as a function of distance from the ice margin, a band exists parallel to, but at a distance from the margin, such that the weight of the glacier is enough to maintain dilatance and deformation. Smalley and Unwin (1968) conceive that drumlins develop in this zone between an inner boundary where glacier loading exceeds that required to initiate dilatance and an outer limit where glacier loading is less than that required to maintain dilatance. Where the weight is less than that necessary to maintain dilatance, the expanded structure in the basal load collapses into a stable, static more tightly packed form. Where the overriding ice sheet imposes a load greater than that necessary to achieve dilatance "everything is carried forward" (Smalley and Unwin, 1968). As material is transported radially it moves into the zone where the glacial loading is sufficient to maintain, but not to initiate dilatance. Similarly, as an ice sheet thins, the zone where once dilatance could be initiated becomes an area where it can only be maintained but no longer initiated.

Dilatance limits (A and B, Fig. 2) depend largely upon the concentration of large rock particles and to a small extent upon the fluidity of the clay-water matrix as it may facilitate readjustments within the dispersed system. Therefore, within a continuously deforming basal debris layer, collapse to the stable, static form takes place first, locally, wherever stress resistance of the system exceeds the stress imposed upon it. If water has been an essential component of the deforming layer, dewatering must occur or its incompressibility will prevent collapse. Where the substrate below the

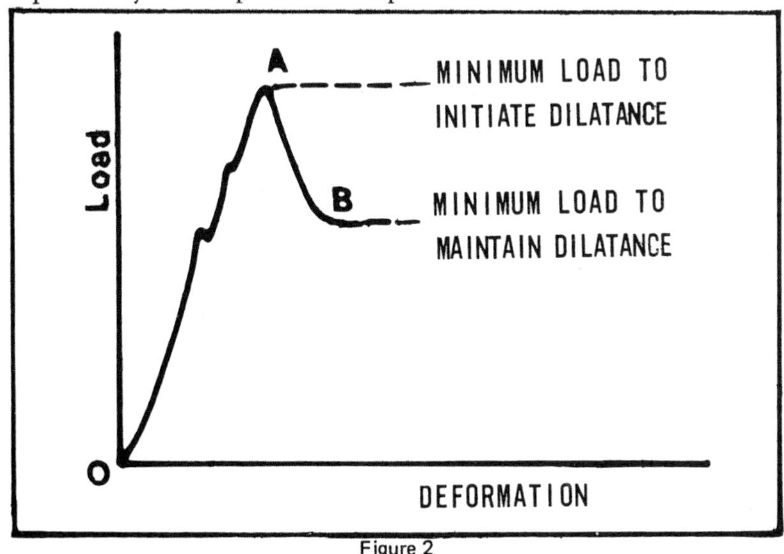

Figure 2

Relation of deformation to load for a representative glacial till, showing dilatance limits (after Smalley and Unwin, 1968).

continuously deforming layer is unbroken rock, the contact is sharp. Where the substrate is drift or brecciated rock, the contact is transitional and transitory. In such cases, a stable, static structure with packing too tight to yield to the ambient pressure is locally determined by variations in dilatance properties of the material. Under such conditions the potential for initiation of nuclei for subsequent drumlin development may be more or less continuously present.

Lundqvist's (1969, p. 27) recognition of a genetic relationship between drumlins and "rib" or Rogen moraine indicates that in some drumlin fields much of the relief may be achieved in this phase of development. In fact, the continuously deforming basal debris layer may be necessary only for initiation of the nucleus upon which accretion will subsequently occur, a process demonstrable on the basis of till fabric analysis.

Probably neither boulder concentration nor fabric will afford adequate criteria by which to distinguish lodgment till from a till core emplaced by collapse of the deforming basal load to a stable static state, too densely packed for further deformation without renewed dilatance. Alternatively, the absence of distinctive cores in some drumlins might suggest the possibility of drumlin migration during development. In such case, the initiating obstruction might be left behind, or might even cease to exist as the drumlin develops beyond it. In the absence of supporting evidence, such an hypothesis seems conjectural and unsatisfying.

Although Smalley and Unwin (1968) consider drumlins to be primarily accretional, they point out that dilatance theory accounts equally for selective mobilization and erosional origin of drumlins. Parts of the former drift cover with low boulder concentration possess low dilatance limits and are subject to selective erosion. So also is severely brecciated rock, and the shale fissility gives it a particularly low dilatance limit.

Dilatance theory affords limited insight into subsequent processes of drumlin development but till fabric data begin to confirm the pattern of deformation by glacial ice around a basal obstruction which is predictable on the basis of plastic flow theory.

TILL FABRIC DATA

Statistical treatment of pebble orientation data in lodgment tills has become a standard technique in glacial geology, commonly applied without regard for possible local variation in fabric. Holmes (1941) in his benchmark study of till fabric cautioned that variation may be encountered within closely related till exposures, a fact subsequently documented by Andrews and Smith (1966); Young (1969); Boulton (1971); Hill (1971) and others. Although abundant theoretical reason exists for variation of till fabric in drumlins, it is only since Wright's (1957) study of pebble orientation in the Wadena drumlin field, Minnesota, that fabric variations within a single drumlin have been analyzed in detail (Andrews and King, 1968; Savage, 1968; Hill, 1971; Shaw and Freschauf, 1973).

Andrews and King (1968) describe variation among till fabric analyses at ten positions in the face of a Yorkshire drumlin obliquely truncated by stream erosion. One analysis was made near river level, about 18m below the drumlin crest; the other nine were measured within 3m of the drumlin surface. The published information, however, does not establish the position of the oblique profile relative to original length. Mean vector orientations as determined for eight of these samples are south of the axis of drumlin elongation. The two samples collected nearest to the drumlin base show the least departure from drumlin axis orientation, 25° and 29°, respectively. At the next higher site the mean fabric vector departs by 59° from drumlin axis orientation, "suggesting that the drumlin grew by the accretion of material added by an increasingly deflected" flow which turned "the stress into a more lateral force" (Andrews and King, 1968, p. 456). They hypothesize that "some element of lateral stress" may be necessary for the formation of a drumlin, but offer no suggestion as to the cause of such deflection. Vector magnitudes are least for fabrics analyzed at the summit of the drumlin, leading to the suggestion that lateral pressure may have been exerted from both sides at the drumlin top. The two vectors with northward departure from orientation of the drumlin axis represent positions north of and near the drumlin crest. In summary, (a) fabrics in lodgment till of a single drumlin were shown to vary by 90° (b) deviation of mean fabric vector from drumlin orientation is least at the base, but increases upward and presumably outward, and (c) the pattern is such as would be produced by deflection at the stoss end of a growing subglacial obstruction.

A more adequately three-dimensional sample-site distribution in a drumlin excavated for construction in southeastern Syracuse, New York, demonstrates fabric variation which conforms to the theoretical pattern of slip lines in plastic deformation around a basal obstruction (Savage, 1968). Such a pattern should show maximum divergence of 90° at the stoss end of the obstruction and similar convergence in the lee, with fabric nearly parallel to drumlin elongation along the flanks away from either end of the drumlin. The sixteen fabric roses obtained at various positions on this drumlin (Fig. 3) show remarkable accordance with the predicted pattern and support a theory of accretional growth by lodgment on a nucleus. Accordingly, Savage (1968, p. 29) proposed that "formation of drumlins begins along slip lines in basal ice bounding an obstacle. . . initial deposition at this slip line will produce a larger obstacle and hence a new set of bounding slip lines on which further deposition occurs".

Glacial flutings are particularly elongate streamline features which maintain maximum crest elevation for long distances in the direction of iceflow. In six transverse sections across glacial flutings near Athabasca, Alberta, Shaw and Freschauf (1973) demonstrated consistent convergence of 20° and 60° between fabrics measured on opposite flanks of flutings at 1/3 and 2/3 of the distance from their axes to their edges respectively. Shaw and Freschauf

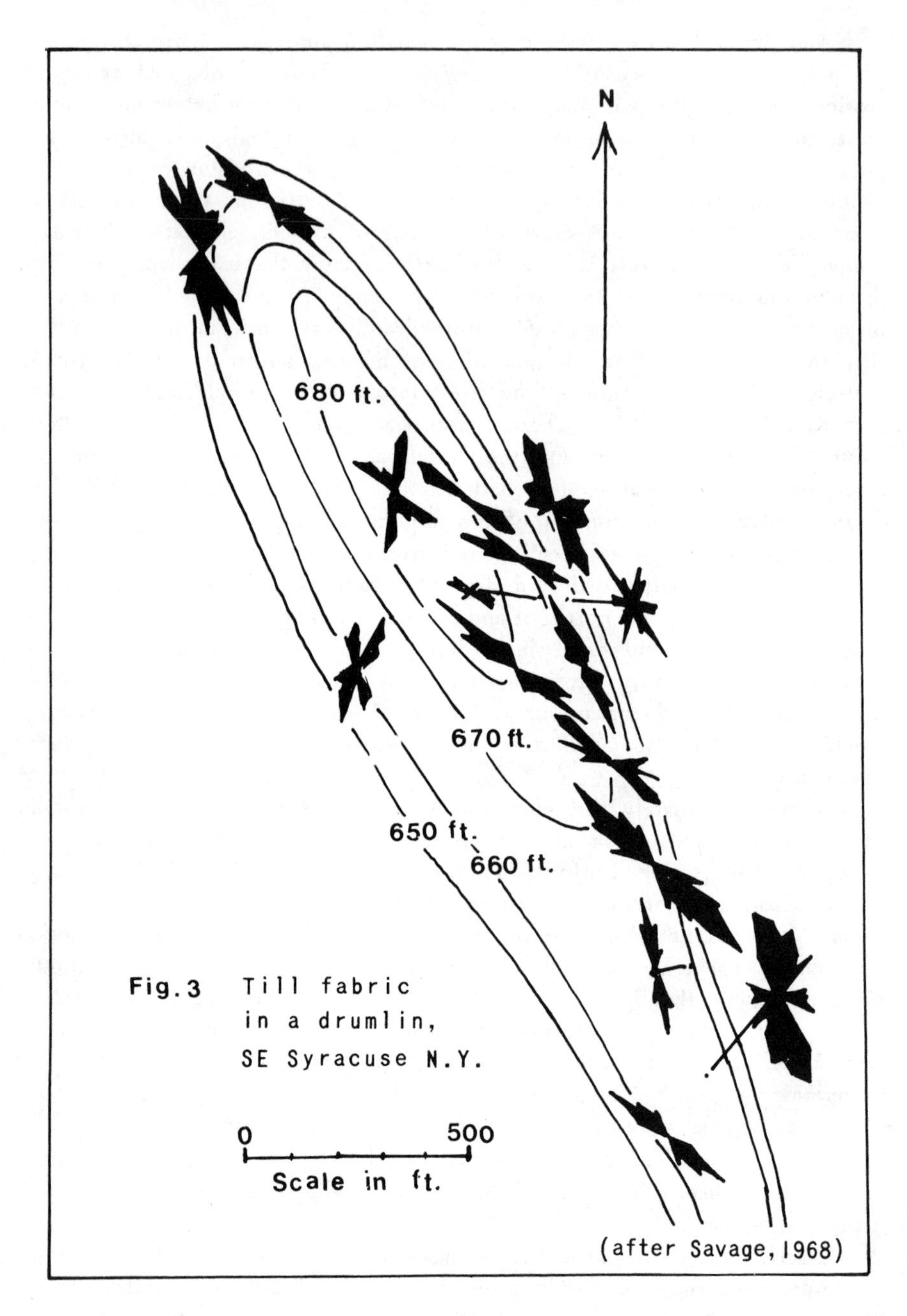

Figure 3
Till fabric in a drumlin, SE Syracuse, N. Y. (after Savage, 1968).

conclude that glacial fluting results from till transport with a component lateral to the principal iceflow direction. Further investigation is necessary to determine the conditions which cause a consistent and continuous component of lateral flow for long enough distances to produce glacial fluting rather than typical drumlin development.

STRATIFIED DRIFT IN DRUMLINS

Drumlins are composed primarily of lodgment till, but in many stratified drift also is a component. Stratified drift occurs not only in drumlin cores, but also intimately associated with lodgment till in a manner that compels recognition of the role of basal meltwater in the processes of drumlin development. Drumlin cores may as readily consist of stratified drift as of non-stratified drift or lithified material. Indeed, in the light of dilatance theory, coarse clastic sediment is a very plausible core. Upham (1894) referred to gravel-cored drumlins as "Madison-type drumlins". Although the core fills an essential function, its composition is more or less incidental where till accretion accounts for drumlin form.

Evidence of basal meltwater activity during drumlin development ranges from microstructures to development of drift tails. On microscopic scale, Sitler and Chapman (1955) described microfoliation and veining in lodgment till. Microfoliation is a crudely parallel arrangement of clay particles attributed to the lodgment process. Veining consists of thin bands rich in silt and clay-sized mineral flakes with high perfection of parallel orientation commonly at a small angle with the vein wall. Veins typically cut across microfoliation, therefore postdate the initial lodgment process and are attributed to shear. The segregation of silt flakes into veins was attributed to purely mechanical differentiation -- shear being favored by orientation and concentration of mineral flakes whereas non-flaky minerals tend to be excluded by rotation out of the shear zone. As illustrated (Sitler and Chapman, 1955, p. 267) vein borders are not characterized by concentration of silt grains greater than that in the intervening areas, as might be expected if segregation were solely a product of mechanical differentiation. It seems plausible instead to attribute the sorting to grain transport by water moving out of the system as it was compacted.

Fissility is characteristic of, but by no means restricted to, lodgment till of drumlins. Such fissility has been attributed to "accretion" of successive layers of till in a kind of layer-by-layer lodgment process. According to Virkkala (1952) "these layers are made by the successive stagnation of thin, overloaded basal layers of ice in the terminal zone of the glacier. The base of the moving glacier shifted upward in jumps, leaving beneath it drift with interstitial ice" (Flint, 1971, p. 159). Boulton (1970, p. 235) has defined "tills produced either by top or bottom-melting of a block of buried debris rich ice, and which are not deformed by subsequent creep or flow" as melt-out tills.

Such tills inherit a fabric and structure modified from the pattern of debris distribution in the stagnant ice.

Some of the fissility observed in New York drumlins, however, appears to have involved actively flowing ice and lodgment of till layers separated by better sorted laminae which because they are more cohesive tend to stand in slight relief on a weathered till exposure. These interbeds are considered to mark intervals of meltwater activity at the glacier sole. Presumably they represent dynamically – rather than seasonally – controlled intervals when the glacier sole was effectively bathed in a water film.

Roadcuts longitudinally and transversely across the stoss end of a drumlin off Route 48 in Van Buren Town, north of Syracuse, New York (Fig. 4) show sparsely stony red till intimately intercalated with fine to medium sand exposed in the drumlin nose. The stoss end exposures are faulted, firmly compacted and sheared. The intimate structural relationships suggest that this is not a pre-existing mass serving as a drumlin core, but rather that its accumulation was an integral part of the drumlin-forming process.

Washed drift comprises variable proportions of the sediment in the tails of crag-and-tail topography and at the lee ends of drumlins. Composition of such drift tails ranges from sand to pebble gravel. Such occurrences of stratified drift may be viewed as having been washed from the subglacial load, presumably largely by undermelting of the glacier at the upstream end of the growing subglacial obstruction. The sediment tends to be washed into the area of reduced pressure in the lee of the obstruction responsible for or provided by the accreting drumlin.

Figure 4
Intimately intercalated sand and till in stoss end of drumlin in Van Buren Town, northwest of Syracuse, N. Y.

Although tails are not characteristic of most New York drumlins, in some areas they are sufficiently typical to indicate local conditions of development. Drumlins with significant stratified drift at their lee ends may be related to the incomplete drumlinoids with lee-end embayment described by Lundqvist (1969). They afford clear evidence of the role of water in developing subjacent topography to match the streamline configuration of the glacier sole.

DRUMLINS AND GLACIER FLOW THEORY

Theoretical analyses by Weertman (1957, 1964), Lliboutry (1964-1965), and Paterson (1969, p. 110-127) afford insight on the role of subglacial meltwater during drumlin formation.

Weertman (1957) postulated two primary mechanisms for glacier slip over basal obstructions. The first of these bears analogy to the classic experiment in which a block of ice "flows" around a wire on which it is supported, refreezing below as heat is transferred across the wire to the continuously melting ice surface. This mechanism involves pressure melting at the upstream surface of an obstruction and refreezing at the downstream end. The amount of ice subjected to this kind of pressure melting upflow from the obstruction is a function of the upstream area of the obstruction and velocity of the basal ice. The process can be maintained continuously only at a rate limited by return heat flow. Accordingly this mechanism is adequate alone only in flow over small obstacles and has limited direct significance on the scale involved in drumlin development.

A second mechanism mentioned by Weertman involves enhanced plasticity of ice due to the local increase in stress adjacent to an obstacle. Rather than being primarily a mechanism of basal slip, as implied by Weertman, this relates to flow velocity distribution within the basal zone of an overriding ice sheet. The stress differential imposed in flow around an obstruction involves stress increase upstream from the obstruction accompanied by stress decrease downstream from it.

Because velocity of ice is proportional to strain rate times the distance through which the stress is increased, this mechanism is more effective for large than for small obstructions. The larger the basal obstruction, other factors being constant, the greater will be the zone of enhanced plasticity, and the more intense the shear within the ice around the obstruction. Thus in a zone adjacent to an existing obstruction development of a competing core is opposed. Initiation of drift cores may be random, but survival and favored development of cores most suitably located in terms of distance from competing nuclei accounts for regularity in drumlin spacing and form.

Where subglacial melting is effective, complimentary processes invoked in Lliboutry's theory of basal sliding (1964-1965) come into play. Basal melting occurs most effectively where pressure is greatest. In addition to pressure melting, frictional heat release in the region of most rapid deformation favors melting at the upstream edges of subglacial obstructions followed by escape of the resulting basal meltwater to regions of minimum confining pressure - for example, in the lee of subglacial obstructions. Assuming both saturation and limited permeability throughout basal ice and substrate, the favored escape routes will be those which require least heat transfer to open and maintain, hence will be most apt to occur at the ice-rock interface.

The consequences of development of a subglacial water envelope, however discontinuous and concentrated in leeside pockets, are (a) diminished friction between glacier sole and substrate, thus facilitating slip, and (b) delayed closure of confining ice walls, in the lee of basal obstructions. Refreezing of basal meltwater can only occur to the extent that heat can be removed from the meltwater, which at the base of a warm glacier is very limited. Hence, as pointed out by Lliboutry (1964-1965, 2, p. 647-652), the larger the water-filled cavities, the less the frictional resistance to sliding, hence the greater the rate of flow which still further enlarges the water-filled cavities. A drumlin field may well facilitate a surge response by such a feedback mechanism.

The occurrence of stratified drift as an integral though typically subordinate component of drumlins, the critical role of dewatering in collapse and deposition of a dilatant subglacial debris-water system, and the importance of subglacial water in facilitating glacier flow over basal obstructions, all suggest that drumlins develop beneath a wet glacier sole. This suggestion is made in full awareness of the superb development of drumlins, for instance, near the Quebec-Labrador border east of Schefferville, the last center of outflow during disappearance of the Laurentian ice sheet. Polar though the ice sheet must have been at its maximum extent, there came a time during thinning when the bulk of the glacier including its sole was at the critical temperature. It is the dispersal pattern of this much diminished ice sheet rather than of its maximum extent which is reflected in the drumlin alignments.

CONCLUSIONS

The following conclusions are developed to supplement the generalizations at the beginning of this chapter which were derived primarily from previous literature:

1. Drumlins are genetically related to such diverse forms as roches moutonnées, crag and tail, and rocdrumlins, in that all are products of shear beneath actively moving ice, hence share a tendency toward streamline form and elongation parallel to former ice flow.
2. Progressive change in form of streamline glacial features in the direction of ice flow affords a basis for inductive reasoning as to transitional and intermediate forms in the process of drumlinization.

3. The lemniscate loop model affords best fit to drumlin form and possesses advantages over the ellipsoidal model in its sound physical basis as the streamline form affording least resistance to flow.
4. Till is a dilatant system, undergoing expansion in the process of mobilization under load. Once dilatant, a subglacial layer of such material continues to deform even under reduced stress until it collapses into a stable, static state, too tightly packed for ready deformation.
5. The relationship between "rib" moraines and drumlins suggests that both may be initiated by massive collapse of the once dilatant and continuously deforming subglacial debris layer. This origin is proposed also for the nuclei of drumlins which lack distinctive cores.
6. Till fabric data indicate growth of drumlins through accretion by lodgment of till along the slip lines of maximum shear stress of a medium deforming plastically around a basal obstruction.
7. The greater the area of an obstruction normal to glacier flow, the more plasticity enhanced and shear intensified around it creating a field hostile for development of competing cores.
8. A selective development of cores most favorably located beyond the sphere of competing nuclei transforms an initially random field of potential cores, developing a degree of regularity of form and spacing. This is expressed, for instance, by modal and minimal values for perpendicular spacing, and by the inverse relationship of drumlin length to drumlin height and spacing.
9. Stratified drift is an integral but not an essential component of drumlins, recording the role of basal meltwater in drumlin development. Most, if not all, drumlins were produced beneath basal ice at the pressure-melting temperature, in equilibrium with subjacent meltwater.

ACKNOWLEDGEMENTS

Thanks are due to William Z. Savage for suggestions and discussion during the preparation of his M.S. thesis at Syracuse University; to William A. White, University of North Carolina, for suggestions in correspondence; and to Max Gage of the Canterbury Museum, Christchurch, New Zealand, and Stephen Porter, University of Washington, for discussion and critical reading of this manuscript.

REFERENCES

Andrews, J. T. and C. A. M. King. 1968. Comparative till fabric variability in a till sheet and a drumlin: a small-scale study: Yorkshire Geol. Soc. Proc. 36:435-461.

Andrews, J. T. and D. I. Smith. 1966. The variability of till fabric: Occl. Pap. Brit. Geomorph. Res. Gp. 3:33-37.

Boulton, G. S. 1970. On the deposition of subglacial and melt-out tills at the margins of certain Svalbard glaciers; Jour. Glac. 7:391-412.

Boulton, G. S. 1971. Till genesis and fabric in Svalbard, Spitzbergen, p. 41-72 in Goldthwait, R. P., ed. *Till: a Symposium,* Ohio State Univ. Press, Columbus, 1402 p.

Charlesworth, J. K. 1957. *The Quarternary Era, with special reference to its glaciation.* Edward Arnold, London, 2 v, 1700 p.

Chorley, R. J. 1959. The shape of drumlins: Jour. Glac. 3:339-344.

Ebers, F. 1926. Die bisherige Ergebnisse der Drumlinforschung. Eine Monographie der Drumlins: Neues Jahrb. Min. Geol., u Palaont. 53B:153-270.

Embleton, C. and C. A. M. King. 1968. *Glacial and periglacial geomorphology.* St. Martin's Press, N. Y., 608 p.

Fairchild, H. L. 1907a. Drumlins of central western New York. N. Y. State Mus. Bull. 111:391-443.

Fairchild, H. L. 1907b. Drumlin structure and origin: Geol. Soc. Am. Bull. 17:702-706.

Fairchild, H. L. 1911. Radiation of glacial flow as a factor in drumlin formation (abstract):Geol. Soc. Am. Bull. 22:734.

Fairchild, H. L. 1929. New York drumlins: Rochester Acad. Sci. Proc. 7:1-37.

Flint, R. F. 1971. *Glacial and Quaternary Geology:* John Wiley, N. Y., 892 p.

Gravenor, C. P. 1953. The origin of drumlins: Am. Jour. Sci. 251:674-681.

Heidenreich, C. 1964. Some observations on the shapes of drumlins: Can. Geog. 8:101-107.

Henderson, Eric. 1959. A glacial study of central Quebec-Labrador: Geol. Surv. Canada Bull. 50, 94 p.

Hill, A. R. 1971. The internal composition and structures of drumlins in North Down and South Antrim, northern Ireland: Geog. Annaler 53A:14-31.

Holmes, C. D. 1941. Till fabric: Geol. Soc. Am. Bull. 52:1299-1354.

Kupsch, W. O. 1955. Drumlins with jointed boulders near Dollard, Saskatchewan: Geol. Soc. Am. Bull. 66:327-338.

Lliboutry, Louis. 1964-1965. *Traite de glaciologie.* Masson et Cie, Paris, 2 v, 1040 p.

Lundqvist, Jan. 1969. Problems of the so-called Rogen moraine: Sveriges Geol. Undersoknung Ser. C., 648, Arsbok 64, p. 1-32.

Miller, J. W., Jr. 1972. Variations in New York drumlins: Ann. Assoc. Am. Geog. 62:418-423.

Muller, F. H. 1963. Geology of Chautauqua County, New York. Part II, Pleistocene Geology. N. Y. State Mus. Bull. 392, 60 p.

Paterson, W. S. B. 1969. *The physics of glaciers.* Pergamon Press, N. Y., 250 p.

Reed, Bruce, C. J. Galvin and J. P. Miller. 1962. Some aspects of drumlin geometry: Am. Jour. Sci. 260:200-210.

Russell, I. C. 1895. The influence of debris on the flow of glaciers: Jour. Geol. 3:823-832.

Savage, W. Z. 1968. Application of plastic flow analysis to drumlin formation: M. S. thesis, Syracuse University, 60 p.

Shaw, John and R. C. Freschauf. 1973. A kinematic discussion of the formation of glacial flutings: Can. Geog. 17:19-35.

Sitler, Robert and C. A. Chapman. 1955. Microfabrics of tills: Jour. Sed. Pet. 25:267-269.

Slater, George. 1928. Structure of drumlins on southern shore of Lake Ontario (abstract): Pan-Am. Geologist 50:232.

Slater, George. 1929. Structure of drumlins on southern shore of Lake Ontario: N. Y. State Mus. Bull. 281:3-19.

Smalley, I. J. and D. J. Unwin. 1968. The formation and shape of drumlins and their distribution and orientation in drumlin fields: Jour. Glac. 7:377-390.

Upham, Warren. 1894. The Madison type of drumlins: Am. Geol. 14:69-83.

Vernon, Peter. 1966. Drumlins and Pleistocene ice flow over the Ards/Strangford Lough area, County Down, Ireland: Jour. Glac. 6:401-409.

Virkkala, Kalevi. 1952. On the bed structure of till in eastern Finland: Comm. Geol. de Finlande Bull. 157:97-109.

Weertman, J. 1957. On the sliding of glaciers: Jour. Glac. 3:33-38.

Weertman, J. 1964. The theory of glacial sliding: Jour. Glac., 3:287-303.

Wright, H. E. 1957. Stone orientation in the Wadena drumlin field, Minnesota: Geog. Ann. 39:19-31.

Young, A. 1969. Variations in till microfabric over very short distances: Geol. Soc. Am. Bull. 80:2343-2352.

CHAPTER 8

REAPPRAISAL OF THE GLACIATED APPALACHIAN PLATEAU

Donald R. Coates

ABSTRACT

This chapter provides a partial literature review and an updating of ideas of the Glaciated Appalachian Plateau. This is a very diverse region containing many differences when compared with non-glaciated areas to the south. For clarity the Plateau is divided into 11 sections, each with its own characteristic suite of geomorphic features. In the United States the Plateau is unsurpassed in many landforms such as the Finger Lakes, through valleys, glens, sluiceways, hanging deltas etc. There is a variety of depositional features, but a surprising absence of well-defined morainic systems except for the western part.

The Plateau contains four major rivers and reaches elevations of 4,200 ft in the Catskill Mountains. Contrary to some views the region is not homogeneous, but contains many areas of non-dendritic terrain. Glaciation had a profound impact, leaving thick drift deposits, selectively eroding troughs 1,000 ft, reducing uplands and notching cols,. and greatly altered many drainage patterns. When considering the total topographic fabric of the Plateau it is important to use a multicyclic model of landscape denudation. Not only are glacial episodes of erosion-deposition important, but the remodeling that occurs by the fluvial-gravity system during inter and intra-glacial events is also of major importance.

INTRODUCTION

The Glaciated Appalachian Plateau is a 30,000 sq mi area that covers parts of New York, Pennsylvania, and Ohio. This chapter emphasizes the New York part, and space does not permit treatment of Ohio. Obviously such a large and diverse region cannot be discussed in great depth within a few pages. So much recent work has been done, however, that the time is now appropriate to provide a status report and to compare such work with earlier and traditional reports that have become entrenched in the literature of the region. Another rationale for this chapter is that these symposiums have been held in Binghamton, the heart of the glaciated plateaus, and such a review can hopefully provide participants with an increased level of geomorphic understanding in the area of their visit.

It is important that the reader understand the nature of the material covered in the review chapter, as well as the limitations in topical matters. This is primarily a geomorphic study with concentration on landforms. For example the stratigraphy and correlation of various glacial units is only minimally referenced and the dynamics of the glacial ice and associated processes is discussed only briefly. Very little attention is devoted to the usual and more or less ubiquitous glacial forms such as kames, valley trains, deltas, etc. Instead greater coverage is given to those features that are more unusual and for which the Plateau is known . . . the Finger Lakes. through valleys etc. (Hereafter the word "Plateau" will always mean the Glaciated Appalachian or Allegheny Plateau. When discussing the non- glaciated part of the Appalachian Plateau the terms "Unglaciated Plateau" will be used.) Such large questions as that of possible peneplanation, and pre-Quaternary drainage evolution will be evaluated in subsequent investigations.

One purpose of this chapter is to disspell some of the misconceptions about the Plateau. A casual reader could easily obtain the view that glaciation produced only minimal terrain changes:

> "Were it not for the accident of glaciation the northern section of the Allegheny Plateau might be included in the description . . . . for the southern section . . . . the extent of dissection is much the same and the relief similar in amount though qualitative differences are noteworthy (Fenneman, 1938, p. 304)."
>
> "The topography as it was before glaciation has not been greatly changed (Fenneman, 1938, p. 313."
>
> "Only the fact that it has been glaciated justifies setting this section apart from the Unglaciated Allegheny Plateau section; despite this, the topographies of the sections have much in common (Thornbury, 1965, p. 132)."

Other erroneous ideas commonly held are that the Plateau is a rather homogeneous and featureless terrain, and that all drainage systems can be classified as dendritic. To illustrate the dramatic differences within the Plateau it is necessary to subdivide the region into at least 11 sections, each with its individual topographic flavor (Fig. 1). These sections will be discussed but will not be given uniform or equal analysis.

It is also the aim of this chapter to provide some of the more recent ideas and literature sources, especially those materials that have been produced in the last 10 years. This will also include references to unpublished theses. Such information should provide a useful supplement to the very helpful bibliography published by Muller (1965a). Armed with such a compilation of materials it becomes possible for a new geomorphologist to comprise his own assessment of the Glaciated Appalachian Plateau.

Previous Work

Justice cannot be done in this brief summary to the vast amount of work and literature that has been written about the Plateau. The best possible beginning for one interested in the region would be a review of the information sources provided by Muller (1965a). Since this only covers New York and the period prior to 1965, I will update the New York area and also point the way to some of the most important Pennsylvania publications. Early work in New York will only be mentioned when relevant to the particular topic under discussion.

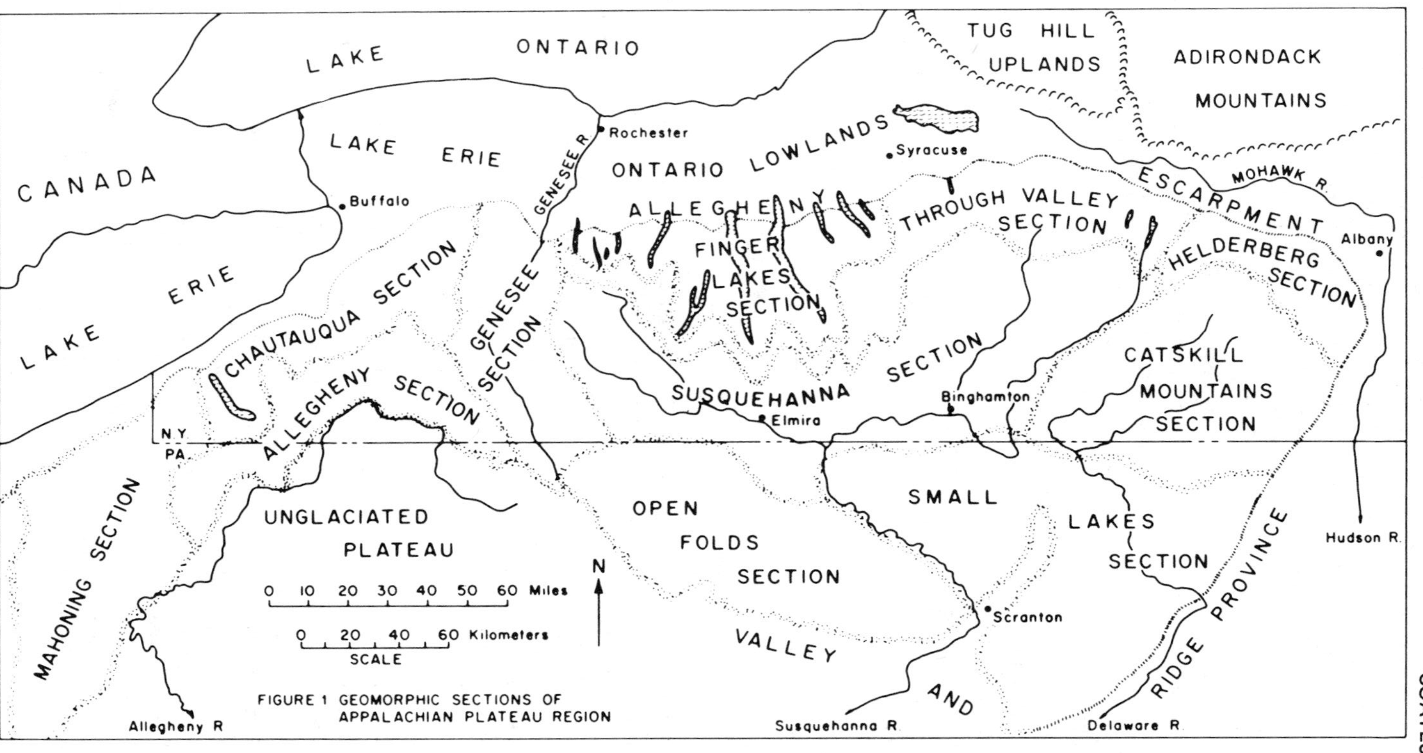

Figure 1. Map of geomorphic sections of the Glaciated Appalachian Plateau with adjacent provinces.

It is important to acknowledge work that specific scientists have accomplished in the Plateau as well as the role different governmental agencies have played in contributing their support and funding for various investigations. For example the most prolific investigator has been H.L. Fairchild and typical of his work is the Susquehanna study (1925). Many of his studies were financed by the New York State Science and Museum Service, as was the work of others such as J.L. Rich in the Catskills (1935). The U. S. Geological Survey and its personnel have aided in assembling significant information of the area in a variety of reports, most of which are aerial in nature. These include the folio series Fuller and Alden (1903a, 1903b), and Tarr (1909); the professional paper series Denny (1956), and Denny and Lyford (1963), and ; groundwater reports Wetterhall (1959), Soren (1963) etc. The U. S. Department of Agriculture has produced soil maps for most counties in the Plateau, and the surveys that date after 1955 are especially helpful for such counties in New York as Broome, Cortland, Tioga, and Tompkins. The Pennsylvania Geological Survey has been especially active in that State and instrumental in publishing a large variety of materials. These include the helpful regional groundwater studies of Leggette (1936) and Lohman (1937, 1939); the regional glacial studies of White, Shepps and others (1959, 1969), and; the special project studies of Leverett (1934), Peltier (1949), Epstein (1969), and Crowl (1971). Some of the earliest reports occur in Lesley (1876) and Lewis (1884).

There is also a voluminous literature in the Plateau that occurs in journals of various science societies, the annual meetings of the Friends of the Pleistocene, and the New York State Geological Association. These publications will be mentioned throughout this chapter, but several are deserving of special attention. Those by Holmes (1937, 1941, 1952, 1960) and von Engeln (1921, 1929, 1961) set the stage and helped paved the way for others. Muller (1965b) provided the most recent overview of the New York region. Many unpublished graduate theses also provide important data.

GENERAL SETTING

The Plateau contains elements of four major drainage systems and other rivers of subordinant size. The principal rivers are the Allegheny, Delaware, Genesee, and Susquehanna (Fig.2) Other large rivers are the Neversink, and Schoharie. Glaciation produced a profound impact on the drainage networks. Ice margins and preglacial meltwaters have altered the courses of many rivers. and the glacially-derived deposits have partially filled valleys giving them a facade of "maturity". The Plateau is entirely underlain by sedimentary rocks of Paleozoic age. The great majority of rocks are Devonian clastics and are deposits that originated in the Catskill delta-alluvial plain type of sedimentation. Limestones occur in the northern part of the Plateau and some coal deposits are present in small parts of Pennsylvania. Structurally the regions contains very gentle folds that trend east to

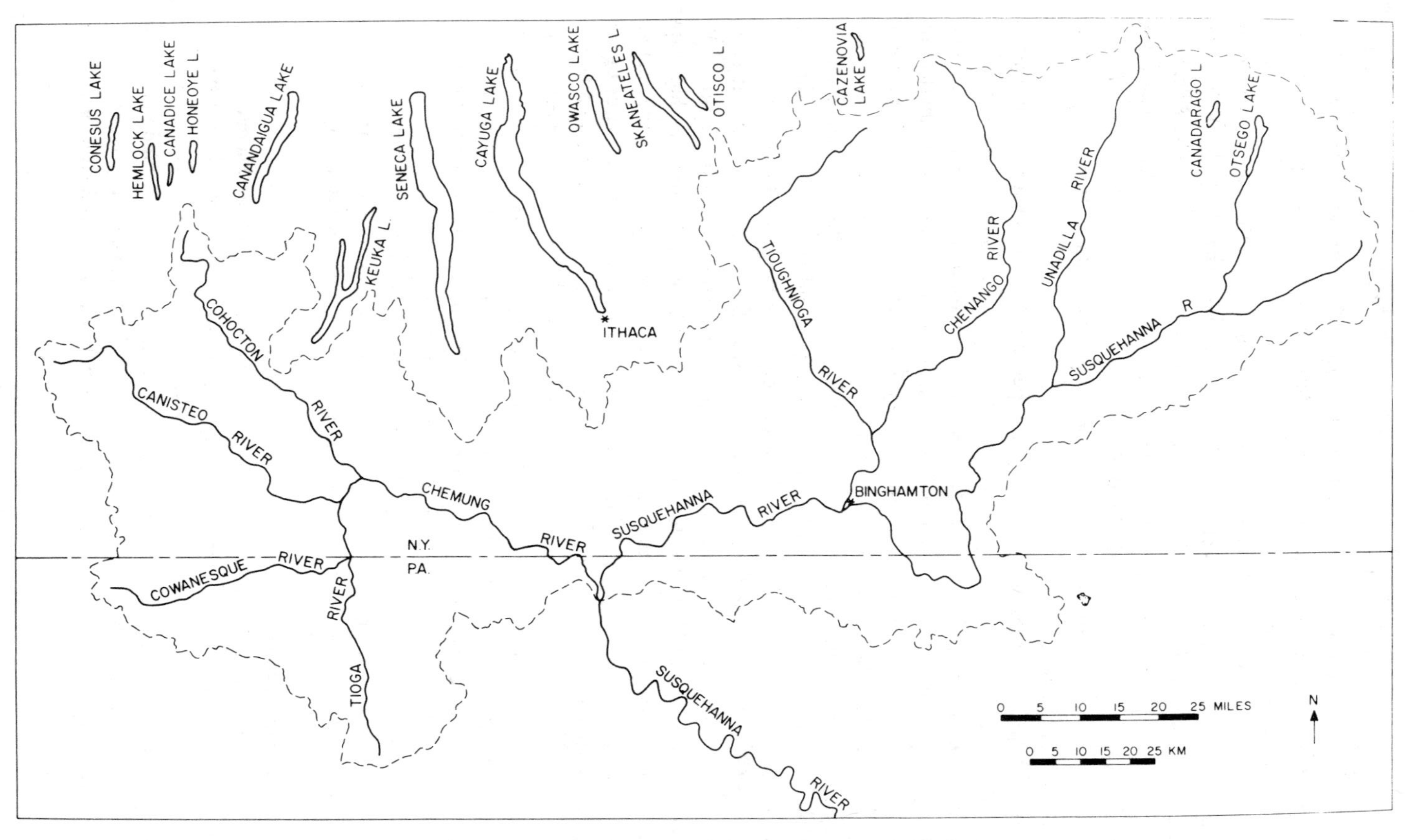

Figure 2. Map of Susquehanna River Basin hydrographic features.

northeast with dips that are generally in the 30–100 ft/mi range. The exception occurs in the Open Folds Section where dips can become a few degrees, and produce a trellis drainage pattern.

The study region (Fig. 1) is bounded on the north by the Lake Erie-Ontario Lowlands and the Mohawk Valley, on the east and southeast by the Valley and Ridge Province, and by the Unglaciated Plateau in the southwest. The Allegheny Escarpment provides an especially sharp demarcation in many places where the relief ranges to as much as 2000 -3000 ft as in the Catskills. However, many parts of the escarpment are irregular and serrate, and it gradually dies out westward of the Finger Lakes Section.

Several quantitative studies were done especially for this chapter to provide a basis of comparison and some generalities about the topographic flavor for much of the Plateau. In one study 121 U. S. Geological Survey topographic maps (1:62,500) scale) were evaluated by random point analysis. 100 random points were marked on each map and the elevation and hillslope amount determined for each position. Using these data the following calculations were made and regional contour maps drawn for each of the parameters: (1) maximum elevation, (2) minimum elevation, (3) mean elevation, (4) topographic slope, (5) relief, and (6) hypsometric integral. Figures 3–5 depict some of these maps which illustrate the general variations of the region. Only the average for each map was used for contouring. A different approach that quantifies elevation factors led to the construction of Figure 6. To produce this map 137 U.S. Geological Survey (1:62,5000 scale) topographic maps were used. Each map was subdivided into either 6 or 9 parts and the highest elevation within each part was located on the base map for contouring. Thus Figure 6 is a contour map showing maximum elevations for the study area. Several factors emerge from analysis of these data and maps, but the most important is the variability that occurs throughout the Plateau. The highest parts are in the east and the west parts of the region with unusually low elevations in the center. There are also differences in relief throughout the Plateau but the average local relief exceeds 1,000 feet. The region has a mean topographic slope of 13 percent, and a hypsometric integral of 57%. The complete analysis of all this information must await further refinement, but it forms the basis for judgments about earlier Plateau surfaces and drainage evolution of the region to be described in future papers.

A Topographic Overview

Previous quantitative geomorphic analysis of a large part of the Plateau (Coates, 1972) provides a basis for portrayal of the region. In one part of this investigation 13 Delaware basin rivers were compared with 12 Susquehanna basin rivers. This group of 25 rivers ranged in size from 14-282 sq mi. A second part of the investigation studied the hydrologic properties of 97 Susquehanna basin rivers, up to those of 400 sq mi. It is repeatedly shown that glaciation and river orientation

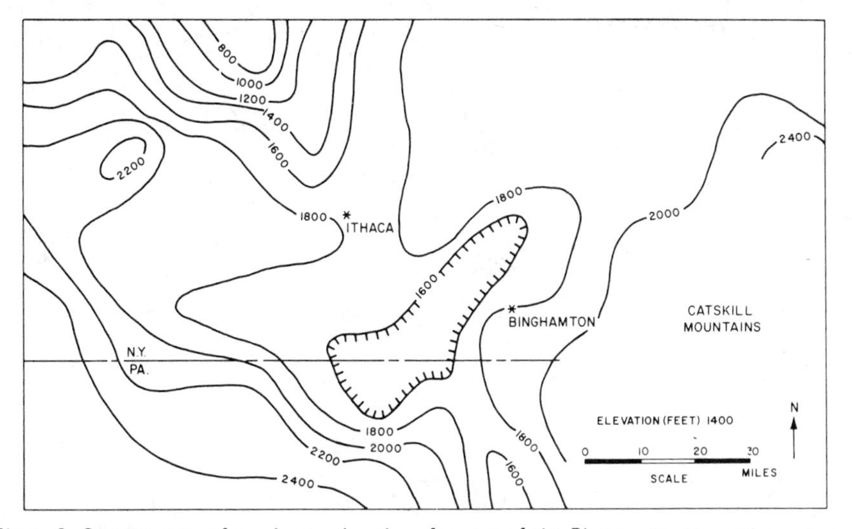

Figure 3. Contour map of maximum elevations for part of the Plateau. Contoured from single point sources of 1:62,500 scale topographic maps (U.S.G.S.)

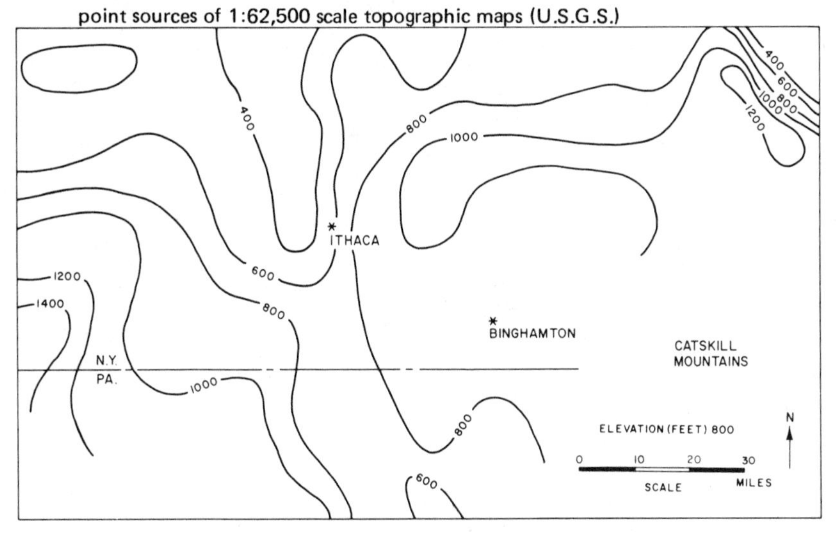

Figure 4. Contour map of minimum elevations for part of the Plateau. Contoured from single point sources of 1:62,500 scale topographic maps (U.S.G.S.)

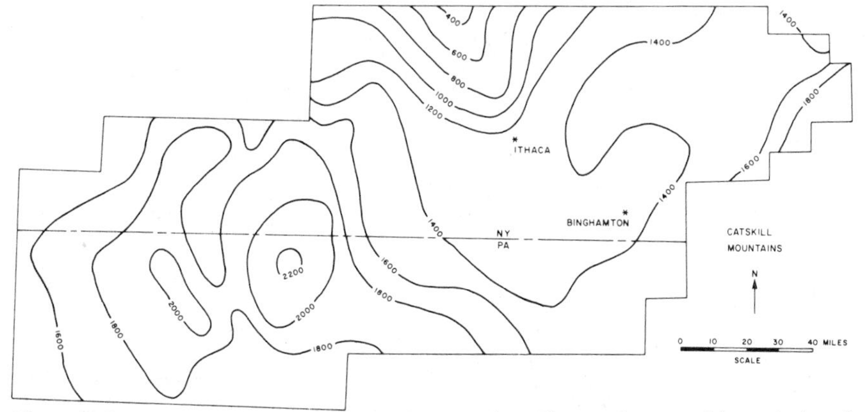

Figure 5. Contour map of mean elevations for part of the Plateau. Contoured from single point sources of 1:62,500 scale topographic maps (U.S.G.S.)

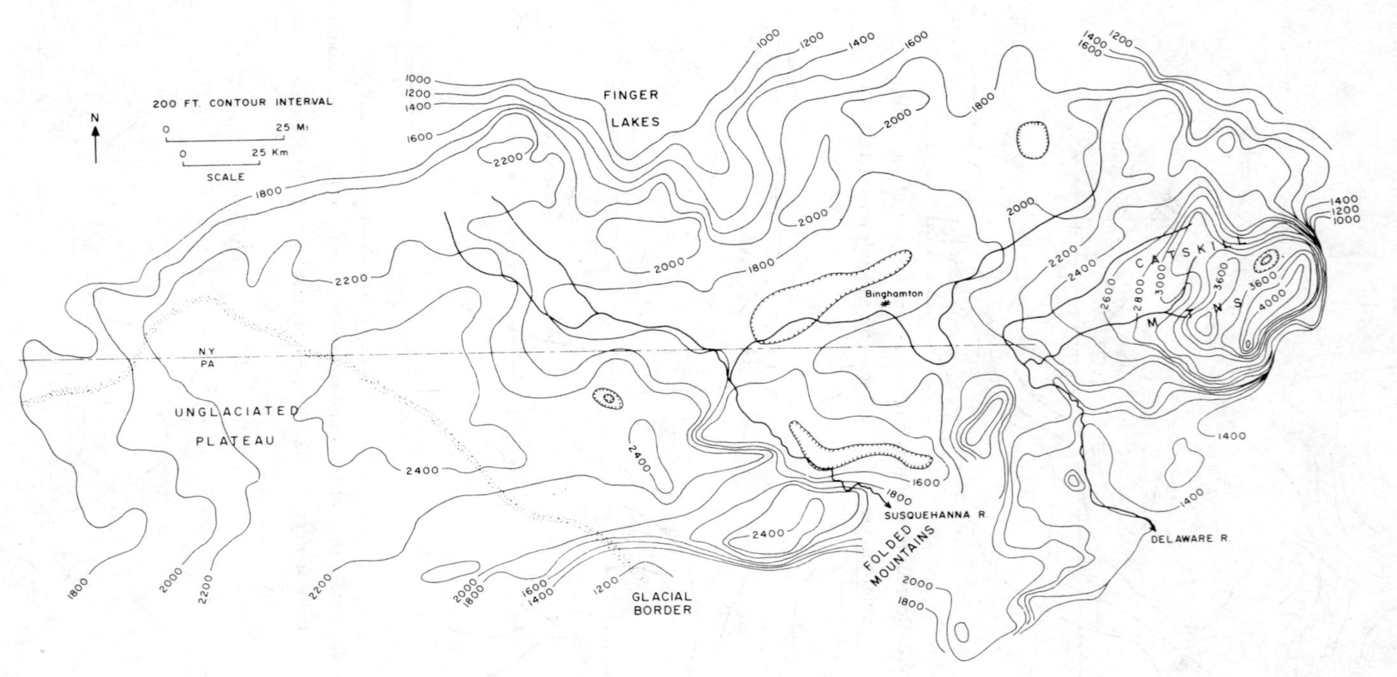

Figure 6. Maximum elevation contour map of Plateau and adjacent area. Contoured from multiple sources of 1:62,500 scale topographic maps (U.S.G.S.)

effect streamflow regimes. Figure 7 depicts 8 generalized cases wherein each river handles its water budget differently than the others. For example river no.1 has more sustained flow and much higher base flow than river no.5 which is very flashy and possesses extremely low discharge during base flow conditions. Other rivers form a continuous hierarchy between these extremes.

Map depiction of terrain differences have already been referred to and by comparison of Figures 3-6 with an overlay onto the various geomorphic sections such differences can be identified section by section. Much additional map work has been done in preparation for a companion article to this chapter, but a quick review of some of this information is in order. (Table 1).

TABLE 1

GEOMORPHIC CHARACTERISTICS

GEOMORPHIC SECTION	ELEVATION (ft)	AVERAGE MAXIMUM ELEVATION (ft)	SLOPE (%)	HYPSOMETRIC INTEGRAL (%)
FINGER LAKES	1000	1650	6	43
THROUGH VALLEYS	1490	2110	9	55
SUSQUEHANNA	1520	2120	11	57
CATSKILL MOUNTAINS	1710	3290	19	55
OPEN FOLDS	1740	2370	14	57
SMALL LAKES	1380	2050	13	55
UNGLACIATED PLATEAU	1730	2250	17	60

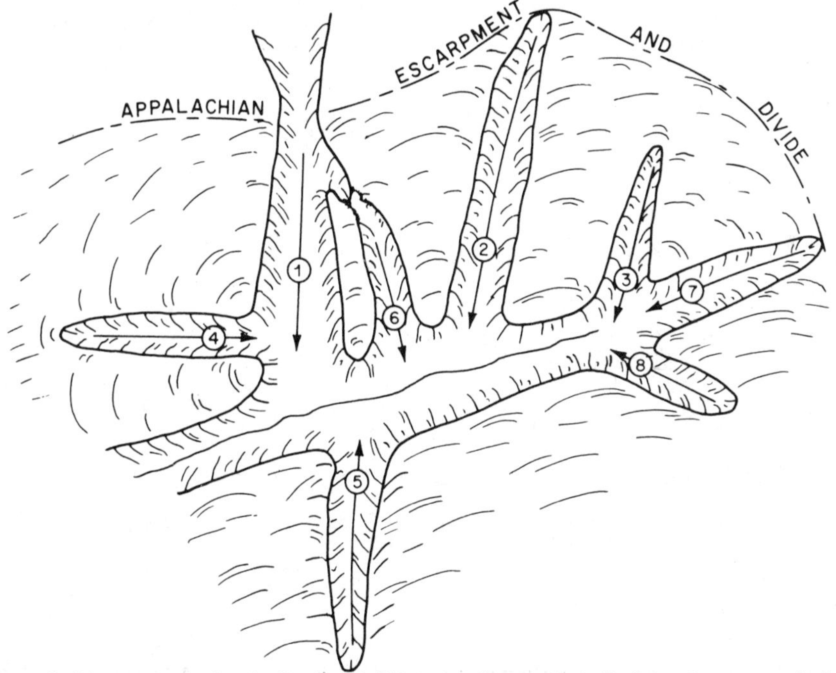

Figure 7. Diagrammatic sketch showing 8 different valley settings. Each has its own particular hydrologic regime. (From Coates, 1972).

Other studies include measurement of geomorphic properties of each river that is 5 mi long throughout most of the Plateau. Such data is revealing much information on drainage anomalies, developmental history, and serve as a basis for comparison both inter and intra-sections. For example stream junction angles are 44° for the eastern Susquehanna Section and 71° for the western part; 65° for the Delaware River; 76° for the Genesee Section; 80° for the Open Folds Section, and ; 72° for the Unglaciated Plateau. The average circularity for all rivers > 200 sq mi is 40%. Twenty-two sluiceways showed an average sinuousity index of 1.04 and a length/width ration of 4.08 (nearly double the usual figure for normal river systems). Other quantitative data occur in such works as Rideg(1970); and Donahue (1972).

There is a large variety of erosional landforms throughout the Plateau. Several of these will be specifically discussed under the different geomorphic sections, but many lanuforms are common to several of the sections and can be discussed in general terms. Coates and Kirkland (in press) provide a synthesis for many of these features. They take into consideration landforms developed parallel to ice margins and those formed parallel to ice-flow direction. The thesis is presented that the arcuate pattern of many rivers (Fig.8) is a reflection of ice marginal positions . . . . such as parts of the Allegheny, Canisteo, Cohocton, and

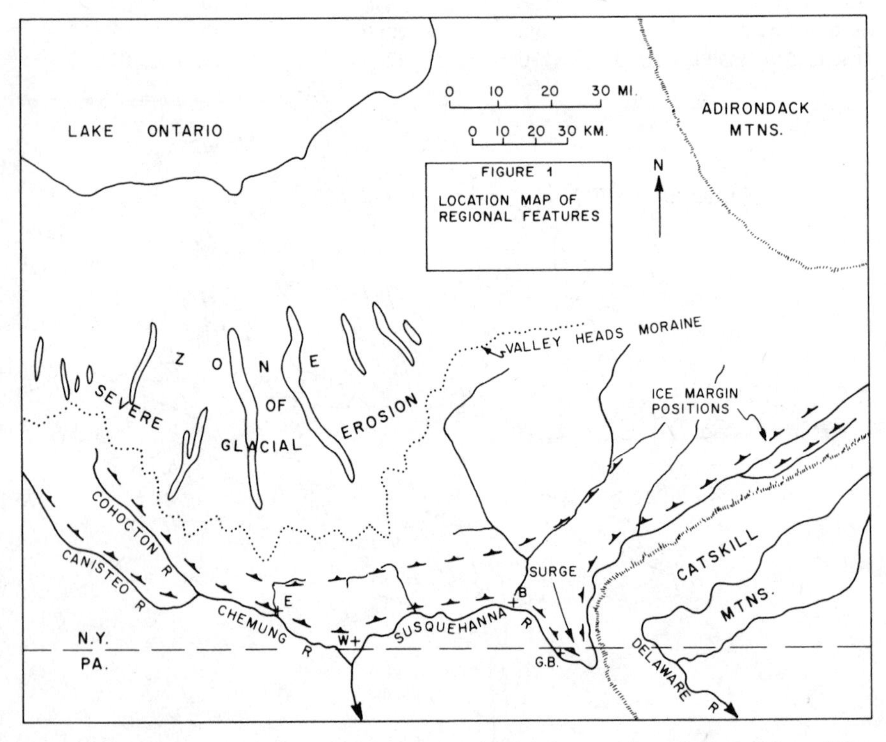

Figure 8. Map showing regional features and arcuate drainages and ice margin positions. (From Coates and King, 1973).

Susquehanna Rivers. In addition a hierarchy model is developed for that series of landforms that are initiated beyond the ice margin and which become increasingly eroded when ice moves through and parallel with the feature. Thus an entire sequential series of landforms are created and include the following evolutionary stages: (1)notch stage, (2)col stage, (3) overflow channel stage, (4)sluiceway stage, (5)through valley stage, (6)finger lake stage, and (7)composite valley stage. The Notch Stage forms when proglacial meltwaters create a V-shaped entrenchment at a divide. Subsequent erosion, often aided by ice action, hollows the notch producing a gentler and flatter floor during the Col Stage. Depending upon the vigor of the processes and the pre-glacial character of the terrain overflow channels, or spillways are formed. The Overflow Channel Stage occurs when a flat-floored channel extends immediately down-gradient from the drainage divide. The Sluiceway Stage evolves when meltwaters are moving some distance from the divide and are deeply entrenching the terrain. Gradients are steeper, incisement is greater, the chute valley is longer, and deposits generally thinner than in the overflow Channel Stage. The Through Valley (terms coined by William Morris Davis as reported by Tarr,1909) Stage is reached when active glacial ice has exploited the previously 'softened'gaps in the divide area and unites oppositely oriented streams into a singly and continuous valley. The Finger Lake Stage can evolve from a through valley when glacial deposits effectively dam outlets in the trough. The Composite Valley Stage is reached by valleys some distance from the divide area where sufficient ice erosion has aided in truncating many spurs, straightened the pre-glacial valley, allowed for fluvial-gravity reshaping of hillslopes, and in which there presently resides thick valley fill deposits (largely formed of meltwater sediments).

A special landform, with many examples in the Plateau is the umlaufberg (a term introduced into the English literature by von Engeln). An umlaufberg is a bedrock outlier in a valley surrounded by glacial drift that is usually stratified. As shown in Figure 9 unlaufbergs come in a wide assortment of sizes and shapes. The common bond is their origin. They formed when an ice margin forced meltwater drainage to flow over what had pre-glacially been part of the bedrock spur. Such fluvial superposition thus entrenched a new channel, and when ice retreats from the area a beheaded spur remains paying testimony to this succession of events. As will be pointed out in subsequent papers, umlaufbergs are important in proving multiple glaciation of the Plateau because they can be classified into groups that are single, double, or multi-cycle features. The latter types document the necessity for having several different glacial events for their development. Thus they can be added to the other types of evidence which can be used on the Plateau to show there were earlier glaciations in the region. These include such features as (l) interglacial deposits (von Engeln,1929), (2) preglacial or interglacial valleys (von Engeln,1931),(3) interglacial conglomerates (Schmidt, 1947), and the possibility for restricted occurrences of pre-Wisconsin drift, differences in surficial changes of drift, and unusual drift changes in some well logs.

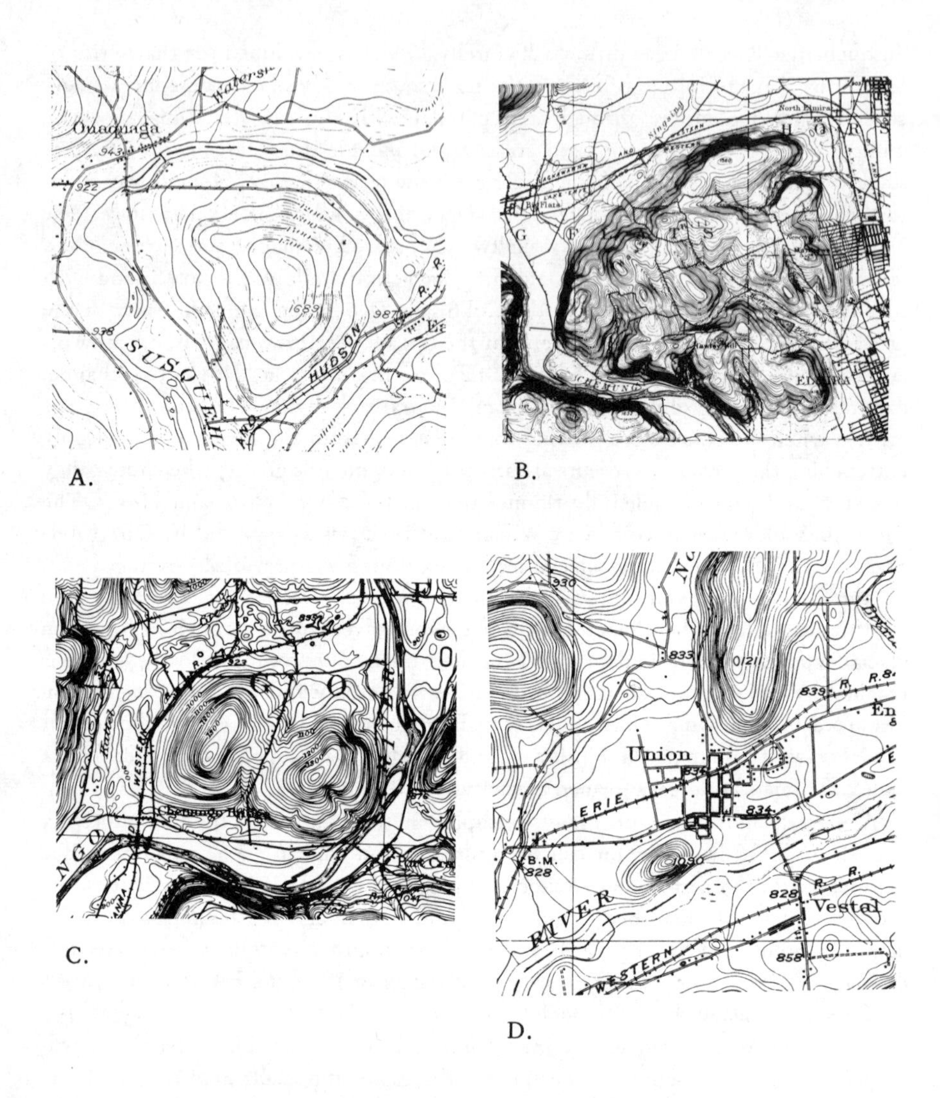

Figure 9. Maps showing variations in umlaufberg size and perfection in development. Copied from 1:62,500 scale U.S.G.S. New York topographic maps.

A. Single cycle. Susquehanna River near Windsor (Ninevah Quad.). Umlaufberg is l.6 mi. long.

B. Double cycle, Chemung River near Elmira (Elmira Quad.) Umlaufberg is 5 mi long.

C. Multicycle. Chenango River near Binghamton (Binghamton Quad.) Umlaufberg is 2 mi long. Note it is double crested.

D. Multicycle end member. Susquehanna River at Union. (Apalachin Quad.) For scale the "Union" lettering is 0.4. mi. long.

GLACIATION

This part discusses those aspects of glaciation that serve to unify the features of the Plateau. Emphasis is placed on a generalized description of the glacial process, and the erosion, deposits, and landforms that are created. A more detailed and localized selection of certain characterisitics is reserved for analyses in the last parts of this chapter.

Glacier Characteristics

The glacier that covered the Plateau was the Laurentide ice sheet. Its thickness was several thousand feet since it completely covered the Adirondacks(>5,300 ft) and the Catskills (>4,200 ft). Using many diverse types of evidence. . . . . striae, indicator rocks, till shadow hills and other streamlined forms, ice marginal features etc. . . . . the nature of ice advance and retreat during the late Wisconsinan glaciation can be unravelled. Coates and Kirkland (in press) provide a sequential series for these events and show compatibility when a 1 Bar basal shear-stress model is used in the reconstruction. Ice advanced into the region flowing down and out of the Ontario Basin. A second lobe flowed around the eastern Adirondacks using the Hudson valley as its avenue. Local glaciers may have occurred on the Adirondacks but as ice sheet thickness increased it topped the mountains at which point the Adirondacks became an accumulation and outflow center which shifted direction of ice flow to the southwest. In turn the Catskills were topped and ice reached its maximum position in Pennsylvania. An arc drawn through the High Peaks area of the Adirondacks intersects very conformably with the Wisconsinan ice margin in New York, Pennsylvania, and New Jersey. Other ice margin loci mirror positions of several rivers. Deglaciation was complex with either downwasting or backwasting depending upon terrain aspect and ice thickness. With the exception of western New York and Pennsylvania the only significant morainal margin in the region was formed by Valley Heads ice. Thus upland moraines are rare and valley ice positions can be more easily deciphered. With climate amelioration ice sheet profiles were lower than the main Adirondacks so that the Ontario and Hudson lobes became reestablished and invaded such areas as the Mohawk Valley.

Erosion

Quaternary geomorphic processes have produced far greater erosion in the Plateau than in the Unglaciated Plateau Section. Clayton (1965) was the first to call attention to the magnitude of denudation throughout the Finger Lakes Section (see also W. White, 1972) although earlier workers had shown that certain lakes such as Seneca and Cayuga had been glacially deepened more than 1,000 ft. Thus pre-Quaternary surfaces in this northern part of the Plateau were reduced 100's of ft. In the southern part of the Plateau Muller (1965) believed glacial reduction was 50–100 ft in the Chautauqua and Allegheny Sections and Denny (1956) shows lowering of surfaces as much as 200 ft in Potter Co., Penn. (headwaters area of the Genesee River). Coates and Kirkland (in press) show that extensive Quaternary erosion occurred not only in the boundary areas of the Plateau but throughout the central part as well. When all aspects are considered. . . . . the lowering of uplands and the preferential erosion cols, valleys, and troughs. . . . . the averaged erosion in the Plateau from processes associated with glaciation and those of interglacial times is more than 200 ft

Much work remains undone concerning complexities of the glacial erosion -deposition regime in a rugged terrain. This problem will be addressed in forthcoming papers, but a brief reference is in order. Previous studies in the field of glacial dynamics have treated the two end members in the series, namely, processes in an alpine-mountainous setting, or in a plains-type setting. This has left unresolved the manner of activitities in moderately rugged topography as exists in the Plateau. The development of landforms parallel to ice transport and those transverse to flow needs to be placed into a unified model. For example ice greatly erodes the former while the latter receives minimal sculpture. The opposite is true for the uplands . . . . . . divides parallel to ice motion receive smaller reduction than those at right angles to flow. It is believed that the relief of the Plateau provides the appropriate threshold so that much erosion can occur. This is possible because the terrain maximizes the number of localities where accelerating and decelerating ice currents occur, thus facilitating the differentials that are necessary to initiate erosion (see Boulton, and Clayton and Moran, this volume). These inequities that are caused during a glacial cycle are transformed into still differing terranes when the inter and intra-glacial fluvial-gravity erosion cycles are reestablished. This matrix of events is what has produced the unusual sculptured features of the Plateau.

Glacial Deposits

Several generalities can be applied to glacial deposits and the landforms they create, and these will be discussed in this section. Specific features associated with glacially-derived sediments will be reserved for analysis within the various geomorphic sections of the Plateau.

Age

The reader is referred to Muller (1965b) for an excellent review of the literature on age of glacial deposits, including their stratigraphy and correlation for the New York part of the Plateau. G. White (1969a) provides the most complete stratigraphic appraisal of deposits in Pennsylvania. For recent work in the glaciated part of eastern Ohio one should review Totten (1969, 1973) and White (1969b).

Most workers who have studied glacial deposits in ice terminal positions in Pennsylvania have concluded they represent two or more periods of glaciation (Leverett,1934). This is an enigma for Plateau glacial stratigraphers because all drift throughout the Plateau has been assigned a Wisconsinan age. The question has never been satifactorily answered concerning the absence of pre-Wisconsinan glacial deposits which should underlie Wisconsinan drift. Since deposits dated as Illinoian occur south of Wisconsinan deposits in Pennsylvania, it is assumed they originally covered the Plateau and that Illinoian glaciation was probably even more vigorous than Wisconsinan ice sheets. Therefore, why are the earlier deposits missing? Will underlying Illinoian deposits be discovered sometime? If not, why and how were they all removed? Was fluvial erosion of Sangamon time sufficiently intense to obliterate all traces of drift? Do the deposits now known as Wisconsinan contain appreciable amounts of reincorporated Illinoian sediments?

The greatest area of the Plateau is covered with drift described as "Olean" but there is no agreement about its exactWisconsinanage. Valley Heads drift covers a much smaller area in the northern part of the Plateau and can be dated as very late Wisconsinan (12,000–14,000 yr B.P.). The unique non-glacial sediments at Otto in western New York have been assigned an intraglacial position by Muller (1965) and age of 63,000 ± 1700 yrs B.P. The equally unique locality of Fernbank on the west side of Cayuga Lake contains non-glacial sediments believed to be Sangmon age because of their resemblance to the Don Beds at Toronto, and a radiocarbon date older than the method will allow.

Till

Till is nearly ubiquitous on all uplands and hillslopes of the Plateau. Of course on hillslopes the original *in situ* till has in places been translocated by gravity movement (King and Coates, 1973) and become part of the colluvial slope cover. The great majority of till is composed of local rocks with length of transport restricted to a few miles (Holmes, 1952). The percentage of far-travelled exotics ranges from 1-3% in the western Plateau to much less than 1% in such areas as the Catskills.

With the notable classic exception of Holmes' (1941) till fabric study, the topic has received very little attention in the Plateau. Occasional other fabrics have been measured, however, (King and Coates, 1973) to indicate that such studies can yield great dividends.

Another gap in knowledge of Plateau tills is the lack of studies that relate to the dynamics and classification of ice regimes responsible for the deposit. For example the nature of till emplacement and the appropriateness of describing the tills as lodgment or ablation have not been analyzed. Flow till has been described as occurring within ice-contact deposits of the Chenango Valley (Cadwell, 1972).

Thickness of till is variable throughout the Plateau and there are even large differences within a single locality. Till in the Finger Lakes Section is often thin, whereas it is thickest in the Susquehanna and Small Lakes Section where depths of 250 ft to bedrock are encountered (Coates, 1966a).

Stratified Deposits

The Plateau contains a wide assortment of meltwater sediments (glaciofluvial and glaciolacustrine beds). These materials occur in the entire range of ice-contact deposits, and in the outwash strata whose occurrence may be far from the ice margin.

The composition of stratified deposits usually shows higher percentages of non-local material than is found in tills. For example, deposits in many valleys of the Susquehanna system contain several percent foreign rocks and in some size categories 25-35 percent of the constituents are non-local, as in the Chenango Valley. The character of the outwash and its permeability is, of course, important for water supplies because it is these materials that form the best aquifers in the Plateau. Unfortunately in parts of some valleys there are clay thicknesses in excess of 100 ft which prevent groundwater development. Particle size of sediments in many valleys is a function of their orientation. Figure 10 illustrates that

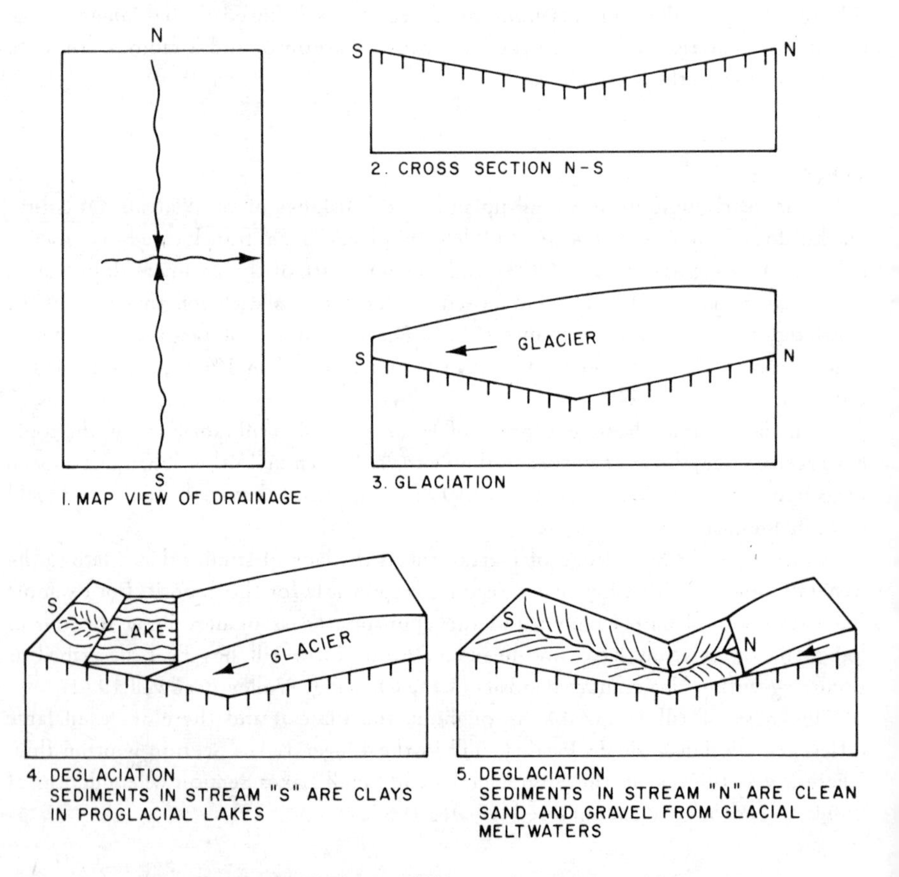

Figure 10. Diagrammatic sketch illustrating differences in meltwater sedimentation in north and south-flowing stream channels.

north-sloping valleys contain more silt and clay because of impoundment in proglacial lakes, whereas south-sloping valleys were self-cleaning and the fines were washed out of the valleys leaving behind (as a type of lag material) coarser grained sand and gravel.

The thickness of stratified deposits is more than 1,000 ft in the deep troughs of the Finger Lakes Section (Coates, 1968), however, not all the sediments are glacial. Indeed the Holocene demarcation of the valley fill materials is imperfectly mapped and understood. Thus, it cannot be assumed that all sands and gravels are of glacial vintage. Coates et al (1971) and Randall and Coates (1973) have shown that in some areas stratified deposits as thick as 45 ft are only 3801 + 60 yrs B.P. The total thickness of valley fill materials is generally more than 100 ft in most Plateau rivers and is more than 200 ft in many of the large rivers. Unusual thicknesses occur in many New York localities (Randall, 1972) such as 434 ft at Oneonta, 373 ft at Norwich, and 398 ft near Winfield. It should be noted that often rock basins in the valleys seem to occur downstream of river junctions with southerly-pointing orientation.

Landforms

The literature of the Plateau is sprinkled with accounts of depositional landforms, but few reports have their study as the primary purpose of the investigation. A notable exception is the terrace study by Peltier (1949). Glacial landforms cover a wide range from those composed entirely of till such as the "drumlinoid" hills near Binghamton to those largely of stratified deposits as the valley trains. The region also abounds with landforms with mixed deposits as kames, kame terraces, kame deltas, eskers etc. Denny and Lyford (1963) show the occurrence of many of these features in a large part of the Plateau.

Moraines have been the topic of several studies such as those by Leverett (1934), and Connally (1964). White and others (1969a) have been especially productive in determining important new guidelines for moraines. For example the surface form is often not coincident with farthest positions in so-called "end moraines". In general the Plateau is not blessed with a series of ice retreatal positions that show up as well-developed upland recessional moraines as is so prominent in the midwest. The closest cousin to such occurrence is the Valley Heads moraine. Many glacial geologists are aghast when they see this feature and discover the frontal position is mostly characterized by massive meltwater deposits, instead of the more textbook type of till with icethrust or icepush features. Recent studies have shown the best indicator of ice margins in valleys (Cadwell, 1972; Kirkland, 1973) is the sequence model largely developed by the U.S. Geological Survey in their New England studies (see Koteff, this volume). Thus in the Susquehanna and Delaware systems it can be demonstrated that a series of ice-contact deposits grade into outwash deposits and that correlation of these units establishes an ice marginal zone of activity during the deglaciation stage. During the formation of this sequence plains and deltas are also produced. Unfortunately this model was unavailable when

Fairchild was doing his study of the Susquehanna (1925). He believed that massive glacial lakes formed in the Elmira and Binghamton areas. He proposed that glacial lake Binghamton extended from Towanda to Great Bend, Pennsylvania . . . . a distance of 75 mi. Reexamination of the data he used for this assertion shows that his supposedly concordant deltas and plains actually consist of a multiple series of individual units in several different sequences.

GEOMORPHIC SECTIONS OF THE GLACIATED APPALACHIAN PLATEAU

I believe it vital for those who would study the Plateau to realize it is extremely diverse and cannot be placed into a simple scheme. Fenneman (1938) considered only two principal subdivisions, with one of them the Catskill Mountains. On the other hand the Physiographic Map of Pennsylvania (Pennsylvania Geological Survey) shows three different divisions of the glaciated part of that State. For convenience I have divided the study area into 11 different sections. I have omitted Ohio from this discussion as that portion commonly referenced as part of the Plateau is more a series of gently undulating plains. It has more in common with midwest style of glaciation and topography than with Appalachian Plateau terrain. Many sectional boundaries are gradational and should be viewed as only approximate in nature. Separation of the various sections is based on one or more of the following criteria:

1. Relief and ruggedness. The Catskill Mountain Section is the best example with local relief in several parts in excess of 2,000 ft.

2. Accepted prior terminology. In such instances as the Finger Lakes Section and the Catskill Mountain Section where well established names are appropriate they have been retained. Others such as the Through Valley Section have also been described in the literature, but formal boundaries were not drawn. In some instances names that have been used, such as the Pocono, were rejected because the term has become geologically confusing. Furthermore the region usually referenced as the "Poconos" (which occur in Pennsylvania southwest of the Delaware River) is not significanly different from the section which is herein labelled the Small Lakes Section.

3. Topographic and drainage fabric. These factors easily aid in delineation of several sections . . . . the Open Folds Section contains a trellis drainage pattern; the Genesee Section delineates a major north-flowing river; the Chautauqua Section is marked by a series of sub-parallel valleys, and; the Small Lakes Section is one dominated by rather irregular topography containing hundreds of lakes and basins.

Some of the sections are somewhat unique and contain topographic features not duplicated elsewhere in the Plateau or perhaps even throughout the country . . . . such as the Finger Lakes, through valleys, and the array of "drumlinoid" hills. Other features, although important, are common to several sections . . . such as cols and sluiceways.

Major topographic features were used whenever possible as section names-Helderberg, Catskill, lakes etc. For other sections the name of the most prominent river was used - Susquehanna, Mahoning, Genesee, Allegheny etc. Such terminology is logical and consistent with guidelines used in the establishment of geographic names

Susquehanna Section

Since this is the largest single section, comprises the heart of the region, and the one I know best I will discuss it first and at greatest length. Muller (1965a, 1965b) provided a bibliography and discussion of this area, and others in New York, prior to 1965, so I will deal primarily with the newer literature for the New York sections.

The glacial writeup by Tarr in the Watkins Glen-Catatonk Folio (1909) is a classic for this area. Merritt and Muller (1959) showed the depth of leaching in the central New York area was inversely proportional to carbonate content of the drift. Denny and Lyford (1963) provide an excellent summation for part of the section and were among the first (see also Moss and Ritter, 1962; Coates, 1963) to cast doubt about the validity of the Binghamton Substage of MacClintock and Apfel (1944). The most recent work on this problem is in Randall and Coates (1973). The present status is that the Binghamton area cannot serve as the type locality for this exotic-rich drift, because the amount of exotic-rich till around Binghamton covers far less than 1% of the hillslopes. In a study of rock composition of valley fill of Susquehanna tributaries between Binghamton and Elmira, Flint (1968) observed higher ratios of foreign rocks in north-flowing streams when compared with south-flowing streams. Glacial drift in uplands and hillslopes of the Susquehanna Section is almost entirely "Olean-type" till with clasts that are largely of local rocks. South-trending valleys that head in through valleys at the Allegheny Escarpment contain deposits with exotics (such as Onondago limestone, Medina sandstone, Potsdam sandstone, Precambrian crystallines etc.) that locally may range to 35 %. Other south-trending valleys that were not in glacial communication with northern rocks by means of low divides contain very few exotics (Moss and Ritter, 1962; Conners, 1969). In all south-trending valleys it is unusual to find appreciable non-local rocks at heights greater than 200 ft above the present channel.

Perhaps one of the greatest errors of early workers in the Susquehanna Section was to ascribe to the area a very thin covering of till. In the literature the following statements can be considered typical:

> " . . . the till cover is only a few feet thick probably averaging less than 10 feet (Tarr, 1909, p. 16)."
>
> " . . . a shallow soil mantle averaging not much more than 3 feet in thickness over the bedrock (Lounsbury, 1932, p. 13-14)."
>
> "The entire area . . . . is covered by a thin layer of glacial till 3 to 10 feet thick (Lounsbury, 1953, p. 5)."

Coates (1966a) in a study of till thickness on uplands of New York southern tier counties showed the average till thickness exceeded 60 ft. Such sediments do not cover hillslopes with a uniform blanket, however. Instead they were differentially deposited so that till thickness on south-facing slopes is generally at least 5 times greater than on north-facing slopes. This condition has led to the development of hill asymmetry throughout the region with the north sides of hills (24%) twice as steep as south sides (12 %). These hills have been termed "till shadow hills" (Coates, 1966; see Fig. 11).

Randall (1972) lists records for 1,990 wells and 725 test borings in the Susquehanna River basin. Examination of these data indicate valley fill sediments are exceedingly variable, and that depth to rock is not uniform. For example valley fill depths in the Susquehanna River between Oneonta and Binghamton, N.Y. range from more than 400 ft to as little as 100 ft. (such as the Great Bend area). Similar disparities occur in other rivers such as the Chenango. Thus thalweg gradients were altered by glaciation processes, rock basins occur under valley fill sediments, and valley fill sediments in major rivers (except for the Tioughnioga) generally range from 150-200 ft in thickness (Fessenden, 1974).

The nature of unconsolidated materials in valleys was discussed by Fairchild (1925) along with other aspects of the glacial heritage of the region. The most recent analysis of valley fill in a large part of the Susquehanna Section is by Cadwell (1972) of the Chenango River. In the 80 mi length of the river Cadwell mapped six different zones, each representing a retreatal phase and ice margin position during deglaciation of the region. The tongue-like character of ice in the valley produced an entire series of glacial deposits and landforms whose resulting mosaic (kames, kame terraces, kame deltas, outwash plains etc.) constitute the type of sequence also described by Koteff (this volume). Radiocarbon analysis from a kettle hole bog in zone 1 (southernmost position) provided a date of 16,650 ± 1800 yrs B.P.

Data are still being collected and computed as part of an investigation by Coates and King (in preparation) for the analysis of a large range of drift hills in valleys of the region. There are several hundred hills in a wide range of sizes and shapes. They

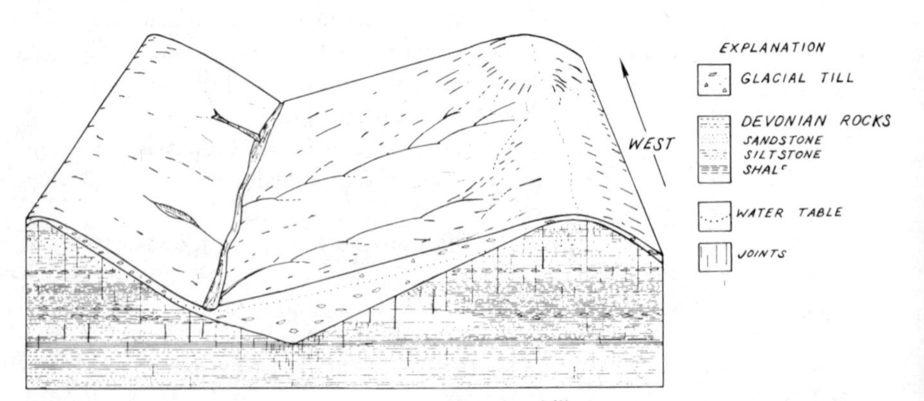

Figure 11. Diagrammatic sketch of a typical till shadow hill.

were first described by Hubbard (1906) and also by Tarr (1909). The hills can be characterized in the following terms: (1) composed almost entirely of till, (2) range from 35 - 200 ft in height, (3) closure may be almost nil to 60 ft, (4) side slopes average 12 %, (5) length/width ratio of 1.15, and (6) diameter average of 1450 ft. The degree of Holocene remodelling of these forms is unresolved. It does seem clear, however, that all such hills cannot be lumped into a single category, but instead are polygenetic. For example King and Coates (1973) in analysis of a rather unique set of hills and mounds in the Great Bend area ascribe them to a glacio-periglacial origin (Fig. 12). The diagnostic feature of these landforms is their linkage to the concave depressions of adjacent uplands. They were formed when the ice retreated leaving truncated spurs and oversteepened hillsides. The till on the slopes was inadequately stabilized by vegetation and massive solifluction resulted with creation of the unusual concavo-convex landforms. Most of the "drumlinoid" features cannot be explained in this manner, however, because they are not associated with corresponding hillside indentations.

Figure 13 illustrates another type of till hill in the Susquehanna Section. The St. Johns Pond area is located immediately north of Binghamton. Within a 20 mile radius there are more than 25 similar features, which consist of a col with a depression on the south and a till mound extended part way across the south-trending valley. The till barrier either ends in another col (as in Fig. 13) or flush with a hillside lacking the concavity forms in the Great Bend area. Other variations of till knobs occur throughout the Susquehanna Section. They are always limited to valleys with largely north-south orientation, and in some valleys, such as Castle Creek and Little Choconut Creek near Binghamton, there may be a series of 5 or 6 hills spread along a distance of several miles. Many hills are quite symmetrical and it is the exception for hills to resemble the streamlined forms of true drumlins.

On the mega-scale the most dramatic aspect of the Susquehanna Section is the arcuate form of the system (Figs. 2,8). This is particularly obvious when the Cohocton and Canisteo Rivers in the western part are linked with the Chenango and Susquehanna Rivers in the eastern part. As a working hypothesis Coates and King (1973) proposed this configuration resulted from ice marginal positions and fluvial incision (Fig. 8), probably forming in somewhat the same way as parts of the Missouri, Ohio, and Allegheny Rivers. This idea has been extended and amplified by Coates and Kirkland (in press) who trace a series of other ice marginal features throughout the Plateau. There is great agreement of such river orientations with marginal positions of the Laurentide ice sheet when a 1 Bar basal shear-stress model is computed. The alinement of other features, especially in the western part, of cols, sluiceways, through valleys etc. lends support to these ideas. The trellis-like character of many drainage patterns, as in the Bath, N.Y. area (Fig. 14) are anomalous since they are not linked to any type of structural control (Coates and Kirkland, in press). The ice marginal theory, however, provides an explanation for such drainage configurations. It must be remembered that such a system did not form overnight, but represents the cumulative effects of several glaciations and additional fluvial erosion that occurred periodically throughout Quaternary time.

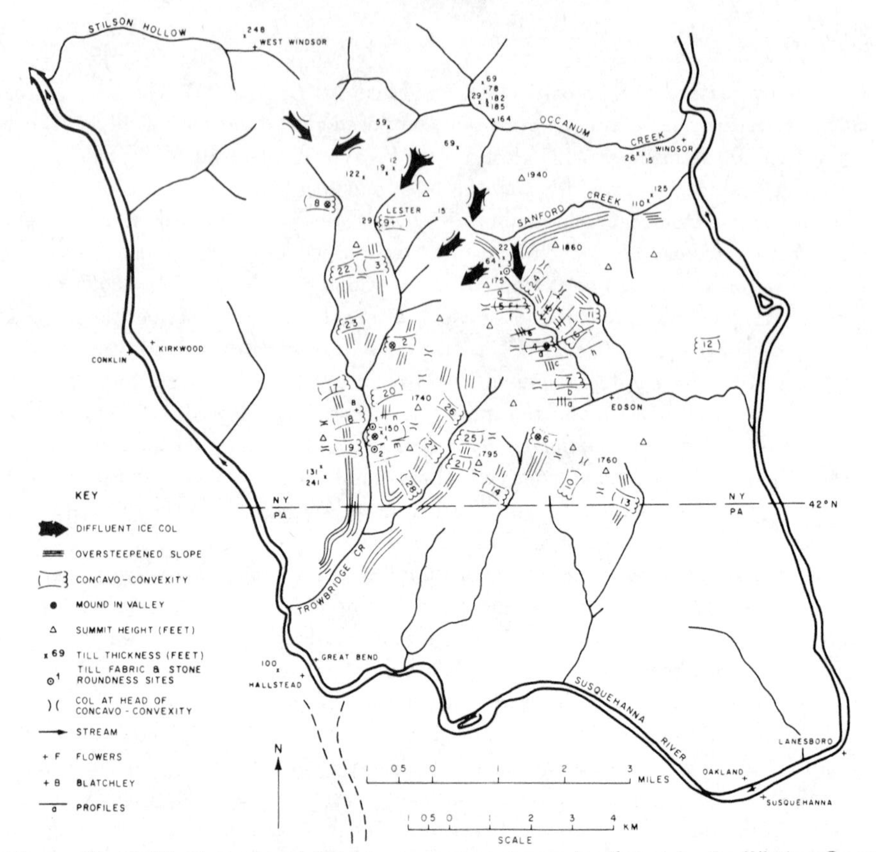

Figure 12. Distribution of periglacial-type concavo-convex landforms in the Windsor-Great Bend area of the Susquehanna. (From King and Coates, 1973).

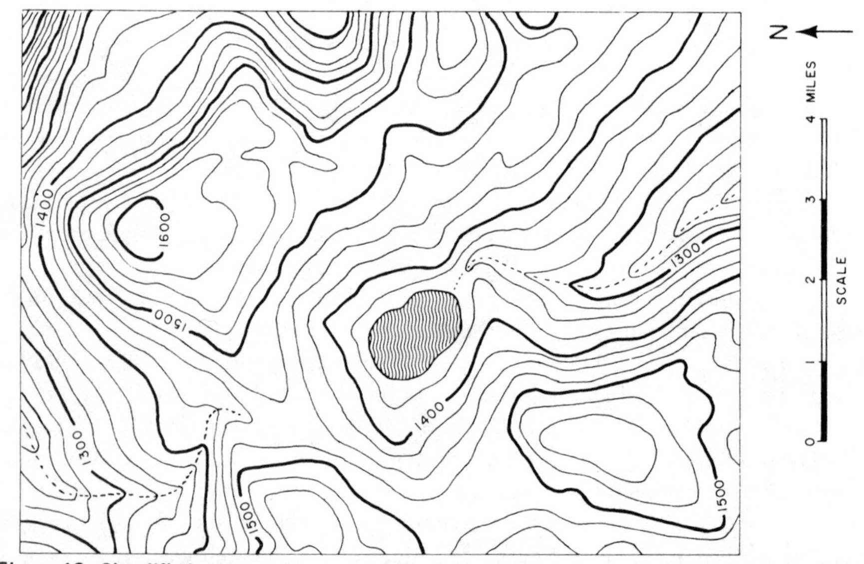

Figure 13. Simplified topographic map of St. Johns Pond area, north of Binghamton, N.Y. Drawn from 1: 24,000 scale U.S.G.S. map, Castle Creek Quadrangle.

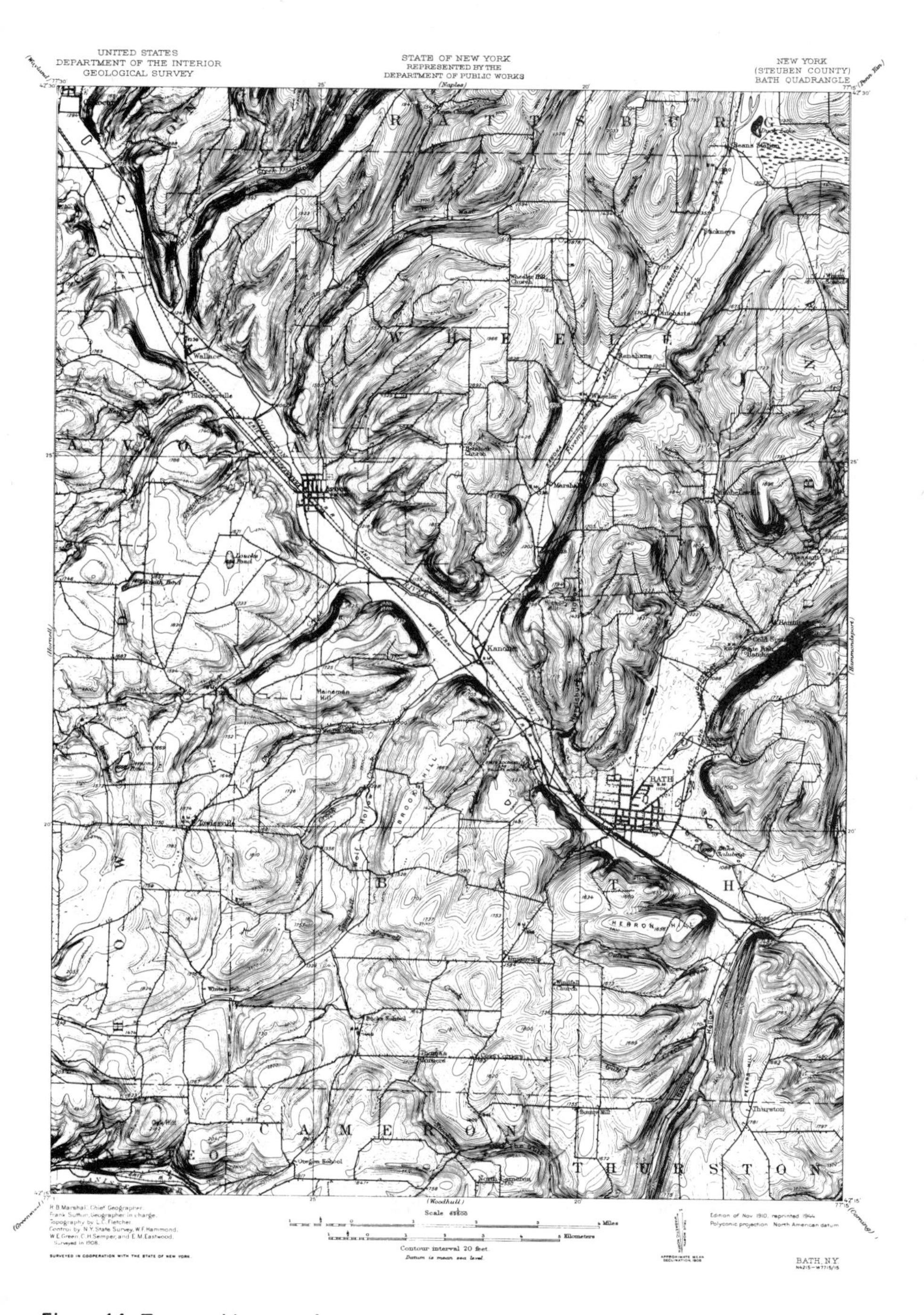

Figure 14. Topographic map of the 1:62,500 scale U.S. Geological Survey Bath, New York Quadrangle. Note pseudo-trellis pattern.

The Susquehanna Section has unusual expression of numerous cols, overflow channels, sluiceways, and composite valleys. For example 100's of cols with erosion 200 ft have been mapped (Coates and Kirkland, in press), and there are l00's more with erosion in the 100-200 ft range. Valley deepening by processes associated with glaciation also amounts to a few hundred feet, especially those valleys that were largely parallel to ice motion.

Catskill Mountain Section

In 1843 Mather assigned the drift and "scratches" on rocks of this region to the Drift division of the Quaternary System, and attributed them to water currents from polar regions. Chamberlin (1883) presented more modern ideas and attributed many features to glacial conditions. He also suggested the possibility of local glaciation, a topic that has become a source of controversy. Rich (1906, 1935) has been the main proponent for local glaciers in the Catskills. Little work on this problem, or other aspects of glaciation in the Catskills was done between the time of Rich's classic 1935 monograph and Kirkland's 1973 dissertation. Kirkland does not support the concept of local glaciation at the end of Wisconsinan time but does not rule out the possibility of local glaciation during earlier episodes. I am skeptical about much of the "evidence" that has been cited as favorable to local valley glaciers. For example many of the "cirques" are on south-facing slopes, a very unlikely orientation. Rich interpreted some of these from appearance on maps of 1:62,500 scale. On the newer large scale 1:24,000 scale maps, such features have almost entirely lost their "cirque-like character". Periglacial and Holocene processes have greatly altered Catskill landforms but few attempts have been made to reconstruct the original mountain configuration. Probably many of the hopper-like features at the head of divides and mountains are instead nivation hollows (see also King, this volume). North-facing hollows in the northern Catskills near Prattsville, N.Y. pose a special problem because of nearby hillocks described by Rich (1906, 1935) as moraines developed by north-flowing valley glaciers. I recently examined these localities and believe such features are subject to other interpretation. For example the lack of fabric in till of the hills suggests they may be ice disintegration features. Several striae occurrences were discovered in bedrock ledges of nearby hillslopes which indicate south-moving ice. The area also abounds with evidence for massive slope movements that would have post-dated glaciation. Additional studies are underway to determine, if indeed, the Catskills contained local glaciers of important magnitude.

Rich (1935) believed drift in the Catskills was formed by two different glaciations. Working mostly in the western Catskills (Fig. 15), but also overlapping Rich's maps, Kirkland (1973) was unable to justify this position. He found no weathering differences in the till, discovered that striae orientation could be explained by topographic factors, and was able to map abundant striae in areas where Rich believed they were rare. Although till composition is nearly 100% of local rock, occasional exotics are found in glaciofluvial deposits. Some of these are so distinctive, such as metanorthosite, that their source area in the Adirondacks can be pinpointed.

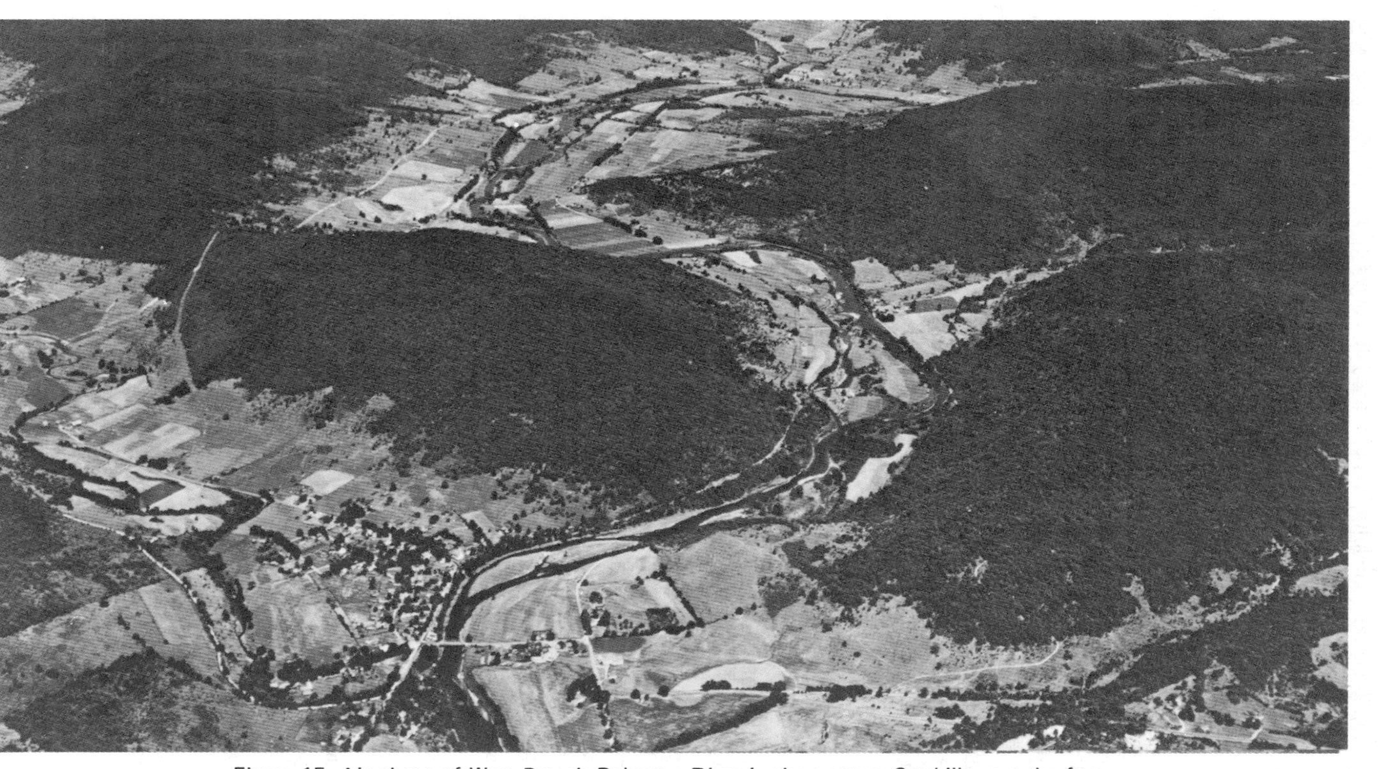

Figure 15. Air photo of West Branch Delaware River in the western Catskills near the former village of Cannonsville. The valleys are now covered with waters of the Cannonsville Reservoir.

Drift thickness in the Catskills is variable. Till ranges from a thin veneer to deposits as thick as 200 ft. The unconsolidated stratified depostis occur in valleys where sands and gravels are invariably interbedded with clay and silt. Thickness more than 200 ft occur in parts of the Delaware River and Beaver Kill valleys. Depth of valley fill exceeds 100 ft in many drainages such as Charlotte Creek, Ouleout Creek, East and West Branches Delaware River etc. (Soren, 1963).

Kirkland (1973) did detailed mapping of deglaciation features in the West Branch Delaware River. He found that the sequence model (Koteff, this volume) applied to the valley fill deposits which he separated into 6 zones that contained a total of 19 stages. At each stage he superimposed basal ice shear-stress models and concluded that the resulting profiles indicate ice stagnation rather than continuous nourishment from the ice sheet.

The height of the Catskills 3,000 ft to 4,000 ft in places, had a profound effect on adjacent sections. The Catskills thus deserve a special place among sections in the Plateau because of their unique height, ruggedness, and relief. Furthermore the "Catskill beds" form the most resistant rocks and all factors combine to produce some unusual glacially- derived markings. For example they contain the only rocks in the Plateau where p-forms have been described (Kirkland, 1973).

Finger Lakes Section

Louis Agassiz visited this area in the 1860's and correctly described their formation as a result of the ice ages. Not everyone has agreed with this viewpoint, and such a renown figure as H.L. Fairchild insisted throughout his lifetime that the dominant processes fashioning the area were fluvial (1905). His opinion should be contrasted with that of Clayton (1965) who was the first to express the idea that ice erosion had been so severe as to remove nearly all traces of any pre-existing landscape. Even the valleys that contain the lakes were viewed as being entirely cut by ice as "intrusive troughs". Furthermore the through valleys were also " . . . . cut into the plateau independently of . . . preexisting water partings"(Clayton, 1965, p. 53). Coates (1966b) in a reply to Clayton proposed a multicyclic theory to explain the features which require a " . . . . combination of a long period of preglacial erosion followed by a series of unusual glacial, interglacial, and proglacial stream diversion channels that were repeatedly exploited by later ice movements". In a 1966 article Clayton modified his earlier opinion by stating, "In my view the ice locally initiated valley development. Far more commonly it occupied, enlarged and joined up existing fluvial valleys". Such discussions should detract nothing from the majesty of this wondersome region.

Eleven elongated and roughly parallel lakes constitute the traditional Finger Lakes Section, and some of their characteristics are listed in Table 2. Inspection of these data indicate there are large physical differences among the lakes. Additional information has been summarized by von Englen (1961) and by Coates (1968). More recently W. White (1972) has discussed the possibility for massive amount of landscape reduction by ice, not only in the Finger Lakes, but regions to the north as well.

The Finger Lakes Section contains a remarkable diversity of features and deposits. It contains bedrock glens such as Watkins Glen and Enfield Glen, that are as spectacular as any in eastern United States. They were initiated during the waning stages of glaciation when upland consequent streams failed to find their preglacial valleys which had been filled by drift (Fig. 16). These earlier valleys, regardless of whether preglacial, interglacial, or interstadial, are always much larger and deeper than the new ones. This indicates the previous fluvial erosion cycle lasted several times longer than elapsed time during the Holocene. Thus as the streams extended downslope they cut into bedrock instead of occupying the position of their former channel. The Section also abounds with hanging deltas, that in some tributaries, such as Coy Glen, contain four and five nested deltas in series that bear testimony to higher lake levels when the Finger Lakes were part of an extensive system of proglacial lakes. These high-level lakes formed when meltwaters were dammed on the south by the drainage divide of the Susquehanna and on the north by the ice margin. Another remarkable feature is Taughannock Falls, with a vertical drop of 210 ft, one of the three highest waterfalls east of the Rockies.

The Valley Heads moraine is the longest and most continuous moraine in New York and marks part of the southern border of the Section. In many places it is massive consisting of waterlaid deposits 100's of ft thick as in the area south of Seneca Lake, and in some areas the deposits are traceable for miles transverse to the strike such as in the valley north of Spencer, N.Y. Southeast of Ithaca in Sixmile Creek are the thickest varves in the Plateau..more than 600 couplets. Valley Heads time is considered to range in the time interval of about 14,000 to 12,000 yrs B.P. (Coates et al, 1971). Valley Heads materials have been described by a number of workers Fairchild (1932), MacClintock and Apfel (1944), Denny and Lyford (19633), Moss and Ritter (1962), Shumaker (1957) and others.

TABLE 2. FINGER LAKES PHYSICAL PROPERTIES

Lake	Surface Elevation (ft)	Greatest Depth (ft)	Mean Depth (ft)	Bottom Elevation (ft)	Water Area (square miles)	Basin Area (square miles)	Length (miles)	Trend
Otisco	788	66	33	722	4	3.5	6	N 38 W
Skaneateles	863	297	143	576	14	81	15	N 35 W
Owasco	711	177	96	534	10	213	11	N 19 W
Cayuga	381	435	179	– 54	66	879	38	N 21 W
Seneca	445	633	290	– 188	67	504	35	N 7 W
Keuka	709	186	—	523	17	248	20	N 20 E
Canandaigua	687	274	128	413	17	175	16	N 18 E
Honeoye	803	30	—	773	2	42	4	N
Canadice	1096	91	54	1005	1	12	3	N 2 E
Hemlock	905	96	—	899	3	40	7	N 6 W
Conesus	818	66	—	752	5	67	8	N 5 E

(Coates, 1968)

Through Valley Section

Many remarks contained in the previous section could also be said about the region of the through valleys. Although the Through Valley Section has been mapped (Fig. 1) with definite boundaries, the actual case is not so simple. Through valleys exist in many other sections such as the Finger Lakes Section, and the Chautauqua Section. However, their perfection and numbers are greatest in the region as labelled. The similarity between the Through Valley and Finger Lakes Section is apparent in the following:

Table 3

Comparison of Finger Lakes and other Troughs

No. of landforms	Description	Greatest Width (mi.)	Ave. Width (mi.)	Length (mi.)	Sinuosity Index	Orientation
26	Finger-lake type through valleys	1.0	0.7	9.2	1.05	180°
11	Finger Lakes	1.1	0.7	16.1	1.04	160°

A further similarity in the two secitons is found in the three large lakes of the through valley region. For example Cazenovia, Canaderage, and Otsego Lakes all have the qualifications of true finger lakes and would be included as such if they

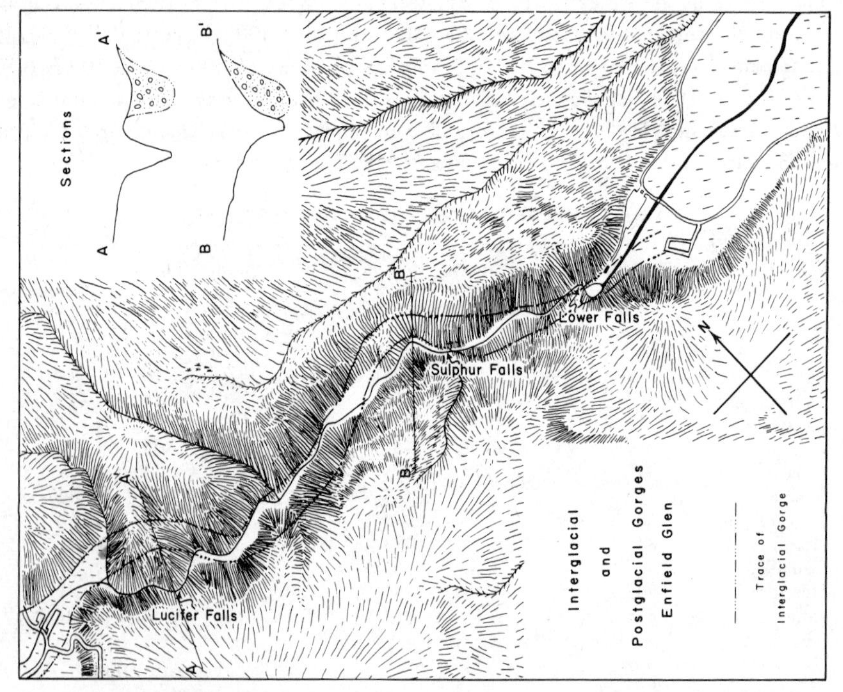

Figure 16. Sketch illustrating differences of post-glacial valleys with earlier valleys formed before Late Wisconsinan time. (From N.Y.S. Geol. Assoc 3lst Ann. Mtg. Guidebook, Cornell Univ., 1959)

were adjacent to the other 11 Finger Lakes. Figure 17 shows the configuration of various drainages for the region. It is important to note that such troughs as those developed in the Otsego Lake area are immediately south of the Adirondack Mountains. More than 1,000 ft of ice deepening occurs here, providing proof of the significance and vigor of glacial movement emanating from the Adirondack outflow center.

Further discussion of through valleys can be found in Tarr (1909), von Englen (1961), and Holmes (1937).

Chautauqua Section

The 17-mi long Chautauqua Lake dominates this section and provides its name. Other features that provide topographic flavor and set it apart from contiguous sections include drumlins and the series of southeat-trending through valleys. Muller (1963) has done the most definitive study of the region and mapped hundreds of drumlins and associated forms. The long axes of these features is southeast and they consist of a large family of landforms that range from drumlins that are not cored to the rock-cored drumlins (rocdrumlin) (see also Muller, this volume). This region also contains several major through valleys and are from east to west the Conewango, Cassadaga, Chautauqua, French Creek, and Findley. The thickness of fill in some valleys of 1,000 ft is the greatest west of the Finger Lakes.

Muller (1963) mapped several morainic systems in the Section and retains the use of earlier terms for them, which are from oldest to youngest Kent, Lavery, Defiance, and Lake Escarpment. The latter are correlative with the Valley Heads moraine and contain a strong topographic expression with well-developed kame and kettle terrain. Such features suggest oscillatory conditions of the ice margin. The ridges rise from about 1,200 ft at the Pennsylvania border to 1,700 ft in the east where ice abutted against the north-facing ridges of the Allegheny Escarpment. This region has been ice-free since about 12,500 yrs B.P. and analysis of pollen stratigraphy can be found in Miller (1973).

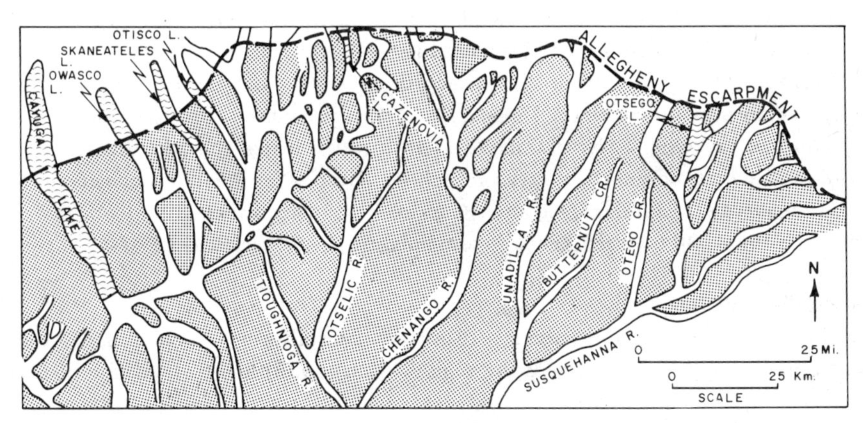

Figure 17. Map showing distribution of through valleys and other main drainages in northeastern part of the Plateau.

Allegheny Section

This Section derives its name from the adjacent Allegheny River which largely forms the southern boundary and receives flow from many drainages in the Section. This general region, which would also include parts of the Unglaciated Plauteau and headwaters of the Genesee River, contains the most rugged and highest elevations in New York east of the Catskill Mountains. The Allegheny Section contains a rather haphazard terrain with absence of an overall fabric, high local relief, and is situated in a geographic position referred to as the "Salamanca re-entrant" (MacClintock and Apfel, 1944). This is the position where ice from the southwestly moving Ontario lobe merged with ice from the southeastly moving Erie lobe. The salient in New York is the only non-glaciated part of the State. It was by a comparison of the topography in this area north and south of the glacial margin, that Muller calculated ice reduction of the upland surface to be on the order of 50-100 ft.

Glacial deposits in the western part of the Section are linked to the stratigraphy established by workers in Ohio and Pennsylvania, whereas eastern Wisconsin-age deposits are largely classified as Olean. Proglacial meltwaters dammed against north-facing slopes (as in the adjacent Genesee River) and cut many spectacular notches several hundreds of feet deep across the divide of the Genesee and Allegheny Rivers (Muller, 1957). A series of seven especially well-developed notches occur at elevations from 1692 to 2111 ft. What been called "Binghamton" by earlier workers, but which should be called "Kent glaciation" was the last glacial event to affect part of the area. The high percentage of foreign rocks distinguish this drift from the Olean, and from the Illinoian which has been described in the southernmost deposits.

The newest work in this Section provides much depth-to-rock information and gives a summary of the geomorphic events:

"Before Pleistocene glaciation, the Allegheny River flowed northward and cut a steep-walled valley in shale and sandstone bedrock. The stream flowed northward through the present upper Conewango and lower Cattaraugus valleys, as part of the St. Lawrence drainage system. Advance of the glacial ice dammed this drainage and formed a long sinuous lake in the valley with a surface altitude of about 1,480 feet above present sea level. An approximately 211-foot thickness of clay and silt was deposited in the lake. To the south the lake spilled over a divide near Kinzua, Pennsylvania; and the outflow eroded a channel to the Ohio River drainage system through which the lake drained. The glacier then advanced into the Allegheny River valley where it deposited large quantities of sand, gravel, and silt to form stratified valley-train deposits of more than 300-foot thickness at some places. (Frimpter, 1974, p. 34-35)."

Helderberg Section

This Section contains several dissected cuesta-form terranes that trend indistinctly southeasterly across the region. Rich (1935) has labelled one of them the Northeastern Escarpment. Their erosional development is largely lithologically controlled by southwesterly dipping Devonian rocks. The lower units in the north part of the area are the Helderberg limestones, whereas clastics form other resistant ridges to the south. Early work in the area was done by Cook (1935). Other pertinent works include LaFleur (1969), and Fullerton (1971), and Krall (1972). These reports are largely concerned with the stratigraphy of deposits and the development of a chronology for glacial events in the general region.

LaFleur discusses several late Wisconsinan ice readvances into the Schoharie River basin and terms them, from old to young, Tannersville, Prattsville, Middleburg, and Yosts and he mapped the margins of each event. He also sheds new information on development of some of the dramatic spillways, such as Grand Gorge, and the sequence of proglacial lakes and glacial meltwaters impounded during ice readvances against the various escarpments.

Genesee Section

This Section is dominated by the north-flowing Genesee River, one of the longest such oriented rivers in northeast United States. Fairchild in many of his writings (see 1928 for example) was mostly responsible for working out its evolution and glacial history. He reported initiation of the Genesee in early Tertiary with the development of the Ontarian River, which partly excavated the pre-glacial Ontario Basin. The Genesee eroded headward capturing several contiguous south-flowing drainages. The pre-glacial Genesee had two branches, the western part in the general vicinity of the present river, and the eastern part flowed in an arcuate pattern from the valley now occupied by Canandaigua Lake. During Quaternary glaciations many parts of the main western valley were clogged with drift and the eastern branch was largely obscured with deposits as thick as 600 ft. In the southern part of the basin proglacial meltwaters formed against the drainage divide and many cols, spillways, and sluiceways were incised into adjacent terrain when the waters overflowed the lakes. Typical chute valleys formed in this manner are Oswayo Creek, Honeoye Creek, Knight Creek, and north-draining equivalents of Hunt Creek, Little Genesee River, Oil Creek etc.

During deglaciation and when north-flowing drainage could become reestablished several new post-glacial gorges formed in the Genesee valley at positions where drift had so clogged the valley that the river was forced to cut a bedrock chasm. For example near the village of Portageville the contrast in preglacial and postglacial valleys is most striking. Within 3 mi the river gradient drops 317 ft over three major waterfalls and a series of rapids. Other deep ravines were cut northward with 500 ft high bedrock walls and one 20-mi reach occurs mainly in the limits of Letchworth State Park.

In the region bordering this Section, Connally (1964) mapped various moraines, such as the Olean, Kent, Almond, and Valley Heads and discovered that in the drift two different provenances occur. He determined this lithology on the basis of garnet ratios, showing that purple garnets were indicative of a Canadian source area, and red garnets came from the Adirondacks.

A BRIEF REVIEW OF GLACIATION IN PENNSYLVANIA

To date the most inclusive works on the Quaternary of Pennsylvania are those by Lewis (1884) and Leverett (1934). Lewis traced the Late Wisconsinan terminal moraine across Pennsylvania and subsequent studies have shown it to be basically correct and needing only slight modification. It is probably Woodfordian in age. Leverett provided an excellent review of earlier works and also looked at deposits beyond the margin. He gave special treatment to drift deposits he thought to be Illinoian. Subsequent work by W. Sevon and others indicates such deposits are probably early Wisconsinan (perhaps Altonian). Although similar in some respects to other Wisconsinan drift they contain an unusual amount of cryoplanation. Although the early glaciation was generally more extensive than the later glaciation, it did not develop any well defined terminal moraine.

Illinoian deposits (Sevon, pers. comm. 1973-1974) occur beyond the limits of Altonian glaciation but are patchy and thin. They do not form a terminal moraine, and appear to be more irregular than indicated by Leverett. The deposits are deeply weathered and were extensively altered by periglacial processes during the Wisconsinan.

Mahoning Section

Under the direction of G.W. White many people have studied the glacial geology of northwestern Pennsylvania. The Mahoning Section, named from the major river, constitutes only part of the area covered by these workers. The principal references for this area are Shepps et al (1959) and White et al (1969; see also White, this volume).

The coal strip mines of the region, up to one mile lengths, have provided unique opportunity to study the detailed stratigraphy of northwest Pennsylvania. Important new principles have also emerged from such studies (White et al, 1969a) that indicate such factors as great variability in drift deposits, and lack of correlation with surface expression and subsurface extent of furthest extent of drift margins.

> ". . . . in early Wisconsinan time the Titusville Till was deposited by a series of advances which laid down a series of till sheets here described for the first time. Pre-Wisconsinan ice advances are less well known, but we can now report a major advance in Illinoian (?) time that deposited the Mapledale Till, a till possibly deposited in two sheets. A still earlier ice advance, which may be early Pleistocene, deposited the now deeply weathered Slippery Rock Till (White et al, 1969a, p. 7)."

White and his co-workers have developed the following geomorphic events for the area:

1. Very Early Plesitocene. Ice blocked north-flowing streams and one relict is silt below Slippery Rock Till.
2. "Pre-Illinoian" Stage. Deposition of Slippery Rock Till followed by long period of weathering.

3. Illinoian (?) Stage. Furthest southeast advance of ice. The Mapledale Till was deposited, perhaps by two advances, and extensive gravel deposits were formed.

4. Sangamonian (?) Interglacial Stage. Weathering produced deep oxidation, deep leaching, and thick soil.

5. Wisconsinan Stage. (a) Titusville (Altonian) ice advances about 40,000 years ago and persisted several thousand years; (b) Kent (Woodfordian) ice advanced about 23,000 years ago and produced a very thin till; (c) Lavery ice deposited a thin till about 16,000 years ago; (d) Hiram ice and Ashtabula ice were not important in the Mahoning Section but did reach the extreme northwest part of Pennsylvania.

Open Folds Section

The diagnostic feature of this Section is the influence of geologic structures on the drainage fabric. The northeastly trending broad folds have produced a trellis pattern for much of the area, as well as arcuate drainages in others (such as occur with tributaries of the Tioga River). The primary information sources for this Section are Fuller, and Alden (1903a, 1903b) Lohman (1939), Denny (1956), and Denny and Lyford (1963).

Some of the most important glacial effects in this Section occur in the northern part where large proglacial lakes were ponded against the divides of north-flowing rivers. Not only were thick glaciolacustrine deposits left throughout many valleys and on some hillslopes, but specatular spillways formed by overflow of the lakes through narrow cols. These are especially well-developed in the Tioga River basin and its major tributaries such as the Cowanesque River. The most awe-inspiring of these features is the Pine Creek Gorge, called the "Grand Canyon of Pennsylvania". Here the nearly vertical 700 ft high walls of the 20 mi chasm were formed by meltwaters impounded in the Tioga and Cowanesque Rivers.

Glacial deposits in the Section range the gamut from till to stratified sediments with abundant development of colluvium and periglacial materials. Coates (1966c) found the till shadow effect to be present in most of the northern area, where till is in excess of 100 ft on many south- facing slopes. The thickest drift that could be verified, mostly stratified, is near the village of Osceola, Pa., where a 245 ft deep well did not reach rock. There is a variety of landforms developed from the glacial drift and they include "choker moraines", kames, kame terraces, and kame deltas. Their abundance is indicative that much of the ice deterioration during the final deglaciation was in terms of downwasting.

The residence time of the ice sheet in and near the area has produced sufficient features to indicate its importance and some of its characteristics. For example Denny and Lyford (1963) were able to restore the glacial profile at ice terminal positions and these data show a range of 100-500 ft/mi.... somewhat greater than the 128 ft report by Peltier (1949) on the eastern border of the Section. Denny (1956) has provided the most extensive account of periglacial effects in the region. He describes a full range of features that include block fields, boulder rings, boulder stripes, and terraces.

Small Lakes Section

Some precedent exists for naming this Section in that the U.S. Geological Survey 1:1,000,000 series topographic map (Hudson River NK 18) labels the region "Numerous Small Lakes". It will be distressing to some diehards that the name "Pocono" does not appear as a geographic place name. The dominant theme that links this region are the hundreds of lakes, marshes, and depressions. For example more than 300 lakes were geomorphically studied in five 1:62,500 scale U.S. Geological Survey maps in the northern part of the Section. They largely occur in south-oriented drainages (Fig. 18) Lakes in the southern part of the Section, often called the Pocono uplands, mountains, or plateau, are even more numerous, larger, and have greater diversity in orientation.

Much of the work by the Pennsylvania Geological Survey in this Section occurs along the borders. Peltier (1949) mapped the terraces of the Susquehanna River and equated their development to the earlier nomenclature as summarized in MacClintock and Apfel (1944) retaining the Binghamton name. In the light of newer data from New York a re-evaluation of the terrace nomenclature is in order. Studies of Wisconsinan drift of eastern Pennsylvania indicate it is very responsive to bedrock, very diverse in character (Epstein, 1969), and not susceptible to the type of rock stratigraphy applied in western Pennsylvania. Deglaciation was complex and has been worked out in some detail by Harrison (1966), Crowl (1971), and Epstein (1969). For example deglaciation sequences formed in the Saylorsburg-Stroudsburg area and the ice fluctuated developing a 1-2 mi wide moraine at the higher elevations. Sevon (pers. comm. 1973-1974) has recognized a significant end moraine, the "Gouldsboro end moraine", north of the classical terminal moraine. He has also called attention to the importance of periglacial deposits in the glacial record of eastern Pennsylvania.

Not only are periglacial features found in the southern border of the section, but they also notably occur in the Great Bend area (Harrison,1966; King and Coates;1973; Coates and King,1973). Here solifluction slopes are prevalent, patterned ground and block fields occasionaly are present, and even tors and rock cities occur on some of the higher hills.

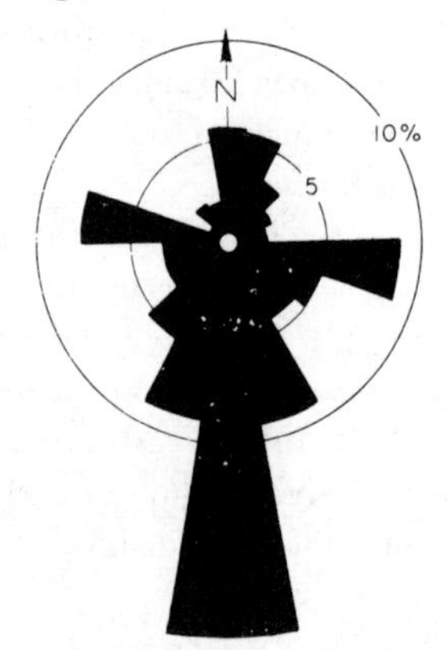

Figure 18. Diagram showing orientation pattern of 350 lakes in north eastern Pennsylvania.

This Section has been little studied in many of the areas that would supply a better picture of Quaternary events. Analysis of the character of the drift is particularly lacking. Lohman (1937) provides sufficient depth to rock data that indicate drift is very thick in many places. Such information is consistent with the view that massive ice stagnation occurred when southerly moving ice thinned to a point during deglaciation that the northern source area was beheaded because of the high elevations of the Catskills. Thus, a massive block of ice on the lee side of the Catskills withered *in situ,* forming a somewhat chaotic assemblage of drift and topographic features. Another topic that needs further elaboration is the meltwater history in the northern part. Although it is known that many cols and overflow channels are present, such as the New Milford Sluiceway (Harrison, 1966), the complete integration of their evolution still poses many problems.

EPILOGUE

This chapter has been devoted to a descriptive analysis of a large glaciated region in New York and Pennsylvania. An attempt has been made to correct some misconceptions, to provide some new insights, to inventory the geomorphic flavor, and to provide a foundation for continuing work. In such broad-scale effort, justice is not done on many matters and prejudices of the author show through. This is a status report which from time to time I will amend, and I am sure others will be only too happy to speak to its flaws. This, however, is the way science advances, and since a specialized treatise on this region has not been done for 36 years its neglect could no longer be tolerated.

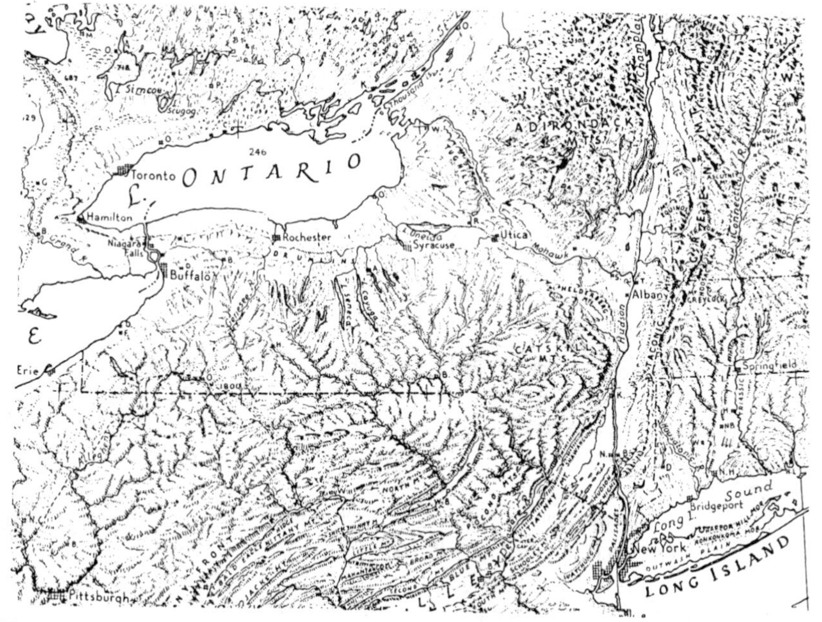

Figure 19. Physiographic diagram of general region. (After Raisz).

The Plateau is blessed with an unusual number of terrain anomalies. I have discussed some in this chapter and spoken to others in previous works . . . . .but many landform problems still remain. What factors have controlled the parallel zig-zag orientation of the Susquehanna andDelaware Rivers(Fig. 19)? Arethey the result of ice marginal drainages in northern Pennsylvania?What caused the meandering valleys of these rivers? Are they relict from different ranges of discharge during theQuaternary? How old are the major geomorphic boundaries of the Plateau and what has been their geomorphic evolution? Pre-Quaternary events obviously marked the region with certain features, but I believe glaciation has made a more profound overprint on the topography than most have suspected.

There is always value in attempting to understand, if only in a reconnaissance, such a large region. I have found that the study of similarities and contrasts of the different sections has provided me with new levels of understanding for glacial processes. This has been beneficial in an academic way, and has proved fruitful in various applied and consulting ventures. For example diagnoses for appropriate landfill sites have been accelerated, location of water wells in "virgin" terrain is more successful, and determination of comparable land values in condemnation cases is more equitable. Thousands of dollars were saved by a construction firm in aggregate exploritory mapping by knowledge that gravels in a large section of the Susquehanna would not pass New York State specifications for concrete.

This effort would have been impossible without the cooperation of Jeanne who spent many solitary months as a "geomorphic widow" during the years I collected information for Plateau articles. Special thanks are also due to my Binghamton students in our many joint exchanges of ideas and data. Ernie Muller and Bill Sevon have been especially kind in helping me round out information sources. I wish to acknowledge the financial sponsorship for varous aspects of this work to the Research Foundation, State University of New York; the Organized Research Fund, Department of Geological Sciences, and; the U.S. Geological Survey.

REFERENCES

Cadwell, D.H, 1972. Late Wisconsinan chronology of the Chenango River valley and vicinity, New York: Ph.D.Thesis, State Univ. of New York at Binghamton, 102 p.

Chamberlin, T.C. 1883. Terminal moraine of the second glacial epoch: U.S. Geol. Survey 3rd Ann. Rept., p. 291-402

Clayton, K.M. 1965. Glacial erosion in the Finger Lakes Region (New York State, U.S.A.): Zeit. Geomorph. NF 9, p. 50-62: and 10, p.475-477

Coates, D.R. 1963b. General geology of south-central New York: in *Geology of South-Central New York:* D.R. Coates, ed., New York State Geol. Assoc. 35th Ann. Meeting, p. 19-57.

------ 1963. Geomorphology of the Binghamton area: in *Geology of South-Central New York:* D.R. Coates, ed., New York State Geol. Assoc. 35th Ann. Meeting, p.97-116.

------ 1966a, Glaciated Appalachian Plateau: till shadows on hills: Science, v.152, p.1617-1619.

——— 1966b. Discussion of K.M. Clayton "Glacial erosion in the Finger Lakes Region": Z. Geomorph., 10, p.469-474.

——— 1966c. Report on the geomorphology of the Cowanesque Basin, Penn. : U.S. Army Corps of Engineers Cowanesque Reservoir Study, Baltimore, 27+ p.

——— 1968. The Finger Lakes: in *Encyclopedia of Geomorphology:* R. Fairbridge, ed., Reinhold Book Corp., p. 351-357.

——— 1972. Hydrogemorphology of Susquehanna and Delaware Basins: in *Quantitative Geomorphology: Some Aspects and Applications:* M. Morisawa, ed., Publications in Geomorphology, State Univ. of New York, Binghamton, p.273-306

Coates, D.R., Landry, S.O. and Lipe, W.D. 1971. Mastodon bone age and geomorphic relations in the Susquehanna Valley: Geol. Soc. Amer. Bull., v.82, p.2005-2010.

Coates, D.R. and King, C.A.M. 1973. Glacial geology of Great Bend and adjacent aadjacent region: in Glacial Geology of the Binghamton-Western Catskill Region, D.R. Coates, ed., Publications in Geomorphology Contribution No.3, p.2-20.

Coates, D.R. and Kirkland, J.T. (in press). Applications of a glacial model for large-scale terrain derangements: Proceedings volume of Symposium on Quaternary Environments, York Univ.,Toronto, May 22, 1974.

Coates, D. R. and King, C.A.M. (in preparation). Drift forms of glaciated regions.

Connally, G.G. 1964. The Almond moraine of the western Finger Lakes Region, New York: Ph.D. Thesis, Michigan State Univ., 102 p.

Conners, J.A. 1969. Geomorphology of the Genegantslet Basin of New York; M.A. Thesis, State Univ. of New York at Binghamton. 154 p.

Cook, J.H. 1935. The glacial geology of the Berne quadrangle: N.Y. State Mus. Bull. 160, 47 p.

Crowl, G.H. 1971. Pleistocene geology of unconsolidated deposits of the Delaware Valley, Matamoras to Shawnee On Delaware, Penn.: Penna. Geol. Survey, General Geology Report G 60, 40+p.l

Denny, C.S. 1956. Surficial geology and geomorphology of Potter County, Pennsylvania: U.S. Geol. Survey Prof. Paper 288, 72p.

——— and Lyford, W.H. 1963. Surficial geology and soils of the Elmira- Williamsport region, New York and Pennsylvania, U.S. Geol. Survey Prof. Paper 379, 60 p.

Donahue, John J. 1972. Drainage intensity in western New York; Ann. Assoc. Am. Geographers, v.62 p. 23-36.

Engeln, O.D. von. 1921. The Tully glacial series: N.Y.S. Mus. Bull. 227-228, p.39-62.

——— 1929. Interglacial deposit in central New York: Geol. Soc. Amer. Bull., v. 40, p. 469-479.

——— 1931. A preglacial or interglacial gorge near Seneca Lake, New York: N;Y.S. Mus. Bull. 286, p.127-131.

——— 1961. *The Finger Lakes Region: Its Origin and Nature:* Cornell Univ. Press, 156 p.

Epstein, J.B. 1969 Surficial geology of the Stroudsburg Quadrangle, Pennsylvania-New Jersey: Penna. Geol. Survey, General Geology Report G 57, 67+ p.

Fairchild, H.L. 1905. Ice erosion theory, a fallacy: Geol. Soc. Amer. Bull., v. 16, p.13-74.

——— 1925. The Susquehanna River in New York and evolution of western New York drainage: N.Y.S. Mus. Bull. 256, 99 p.

——— 1928. Geologic story of the Genesee Valley and western New York; Pub. by author, Rochester, N.Y. 215 p.

——— 1932. New York moraines: Geol. Soc. Amer. Bull., v.43, p.627-662.

Fenneman, N.M. 1938. *Physiography of the Eastern United States:* McGraw-Hill Book Co., New York, 712 p.

Fessenden, R. 1974. Interpretation of Quaternary sediments in the Cortland through valleys from electrical resistivity data: State Univ. of New York at Binghamton, M.A. Thesis, 142 p.

Flint, J.J. 1968. Hydrogeology and geomorphic properties of small basins between Endicott and Elmira, New York: M.A. Thesis, State Univ. of New York at Binghamton, 74 p.

Frimpter, M.H. 1974. Ground-water resources, Allegheny River Basin and part of the Lake Erie Basin, New York: New York State Dept. of Environmental Conservation, Basin Planning Report ARB-2, 98 p.

Fuller, M.L. and Alden, W.C. 1903a. Elkland-Tioga Folio, Penn.: U.S. Geol. Survey Geol. Atlas 93.

——— 1903b. Gaines Folio, Penn.: U.S. Geol. Survey Geol. Atlas 92.

Fullerton, D.S. 1971. The Indian Castle glacial readvance in the Mohawk Lowland, New York, and its regional implications; Ph.D. Thesis, Princeton Univ., 96+p.

Harrison, J.E. 1966. Proglacial drainage evolution and deglaciation of the Great Bend region, Pennsylvania and New York: M.A. Thesis, State Univ. of New York at Binghamton, 71 p.

Holmes, C.D. 1937. Glacial erosion in a dissected plateau: Am. Jour. Sci. (5) v. 33, p. 217-232.

——— 1941. Till fabric: Geol. Soc. Amer. Bull., v. 52, p. 1299-1354

——— 1952. Drift dispersion in west-central New York: Geol. Soc. Amer. Bull., v. 63, p. 993-1010

——— 1960. Evolution of till-stone shapes central New York: Geol. Soc. Amer. Bull., v. 71, p. 1645-1660.

Hubbard, G.D. 1906. Drumlinoids of the Catatonk folio: Am. Geog. Soc. Bull., v. 38, p. 355-365

King, C.A.M. and Coates, D.R. 1973. Glacio-periglacial landforms within the Susquehanna Great Bend area of New York and Pennsylvania: J. Quat. Research, v. 3, n. 4, p. 600-620.

Kirkland, J.T. 1973. Glacial geology of the western Catskills: Ph.D. Thesis, State Univ. of New York at Binghamton, 88 + p.

Krall, D.B. 1972. Till stratigraphy and Olean ice retreat in east-central New York: Ph.D. Thesis, Rutgers Univ., 95 p.

LaFleur, R.G. 1969 Glacial geology of the Schoharie Valley: New England Geological Conference, 61st Ann Mtg., Guidebook for Field Trips, J.M. Bird, ed., 20 p.

Leggette R.M. 1936. Ground water in northwestern Pennsylvania: Penn. Geol. Survey, 4th ser., Bull W 3, 215 p.

Leverett, F. 1934, Glacial deposits outside the Wisconsin Terminal Moraine in Pennsylvania: Penna. Geol. Survey, 4th Ser., Bull. G 7, 123 p.

Lewis, H.C. 1884, Report on the terminal moraine in Pennsylvania and western New York: Penna. Geol Survey, 2nd ser., Rept. 2,299 p.

Lohman. S.W. 1937. Ground water in northeastern Pennsylvania: Penn. Geol. Survey, Fourth Series Bull. W 4, 312+p.

———1939. Ground water in north-central Pennsylvania: Penn. Geol. Survey, 4th Ser., Bull. W 6, 219 p.

Lounsbury, C. et al. 1932, Soil Survey of Broome County, New York: U.S. Dept. Agri., 40 p.

------1953. Soil Survey of Tioga County, New York: U.S. Dept. Agri., 133p.

MacClintock, P. and Apfel, E.T. 1944. Correlation of drifts of the Salamanca re-entrant, New York: Geol. Soc. Amer. Bull., v.55, P. 1143-1164.

Merritt, R.S. and Muller E.H. 1959 Depth of leaching in relation to carbonate content of till in central New York State: Am. J. Sci., v.257, p. 465-480.

Miller, N.G. 1973. Late-glacial and postglacial vegetation change in southwestern New York State: N.Y.S. Museum and Science Service Bull. 420, 102 p.

Moss, J.H. and Ritter, D.F. 1962 New evidence regarding the Binghamton substage in the region between the Finger Lakes and Catskills, New York: Am.J. Sci., v.260, p. 81-106

Muller, E.H. 1957 Physiography and glacial geology of Allegany County and vicinity: N.Y.S. Geol. Assoc. Guidebook for 29th Ann.Mtg., Wellsville N.Y., p. 4-10.

------1963. Geology of Chantanqua County, New York: Part II: New York State Museum and Science Service Bull. No. 392, 60 p.

------1965a. Bibliography of New York Quaternary geology: New York State Museum and Science Service Bull. No. 398, 116 p.

------1965b. Quaternary geology of New York State: in *The Quaternary of the United States*, H.E. Wright, Jr. and D.F. Frey, eds., Princeton Uni. Press, p.99-112.

Peltier, L.C. 1949. Pleistocene terraces of the Susquehanna River, Penn,: Penn Geol Survey, Fourth Series, Bull. G.23, 158 p.

Randall, A.D. 1972. Records of wells and test borings in the Susquehanna River basin, New York: N.Y.S. Dept. of Env. Cons. Bull. 69, 92p.

Randall, A.D. and Coates, D.R. 1973. Stratigraphy of glacial deposits in the Binghamton area: in Glacial Geology of the Binghamton-Western Catskill Region, D.R. Coates, ed., Publications in Geomorphology Contribution No.3, P. 40-55.

Rich, J.L. 1906. Local glaciation in the Catskill Mountains: Jour. Geol., v. 14, p. 113-121

------1935. Glacial geology of the Catskills: New York State Museum Bull. No. 299, 180 p.

Rideg, P. 1970 Quantitative fluvial geomorphology of Catatonk Creek basin, New Yorl: M.A. Thesis, State Univ. of New York at Binghamton, 156 p.

Schmidt, V.E. 1947 Boulders of interglacial conglomerate in central New York: Am.J. Sci., v.245, p. 127-133

Sevon, W.D. Personal Communications. Letters Jan.22, 1974; Febr.1, 1974.

Shepps, V.C., White, G.W. Droste, J.B. and Sitler, R.F. 1959. Glacial geology of northwestern Pennsylvania: Penna. Geol. Survey, Fourth Series, Bull. G 32, 59+p.

Shumaker, R.C. 1957. Till texture variation and Pleistocene deposits of the Union Springs and Scipio Quadrangles, Cayuga County, New York: M.A. Thesis, Cornell Univ., 44 p.

Soren, J. 1963. The ground-water resources of Delaware County, New York: New York Water Resources Comm., Bull. GW-50, 59+p.

Tarr, R.S. er al. 1909 Description of the Watkins Glen-Catatonk district, New York: U.S. Geol. Survey Geol. Atlas Folio, 33 p.

Thornbury, W.D. 1965. *Regional Geomorphology of the United States:* John Wiley, New York, 609 p.

Totten, S.M. 1969. Overridden Recessional moraines of north-central Ohio, Geol. Soc. Amer. Bull., v.80, p. 1931-1946

------1973. Glacial geology of Richland County, Ohio: Ohio Geol. Survey, Report of Investigations No.88, 55+, p.

Wetterhall, W.S. 1959. The ground-water resources of Chemung County, New York: New York Water Resources Comm., Bull. GW-40, 58+p.

White, G.A. 1969. Pleistocene deposits of the north-western Allegheny Plateay, U.S.A.: Quart.J. Geol. Soc. London, v. 124, p. 131-151

White, G.W. Totten, S.M. and Gross, D.L., 1969. Pleistocene stratigraphy of northwestern Pennsylvania: Penna. Geol Survey, 4th ser., Bull. G55, 88 p.

White, W.A. 1972. Deep erosion by continental ice sheets: Geol. Soc. Amer. Bull., v.83, p. 1037-1056.

PART 3

OTHER TERRAIN CONSIDERATIONS

CHAPTER 9. ICE–WEDGE POLYGONS OF NORTHERN ALASKA
Robert F. Black

CHAPTER 10. HISTORY OF GLACIAL LAKE WAWARSING, SOUTHEAST NEW YORK
William B. Heroy

CHAPTER 11. THE RELATIONSHIP OF RIVER TERRACE FORMATION TO GLACIATION IN THE SHOSHONE RIVER BASIN, WESTERN WYOMING
John H. Moss

CHAPTER 12. GLACIAL GROOVES AND PERIGLACIAL FEATURES IN THE SAHARAN ORDOVICIAN
Rhodes W. Fairbridge

There is a large family of features that are attributable to glaciation but which were not the direct product of the ice. Some landforms were developed at such a distance from the ice margin that their linkage to the glacier is highly disputed. Still other features are so geologically old that their heritage, whether from glaciers or some other process, has not received universal agreement. These subjects form the basis for this part, and are only suggestive of the many possible ways that Ice Ages influence the earth's surface. Time and space do not permit a complete exposition of the myriad of ramifications that prove the importance of glaciation, but the reader should be aware of such features and events. It is further suggested that he become acquainted with such texts as that by Embleton and King which provide a fine overview for the subject. For example it would be disastrous for a worker in such desert basins as occur in Nevada or Utah to infer a constant arid climate, because the hydrologic regime was affected by pluvial climates during the Quaternary. In a different manner sedimentologists could obtain spurious results when working in the continental shelf if they failed to realize the magnitude of changes in this environment created by falling and rising sea levels. Of course the understanding of frozen ground phenomena and periglacial features is becoming increasingly important as population increases in arctic terrain, and new resources are obtained ... such as via the Alaskan Pipeline. Thus glaciation and its global effects constitute an exceptionally necessary component for many earth science disciplines.

Chapter 9 by Black draws upon his large experience in arctic terrain to describe and analyze some frozen ground features. Ice-wedge polygons and patterned ground cover large land surfaces in Alaska, Canada, and Siberia, and Black's confirmation of the Leffingwell theory provides a significant contribution on a controversial topic.

Chapter 10 by Heroy blends the use of topographic maps and field observations to unravel the glacial chronology of Lake Wawarsing. This manuscript is an interesting historical document since Heroy's style of work and ideas were closely associated with other glacial geologists four decades ago.

Chapter 11 by Moss is in part a summary of an area he has worked for two decades. The controversial topic of the origin of the terraces is neatly solved with compelling evidence that proves their linkage as glaciofluvial deposits extended from ice margins.

Chapter 12 by Fairbridge is also a topic that is steeped in debate .... the origin of features on the bedrock of the Sahara region. His remarkable photographs, as on the frontispiece, and the personal documentation he marshals along with supporting literature provide a sufficient case to prove glaciation during Ordovician time.

CHAPTER 9

ICE-WEDGE POLYGONS OF NORTHERN ALASKA

Robert F. Black

ABSTRACT

Ice-wedge polygons, commonly 5-30 m in diameter, are strikingly developed over the Arctic Coastal Plain of northern Alaska. Troughs over ice wedges that outline the polygons are a few centimeters to several meters wide. Centers of polygons are flat, high centered, or low centered in a continuum in which relief generally is several decimeters to a meter.

I adopted or modified techniques and equipment of glaciologists and petrologists to study ice wedges in the field and laboratory in northern Alaska. Semi-permanent thin sections aggregating 10,330 cm^2 from 44 wedges were photographed and studied. Numerous observations in the field and laboratory by other techniques revealed grain size and shape, and other parameters that aid in our understanding of the origin of ice wedges.

All ice wedges megascopically and microscopically exhibit marked lineation and foliation of air bubbles and other inclusions, dimensional lineation of ice crystals, and lineation of optic axes with maxima in fabric diagrams generally 10-35%. The fabrics are complicated and not uniform from one wedge to another, nor even necessarily from one part of a wedge to another. Not all fabrics can be explained. Deformation fabrics and growth fabrics are superimposed on each other. Growth of ice occurs in contraction cracks, and later compression produces shear and recrystallization under strong temperature gradients.

Ground-contraction and expansion was measured in five sites and the results correlated with observations and measurements on the ice content of permafrost and its thermal regime. Clearly the permafrost responds like ice to temperature changes. These and other results substantiate the contraction theory for origin of ice wedges outlined decades ago by E. de K. Leffingwell. The annual addition of ice to permafrost may lead in 5,000-7,000 years or more to massive ice wedges, instability, and initiation of thaw lakes. Man's works may initiate a thaw cycle.

INTRODUCTION

Patterned ground (Washburn, 1956, 1973), mostly ice-wedge or tundra polygons, is strikingly developed throughout the Arctic Coastal Plain of northern Alaska (Black, 1952, 1954, 1963; Brown and Johnson, 1966; Lachenbruch, 1962, 1966; Leffingwell, 1919) (Fig. 1). Ice-wedge polygons are polygonal patterns when

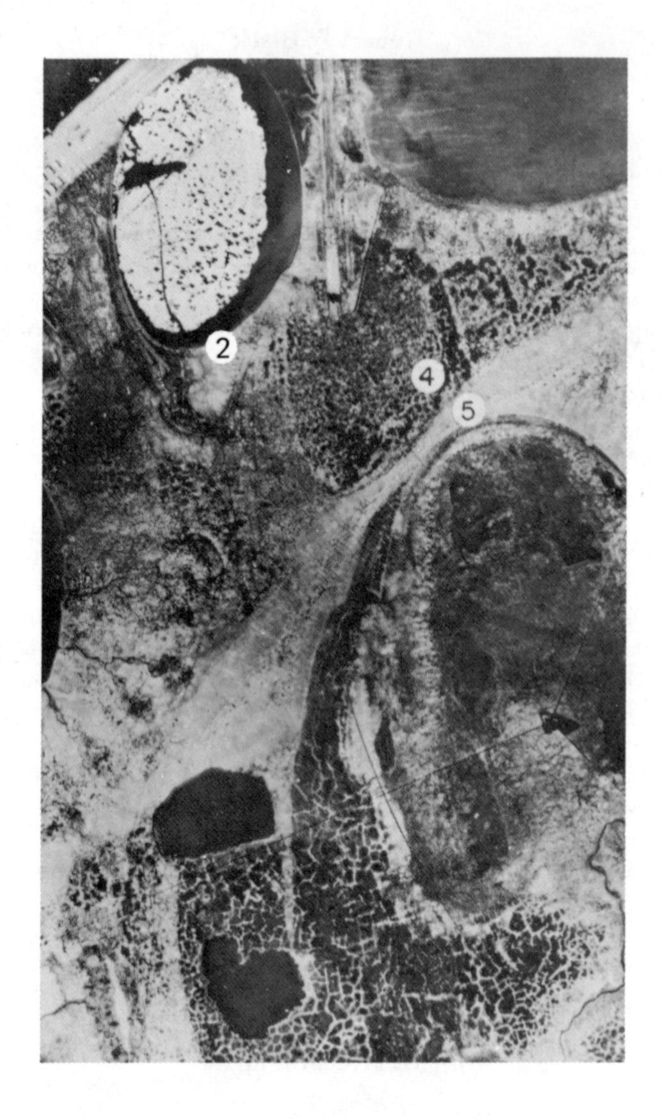

Figure 1.

Air view of the Barrow area, showing oriented lakes and ice-wedge polygons of different ages. Three contraction sites where rate of growth of ice wedges was measured are shown by numbers. The first site is off the photograph to the west; the third is to the northeast. U. S. Coast & Geodetic 9-lens photograph, July 17, 1947. Scale approximately 1/28,000. Cf, Brown and Johnson (1966).

seen in plan. Individual polygons are generally 5-30 m in diameter. Troughs over ice wedges that outline the polygons are a few centimeters to several meters wide. Ice wedges are wedge-shaped masses of ice oriented vertically with their apices downward (Brown and Péwe, 1973) (Fig. 2). Surface wedges are at the top of permafrost, and many are actively widening today. Buried wedges are covered by permafrost and are inactive (Fig. 3). Centers of polygons are (1) flat, (2) high centered, i.e., with rounded mounds, or (3) low centered, commonly with shallow ponds surrounded by raised rims.

Most studies in the Arctic Coastal Plain involving geomorphology, soils, or botany have been concerned with the ice-wedge polygons (Black, 1969a; Brown and Sellmann, 1973; Gunn, 1973; Walker, 1973). Ice wedges add further problems to man's use of the Arctic tundra, especially in construction (Black, 1957; Lachenbruch, 1970; Linell and Johnston, 1973).

Ice wedges grow in the perennially frozen unconsolidated sediments of the Arctic Coastal Plain. The sediments are mostly silt and sand with some clay and gravel, which comprise the primarily marine Gubik Formation of Pleistocene age (Black, 1964; Brown and Sellmann, 1973; and Brown, 1973). The sediments extend northward under the Arctic Ocean with little or no topographic or geologic break. Relief characteristics and a four-fold genetic classification of that coast are covered by Hartwell (1973). The Arctic Coastal Plain, roughly latitude 70–71° N, typifies a periglacial region without glaciers yet with a mean annual ground temperature of -8° to -12° C (Black, 1969a; Wahrhaftig, 1965). Permafrost is continuous and generally 300-600 m thick (Brown and Péwé, 1973; Gold and Lachenbruch, 1973). Tundra vegetation with a paucity of shrubs is zoned in microhabitats with relief in places of only a few centimeters (Britton, 1957; Spetzman, 1959; Wiggins and Thomas, 1962). Drainage is not fully integrated. Eolian deposits and features are widespread (Black, 1951). Regional relief commonly is only a few meters to a few tens of meters. Periglacial microrelief features, such as thermokarst and a variety of mounds, in addition to the patterned ground, are ubiquitous (Hussey and Michelson, 1966). Oriented lakes by the tens of thousands characterize the coastal plain (Black and Barksdale, 1949; Black, 1969b; Sellmann and Brown, 1973). Their drained basins are strikingly and faithfully portrayed by the ice-wedge polygons (Fig. 1).

New ice-wedge polygons are starting today in subdivision of older polygons and as new land becomes perennially frozen, as in drained-lake beds. Annual increments of ice are continually being added to many older ice wedges, increasing their widths. Growth of ice wedges causes flow and shear of permafrost to the surface adjacent to the wedges. These internal changes modify the surface topography and alter the effects of geomorphic processes, especially surface runoff and mass-wasting processes. Buried inactive ice wedges, in places associated with and the cause of thermokarst features, are many thousands of years old (Sellmann and Brown, 1973).

Figure 2.
Cross sections of three ice wedges of increasing age from the smallest to the largest in marine sediments of Pleistocene age about 115 km southeast of Barrow, on the Arctic Ocean. Peat layers, cut by the younger ice wedges, accumulated in low-centered polygon ponds. The bank is 5 m high.

Figure 3.
Buried ice wedges in marine sediments of Pleistocene age about 120 km southeast of Barrow, on the Arctic Ocean. The wedges are rising very slowly like diapirs, distorting formerly horizontal layers of pond peat.

This chapter emphasizes ice-wedge polygons only. Much of it is based on my unpublished field and office studies 1945-1953 in Alaska and Washington, D.C., while I was employed by the U.S. Geological Survey (Black, 1953a, 1953b). Those broad studies included various aspects of the thermal regime of permafrost and its stratigraphy, composition, and role in the geomorphology of the Arctic Coastal Plain. Support at Barrow in 1949-50 by the Office of Naval Research is gratefully acknowledged. I owe a great debt to my wife, Hernelda, for both field and office assistance, at times under very trying circumstances. Many others too numerous to mention have aided in the field or in discussions. Comments by J. Ross Mackay on this paper in draft form were very helpful and much appreciated. He is carrying on similar studies in Arctic Canada (Mackay and Black, 1973).

FIELD AND LABORATORY STUDIES

My study of ice wedges was undertaken to resolve the differences in the opposing theories of Leffingwell (1919) and Taber (1943) regarding their origin. Leffingwell's contraction theory for origin of ice wedges calls for present-day segregation of ice by annual increments after the ground is frozen to considerable depth and for moisture to come from the atmosphere. Taber's concept is that ice wedges and other large masses of ice grew when permafrost formed. Moisture was drawn up from below the descending frost line.

I revisited many of the localities studied by Leffingwell and Taber. At Barrow I undertook quantitative field and laboratory studies, especially on the fabrics of ice wedges, ground contraction and expansion from thermal changes, and composition of the permafrost. These have brought about a clearer understanding of the genesis of the ice wedges along the lines advocated by Leffingwell (Black, 1952, 1953a, 1953b, 1954, 1963; Lachenbruch, 1962, 1966). Only a synopsis of some of my studies can be presented here. The fabrics of ice wedges are emphasized, because little study of them has been done to this day, particularly in North America (cf., Shumskii, 1964).

My general geologic field investigations in northern Alaska were aimed at the delineation of terrain units, determination of their geomorphic history, and their correlation with terrain features, especially the polygonal ground. Microscopic examination of 255 samples and textural classifications of 147 samples were made of soil from permafrost and the active layer. Moisture content of 36 samples was measured. Samples were collected in winter and summer from cutbanks, pits, and drill cores. More than 50 km of continuously exposed cutbanks, 3-20 m high along the Arctic Ocean, were examined; traverses of several thousand km were made by small plane, boats, weasels, and on foot.

Ground-contraction measurements covered one year's cycle near Barrow. Continuous air and undersnow temperatures were recorded at one site. Ground temperatures every 1-2 days were measured every 2.5 cm to a depth of 1.3 m. Additional readings were taken at intervals of 2-7 days in other sites and to greater depths. Thermistors were read to 0.01° C. Depth and density of the snow cover were recorded at intervals throughout the winter at five sites where ground contraction was measured. My meteorologic data were supplemented by data from the U.S. Weather Bureau and others. These and other data (Fig.4) permit an analysis of heat exchange between the ground and atmosphere which can not be discussed here.

FABRICS OF ICE WEDGES

Introduction

The *fabric* of an object refers to the spatial relations of its components (Knopf and Ingerson, 1938; Fairbairn and Chayes, 1949). The components may be megascopic or microscopic grains or particles, irrespective of composition, that are scattered through the object or within certain units or particles, or they may be sub-microscopic crystal lattices. A *deformation fabric* results if the components are oriented by stress; a *growth fabric* results through growth of grains in place by crystallization. The analysis of the fabric includes measurements of scalar properties, i.e., length, width, crystal habit, etc., but no directions; and of vectorial properties which indicate direction. *Dimensional orientation* is expressed by the external shape of particles and grains , whereas *lattice orientation* refers only to the internal structure. The lattice orientation of the ice was determained mostly by orientation of the optic or "c" axis. Secondary axes were located rarely, using Tyndall figures (Nakaya, 1956).

Parallel alignment of dimensional orientations of individual ice grains, air bubbles, and other inclusions and individual lattice orientations determines the degree of development of the *linear fabric* in ice wedges. Layers or zones of air bubbles and other inclusions and ice grains of different size or shape than those in adjacent layers or zones comprise the *planar fabric*. The layers commonly contain linear fabric parallel or normal to the layer. In this paper a parallel alignment of long dimensions of bubbles and other inclusions or ice grains is called *dimensional lineation*; a parallel lattice orientation is *optic axis lineation*; and parallel layers or zones of bubble-rich ice alternating with bubble-poor ice, of other inclusions, and of ice-grain clusters that differ from adjacent zones or layers constitute foliation planes or simply the *foliation*.

During 1949-1950 techniques and equipment of glaciologists and petrologists were adapted or modified for study of ice wedges (Bader, 1951; Black, 1953 a, 1953b). One home-made 4-axis universal stage handled thin sections 5 cm on a side; another took slides to 25 cm. A Bausch and Lomb 5 axis stage and petrographic microscope were also used for thin sections 2.5 cm on a side. Refrigerant piped through the temperature control stage prevented melting of the ice sections.

Of several hundred ice wedges examined 83 samples were taken from 44 wedges, and 99 semi-permanent thin sections of 6.25-62.5 cm^2 were made and studied in a cold room at Barrow. The total slide area was 10,330 cm^2 of which

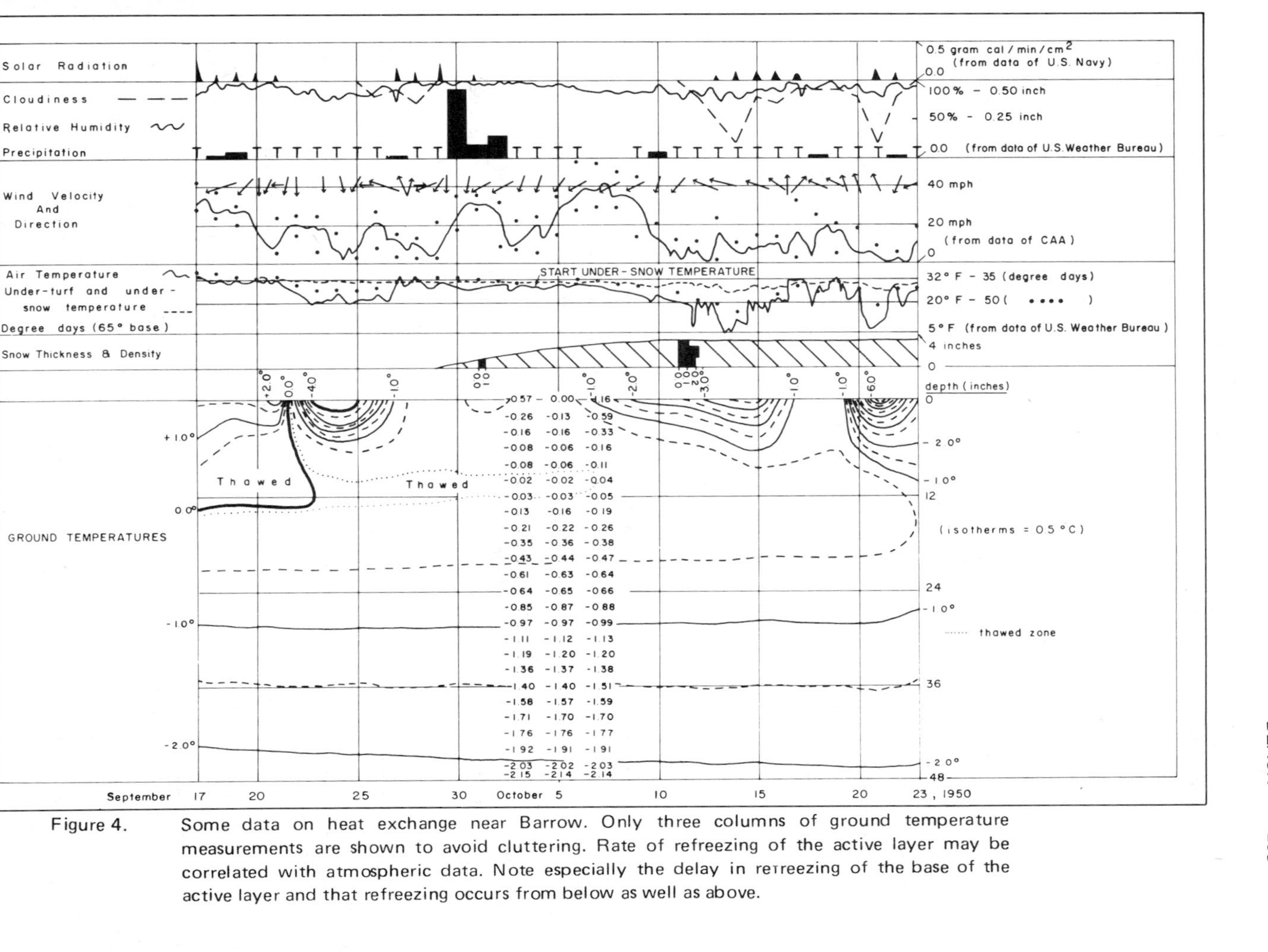

Figure 4. Some data on heat exchange near Barrow. Only three columns of ground temperature measurements are shown to avoid cluttering. Rate of refreezing of the active layer may be correlated with atmospheric data. Note especially the delay in refreezing of the base of the active layer and that refreezing occurs from below as well as above.

most represented active surface wedges (for comparison, 1,825 cm^2 came from ice wedges in the discontinuous and sporadic permafrost of the Fairbanks area). All slides were photographed with plain transmitted light and crossed polaroids, using black and white and color film. Many temporary thin sections and ice "rubbings" were made in the field.

Lower hemisphere equal-area fabric diagrams and grain plots were made of the fabrics data. The grain plots were made from a thin section photograph. Each ice grain was numbered, and optic-axis orientation of each grain was indicated directly on the print. This permits direct observation of each grain's alignment with respect to adjacent grains, bubbles and other inclusions, fractures, or other structures. Ice wedges megascopically and microscopically exhibit marked lineation and foliation of air bubbles and other inclusions (Fig. 5). Dimensional lineation of ice crystals is pronounced in many wedges. Lineation of optic axes, with maxima in fabric diagrams of 10-35% were observed in all thin sections (Fig. 6).

Deformation fabrics and growth fabrics are superimposed on each other in most active ice wedges. Dimensional and lattice orientations of individual ice grains generally are not parallel, and seem to bear little relationship to each other in most wedges. Some striking exceptions are noted.

Lineation and foliation of air bubbles and other inclusions

Lineation and foliation of air bubbles and other inclusions are best seen megascopically on the face of a wedge cut vertical and normal to the horizontal axis (Fig. 5). The patterns are very complex and differ greatly in detail between wedges and within a wedge. All wedges have vertical lineation and lineation normal to a particular structure. Foliation planes lie mostly between the axial plane and the sides, but also crisscross within a wedge.

All wedges examined contained myriads of rounded, elongated, and irregular air bubbles of which most were 0.1-3 mm in diameter (Fig. 7); some rod-like bubbles are as much as 10 mm in length. All types commonly are found in different parts of any particular surface wedge (Fig. 8). The rounder and larger bubbles are in buried wedges.

In surface wedges most air bubbles are scattered indiscriminately within and on the borders of individual ice crystals. Layers of bubbles and other inclusions cut across grains and lie between grains (Fig. 9). In buried wedges, however, more bubbles and other inclusions lie at boundaries of grains. Layers of inclusions are broken and segments offset markedly by single crystals of ice. Bubbles commonly are aligned and flattened in the basal plane of individual ice crystals. These produce a distinctive simultaneous reflection on rotation to the light and permit the identification of the c-axis. On irradiation, Tyndall figures may be produced (like snowflakes) that permit the identification of secondary axes (Nakaya, 1956).

Most air bubbles and other inclusions are in conspicuous layers (Fig. 10). If thick turf overlies a wedge, the layers of bubbles are relatively free of other inclusions. If mineral soil overlies a wedge, the layers commonly contain that soil. If the layers of bubbles are gently inclined and do not reach the top of the wedge, they

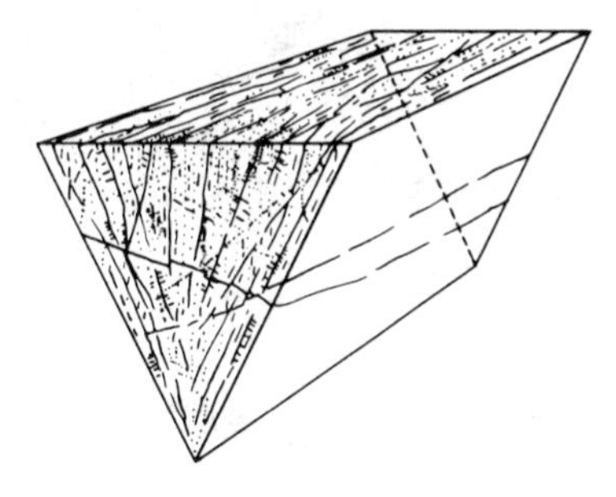

Figure 5.
Sketch of megascopic fabrics of an idealized ice wedge, representing contraction fractures and shear planes, alignment of bubbles and other inclusions, and dimensional orientation of ice crystals.

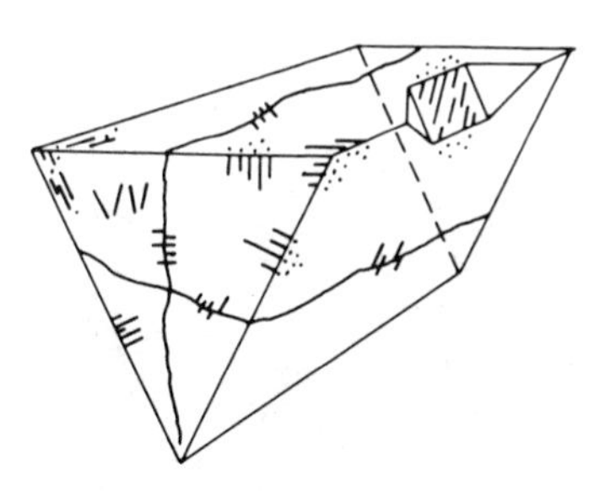

Figure 6.
Sketch of some microscopic fabrics of an idealized ice wedge, representing lineation and foliation of the c-axes of ice crystals and their relationship to contraction fractures and shear planes or to the borders of the wedge.

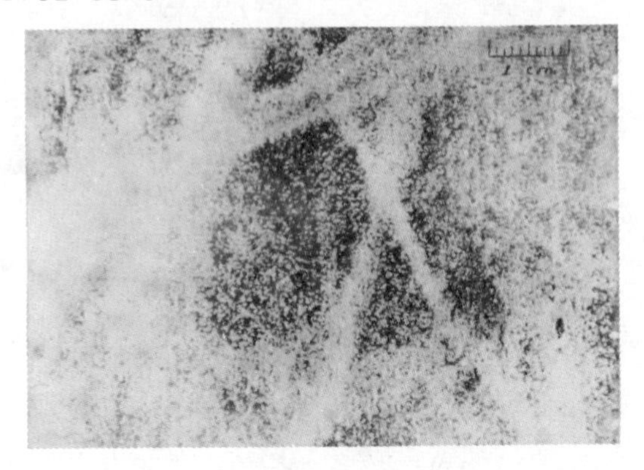

Figure 7.
Air bubbles in a section of ice cut vertical and normal to the horizontal axis of an ice wedge. Reflected light in a section several millimeters thick.

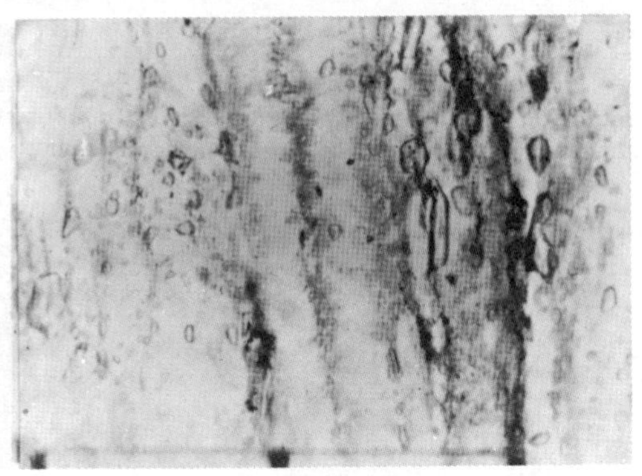

Figure 8.
Details of various kinds of air bubbles in a thin section cut vertical and normal to the horizontal axis of an ice wedge. Photograph is 12 mm wide. Note the horizontally oriented triangular bubbles in a filled contraction crack 2-3 mm wide crossing vertically through the center of the photograph.

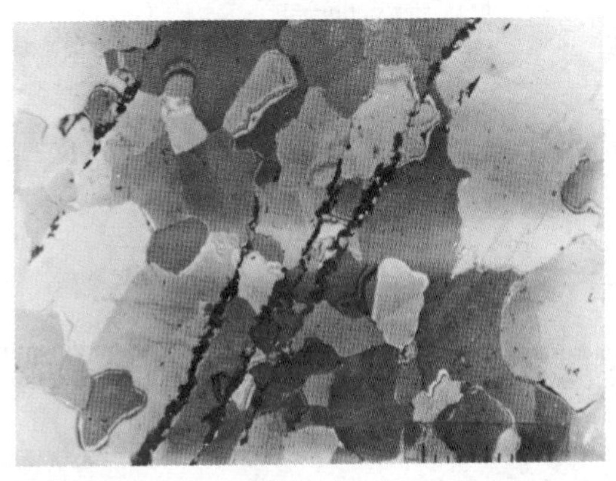

Figure 9.
Dark inclusions in layers parallel to the side of an ice wedge and their relationship to equigranular and irregular ice crystals shown under crossed polaroids. C-axes are oriented mostly normal to the layers. Photograph is 4.6 cm wide.

are generally free of other inclusions. Some layers of inclusions, especially those near and parallel to the sides of a wedge are free of bubbles. Most layers range from thin films to 10 mm in thickness and are several centimeters or meters in extent. Lens-like, elliptical, blocky, rounded, elongated, and irregular inclusions are common. The coarser fragments occur separately and in layers with other inclusions. At the junction of wedges a crisscrossing pattern of air bubbles and other inclusions is best seen on the tops of wedges.

The layers of inclusions commonly extend from the top of a wedge downward most of the vertical range of a wedge. The others, those gently inclined, do not intersect the top of a wedge but extend across the wedge at some depth. Some layers are more irregular than others. Toward the sides of most wedges the layers tend to parallel the sides. In the center of a wedge layers tend to be vertical. Therefore, the traces of layers on the face of a wedge diverge upward forming broken, irregular V's. Conspicuous crisscrossing layers are relatively common.

Most air bubbles and other inclusions are elongated vertically regardless of the orientation of the layer in which they lie. In other layers inclusions parallel the layer or may parallel the sides of the wedge and cross vertical layers of bubbles. In some vertical to inclined layers of air bubbles, individual bubbles tend to be elongated normal to the layer. Most bubbles and other inclusions within crystals are apparently not oriented with respect to the lattice. However, those bubbles forming a layer in the basal plane of a grain are flattened in very thin planes.

Dimensional lineation of ice crystals

Ice in ice wedges is holocrystalline. Individual grains are mostly anhedral (zenomorphic), and about 0.1 mm–10 cm in diameter. In parts of some wedges subhedral (hypautomorphic), and very rarely euhedral (automorphic), grains are seen. These are, respectively, prisms parallel to the c-axis and hexagonal outlines of crystals cut normal to the c-axis. Grain boundaries are straight, curved, irregular, or sutured.

Equidimensional ice grains were observed in most buried wedges and in some surface wedges, but most grains are slightly elongated (Fig. 9). Some have length-width ratios of more than 10 to 1 (Fig. 11). The long dimensions of notably acicular crystals are generally vertical, whereas those of less elongated crystals form girdles normal to the wedge's horizontal axis or one or more maxima normal to that axis.

Dimensional lineations of grains in clusters and layers vary markedly in orientation within a wedge (Fig. 12). Clusters of 5-20 large grains with a pronounced lineation commonly are cut by layers of small grains with different lineation. In some wedges traces of dimensional lineation are fan-shaped and parallel the foliation of silt layers. Lineation within narrow layers mostly is normal or parallel to the layer regardless of its orientation (Fig. 13). Ice crystals also radiate outward from pebbles included in wedges.

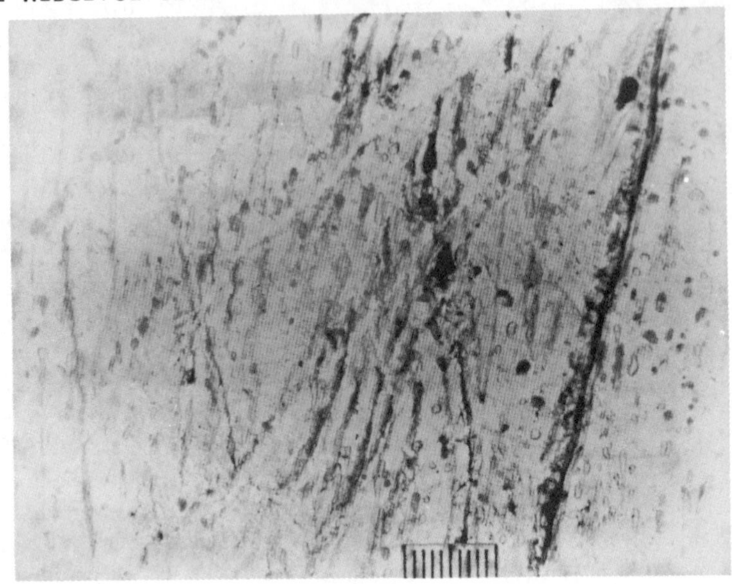

Figure 10.
Layers of dark inclusions and air bubbles in a thin section cut vertical and normal to the horizontal axis of an ice wedge. Scale is 10 mm.

Figure 11.
Thin section cut vertical and normal to the horizontal axis of an ice wedge, showing long prismatic ice crystals and a multiple contraction crack with hoarfrost. C-axes plunged 70–75° degrees to the left. Scale is 10 mm.

Long dimensions of large crystals mostly do not coincide with alignments of optic axes even in those grains with elongation ratios of 10 to 1 (Fig. 12). In small grains, as in whorls around inclusions and in narrow layers or zones normal to a structure, they coincide.

Lineations of optic axes

No thin section of wedge ice lacked noticeable lineation of optic axes. Even buried wedges with the weakest fabric commonly had maxima of 10% or more in the equal area, fabric diagrams; most maxima in fabric diagrams from surface wedges were 10-35%. Some idealized lineations are shown in Figure 6. Most display complicated fabrics (Figs. 12-14).

The most pronounced lineations of optic axes are (1) normal to the wedge top (vertical) (2) normal to the axial plane (normal to the horizontal wedge axis and horizontal) (3) normal to the wedge axis and inclined to one side or to both sides, but commonly normal to only one side of a wedge, and rarely parallel to the side (4) parallel to the wedge axis (and horizontal) (5) normal to a narrow layer in which the grains occur (generally normal to the wedge axis as well) (6) parallel to the layer in which they occur (generally horizontal and parallel to the horizontal wedge axis as well); and (7) some other orientation, generally 30°–60° to the wedge axis and horizontal or at 30°–60° to the horizontal. All directions of lineations were found in most wedges (Fig 15). In some wedges all occurred in a block less than 30 cm on a side. A rare buried wedge may exhibit only one lineation; others only a few.

Vertical lineation and lineations normal to the horizontal wedge axis and horizontal or inclined were generally best developed. Lineation parallel to the wedge axis was persistent and widespread, but rarely of high percentage. In some wedges local maxima at an angle to the wedge axis and inclined were well-defined and of high percentage.

Percentages of maxima in fabric diagrams are less indicative of the degree of optic-axis orientation within a thin section than are grain plots. In many fabric diagrams about equal numbers of c-axes are present in each "area of a maximum" of different percentage; in other diagrams the number of axes in each "area of a maximum" of equal percentage are vastly different. These differences are the result of the degree of linearity as well as of the numbers of axis; e.g., a diffuse orientation of many grains produces a large "area of a maximum" of low percentage, whereas a few axes well-oriented produce a small "area of a maximum" of high percentage.

The degree of linearity of c-axes varies considerably within a wedge. Commonly axes of one cluster of grains are parallel, and axes of grains nearby diverge several degrees or more (Figs. 16-18). In some thin sections alternate grains or clusters of grains are essentially parallel and make an angle, ranging from a few degrees to nearly 90°, to adjacent grains or clusters of grains (Figs. 19, 20). Where they are separated by only a few degrees, the grains make up a single area of a maximum,

Figure 12.

Typical marked variation in dimensional fabrics in a thin section cut vertical and normal to the horizontal axis of an ice wedge. Crossed polaroids. Scale is 10 mm.

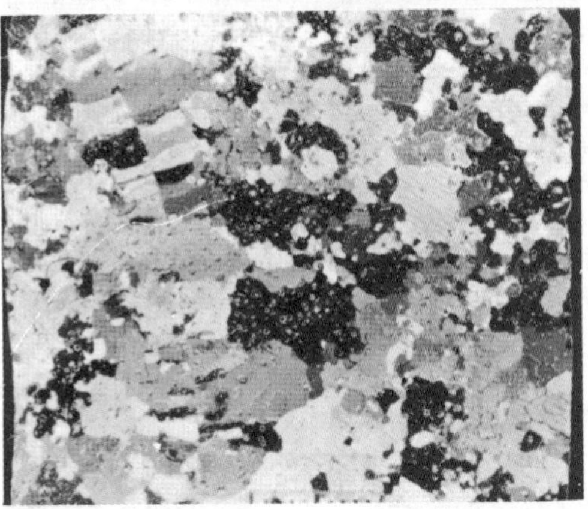

Figure 13.

Dimensional lineation of ice crystals filling a curved contraction crack in the left part of a thin section cut horizontal. The long axis of the ice wedge is parallel to the sides. Scale is 10 mm. Crossed polaroids.

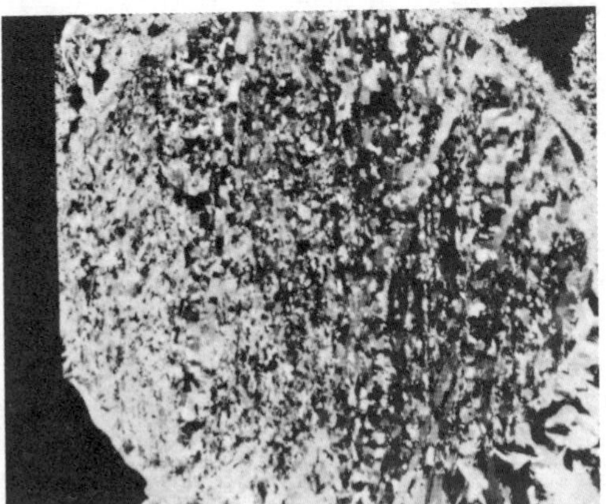

Figure 14.

Typical complicated dimensional fabrics of ice crystals in a thin section cut vertical and normal to the horizontal axis of an ice wedge. Crossed polaroids. Most c-axes were oriented normal to the layers they were in. Those on the extreme left plunged steeply to the right, normal to the dimensional lineation. Scale is 10 mm.

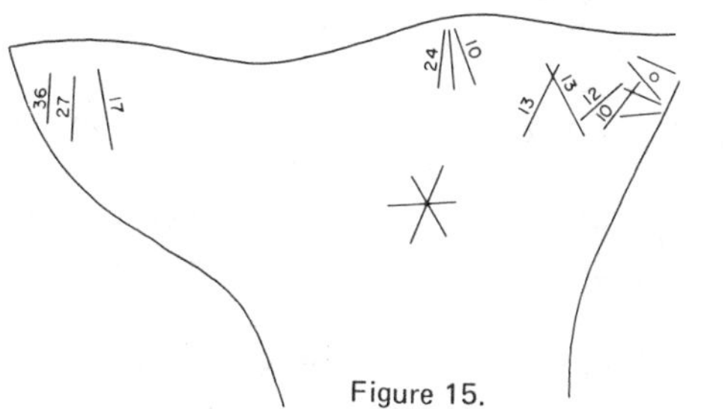

Figure 15.
Sketch of an ice wedge in cross section, showing lineations of c-axes. Numbers indicate maxima in fabric diagrams; others are strong maxima that are obvious in thin sections, but not counted. The small circle indicates a maximum normal to the face.

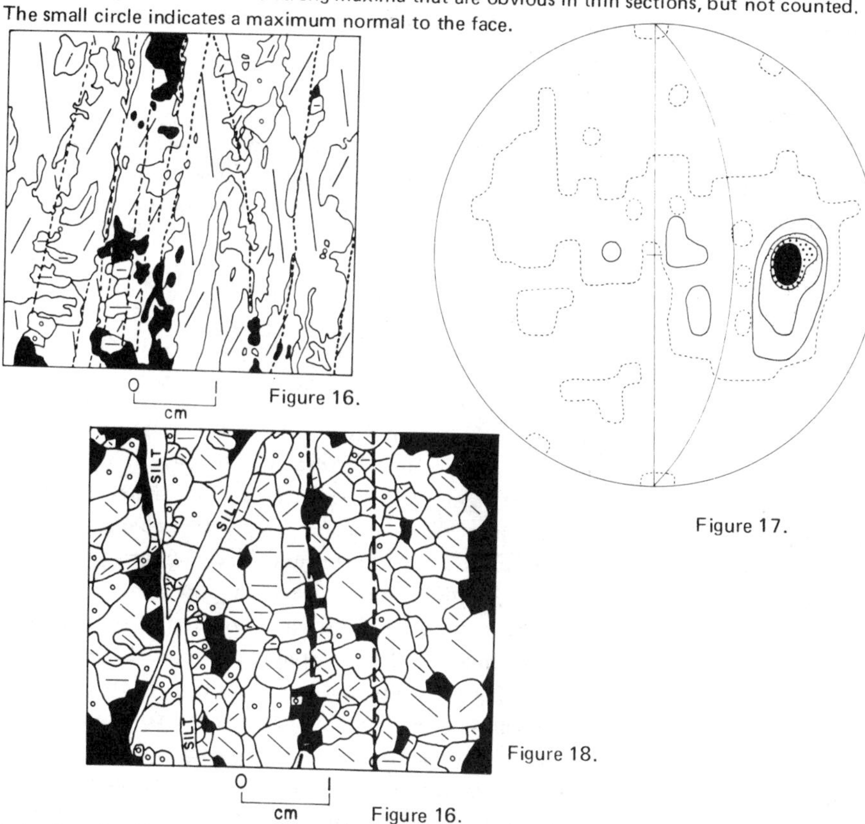

Figure 16.

Figure 17.

Figure 18.

Figure 16.
Grain plot of a thin section cut vertical and normal to the horizontal axis of an ice wedge. Black areas are overlapped grains that were not measured. Dashes are partly healed fractures. Fifteen grains plunge 62 degrees left, 13 plunge 82 degrees right; 7 are horizontal in the thin section; 4 are horizontal and normal to the thin section (circles).

Figure 17.
Fabric diagram of a thin section cut horizontal with the axis of the wedge shown by straight line and the side of the wedge by curved line. Percentages of 162 grains are 12.4-16.1, 9.3-11.8, 6.2-8.7, 3.1-5.6, and 1.2-2.5.

Figure 18.
A grain plot of the same thin section used for figure 17 shows 76 grains plunge 40 degrees right; 46 are within 30 degrees of vertical (circles); 23 plunge 40 degrees left; and 17 are within 30 degrees of horizontal, left-right.

and the angular relationship of alternate grains to adjacent grains is lost unless it is shown in a grain plot (Figs. 19,20). As the angle between c-axes of alternate grains and of adjacent grains becomes larger, distinct maxima appear, but again the alternating alignment of c-axes does not show on a fabric diagram, and a grain plot is necessary. In addition, the relationships of those few grains that are lost on a fabric diagram because of their different orientation and low pecentage are readily discerned in a grain plot. Usually they are the most interesting or important grains in deciphering the history of the ice (Fig. 16). Thus, the grain plots were necessary to correlate the size, shape, optic orientation, and location of any one grain with adjacent grains, layers or structures.

Other lineations

Other lineations from intersections of planar features include those of contraction cracks, contraction cracks with wedge sides, contraction cracks with shear planes, and layers or zones of differently oriented dimensional lineations and lattice lineations. All these lineations occur in some wedges.

ORIGIN OF FABRICS IN ICE WEDGES

Origin of fabrics in ice wedges is only partly understood. This is due in part to insufficient knowledge of ice. The major factors are temperature, pressure, composition, and time. Annual temperature change in the upper 1-3 m of permafrost near Barrow is more than 20° C. Seasonal temperature changes produce both horizontal and vertical stresses in the ground that are greatest where unconsolidated supersaturated (containing more ice than pore space) materials occur. These materials have the same coefficient of thermal expansion as pure ice–about 4-6 times that for most undersaturated soil or rock. A marked temperature drop produces cracking vertically and horizontally; a temperature rise causes expansion with subsequent stresses if the initial cracks change shape or if material has been added to them. Contraction (discussed later) and fabric studies showed that no crack closed completely; material, such as hoar frost, snow, organic matter, or dirt had been added. Thus, stresses sufficient to cause flow of permafrost occur. They result in upturning of horizontal beds and double raised rims adjacent to ice wedges. Horizontal compression produces shear planes adjacent and parallel to the sides of ice wedges, slipping upwards of V–shaped units within wedges. slipping of parallel sheets vertically in *en echelon* arrangement, and bowing of the tops of ice wedges by flow of ice. Apparently all types of movement may occur in some very active wedges, although probably only one type occurs at any one time or place. Thus fracture, granulation, shear, and flow of material are involved on an annual basis accompanied by recrystallization of grains in the ice wedges. Activity decreases with depth as thermal changes are reduced.

Ground temperature changes initiate a chain reaction of differential stresses that are dependent upon the ground composition. The insulating effects of snow and vegetation and the variable thickness of the active layer cause ground temperatures of nearby points in permafrost to vary many degrees Celsius at any one time. Thus, uniform stress can not be expected everywhere, nor can the stresses be repeated in the same direction and amount except fortuitously when like conditions are repeated.

Lineation and foliation of air bubbles and inclusions

Irregular vertical or nearly vertical contraction cracks and horizontal or low-angle contraction cracks within ice wedges are primary planes in which hoar frost produces primary growth fabrics of ice. Such planes once started tend to be self-perpetuating because they are zones of weakness. Fabrics in those planes start in late winter when seedling ice crystals (hoar frost) grow in them, or in late spring and early summer when melt water flows into them and freezes in place. The fabric of the initial ice crystals is dependent on thermal gradients and influence of adjacent ice on which growth may occur. That fabric determines the orientations of initial air bubbles and other inclusions surrounding the hoar crystals or enclosed within the "comb" stucture of the growing crystals in the melt water.

In comb structure, most air bubbles are slightly elongated normal to the sides of the crack regardless of its orientation (Fig. 8). Most cracks roughly parallel the axial plane of the wedge; hence, the bubbles are normal to the horizontal axis and approximately horizontal. If the new ice crystals are near the surface, many bubbles and other inclusions are oriented vertically in the strong vertical thermal gradient. If the hoar crystals grow onto the lattice orientation of an adjacent ice crystal, the lineation of the air bubbles and other inclusions is aligned accordingly.

In an active ice wedge, these initial orientations are modified by shear, flow, or recrystallization shortly after their formation as temperatures rise and compression takes place. The net movement in shear and flow is upward, but the local movement commonly is inclined 30° or more. Preferential transfer of vapor across bubbles in strong thermal gradients causes them to elongate and migrate also. In passive or buried wedges minor temperature gradients and little change in the stress field yield more rounded bubbles and higher concentrations of bubbles and other inclusions at recrystallized grain boundaries.

Many bubble-rich and bubble-poor zones in ice wedges are accountable respectively to contraction cracking which adds air and to shearing which allows it to escape. Bubbles and other inclusions may be aligned and distinctively flattened within the basal planes of individual ice crystals.

Dimensional lineation of ice crystals

Vertical dimensional lineation of notably acicular or prismatic grains of ice presumably results during recrystallization under strong vertical temperature gradients. Horizontal stress possibly aids formation of vertical grains because of the saving in energy of crystallization. Inclined dimensional lineation may result from flow or slight rotation of such elongate grains, although apparently it may also form in that position during recrystallization. The mechanism involved is not known, but probably involves surface energy of the growing crystal in inclined thermal gradients and stress fields. The tendency is for high-energy interfaces to migrate toward lower energy conditions, decreasing the interface area and forming crystals.

I suggest that rapid shear at low temperatures produces fracture and granulation, yielding in places small rectangular grains oriented parallel and normal to the shear planes. However, some small rectangular grains are considered primary growth in contraction cracks that subsequently sheared very slightly, but not enough to destroy the original grains. At higher temperatures, large sutured grains, oriented vertically or within 30°, seem to result from vertical flow and shear accompanied by recrystallization.

Lineations of optic axes

In contraction cracks, seedling crystals may grow with their lattices vertical or at various angles in response to (1) initial orientation of crystals on which they grow or to orientation of nuclei (2) directional feeding (3) directional heat conduction (4) interference of other grains, or (5) some unknown factors. Condensation is easiest at corners or intersections of plane surfaces. Parallel prismatic grains usually grow inward from both walls of a crack. The primary growth fabrics are later altered by recrystallization under shear or compressive stress, and strong thermal gradients. The vertical optic-axis lineation is normal to the cooling surface, presumably under such thermal gradient. Compressive stress directed horizontally probably aids the process.

Horizontal and inclined optic-axis lineations seem to be due to the response of individual crystals to shear stress in which crystal lattices rotate to permit gliding on the basal plane or less commonly to permit rolling around the c-axis. Shear angles within any unit of the ice depend greatly on its temperature and rate of stress.

In passive environments, such as buried wedges, lattice orientations are apparently inherited from a time when the wedges were actively growing. The orientations become more random as recrystallization of equidimensional grains proceeds with a lack of strong thermal gradients and of compressive or shear stress.

GROUND-CONTRACTION STUDIES

Contraction and cracking of the ground during winter are the basis for Leffingwell's (1919) theory of origin of ice wedges, but Taber (1943) holds that these are unimportant. Ice has a linear coefficient of thermal expansion of 50×10^{-6} whereas quartz and most other rock-forming minerals in the soil are $8\text{-}13 \times 10^{-6}$ and average about 10×10^{-6}. The difference of 4-6 times is significant. For example, a block of ice 5 m long on changing temperature 20° C (which is common at Barrow, Alaska) changes length 5 mm, but rock only about 1 mm.

Ground contraction was measured with the aid of vertical rods in five sites where ice-wedge polygons are at different stages of development. They are (1) high-centered polygons in sandy-clayey silt with ice wedges up to 6 m wide (Fig. 21); (2) flat-centered polygons in sandy-clayey silt with ice wedges up to 2 m wide (Fig. 22); (3) flat-centered polygons in sandy gravel with ice wedges up to 1 m wide; (4) low-centered polygons in sandy-clayey silt with ice wedges up to 3 m wide (Fig. 23); and (5) high-centered polygons in sandy gravel with ice wedges up to 3 m wide (Fig. 24). In the five sites, respectively, 27, 13, 13, 27, and 13 solid iron rods, 15.9 mm in diameter and 0.9-1.4 m in length were driven by sledge hammer into the ground during August-September, 1949, when the active layer was at its maximum thickness. The rods penetrated permafrost 20-60 cm. The rods were spaced in rows to enable horizontal contraction to be measured across polygons, across ice wedges, and along ice wedges. Distances between rods was < 1-7 m. All rods were leveled close to the ground surface to support a 30-m tape calibrated by the U.S. Bureau of Standards. Chemical stands supported the longer intervals of the tape. Simple wooden L-shaped "deadmen" supported the ends of the tape which were attached by turnbuckle and spring scale for uniform tension. Readings of intervals between hairline crosses on top of each rod were made to 0.1 mm by means of a 9x magnifier and clip-on scale. All readings were made under calm conditions at night or when thick fog obscured the sun. Air temperature corrections were applied to the tape, and coefficients of thermal expansion of the frozen soil were computed from the intervals, utilizing the contraction from fall to winter accompanying a ground-temperature change of $5-20^{\circ}$ C.

A complete set of readings was made in October-November, 1949, after the active layer was frozen to permafrost. The same configuration was measured in February-March, 1950, after the ground attained its minimum temperature. Measurements of some intervals were made in spring and summer to follow the closing of the contraction cracks and the expansion of the ground. Accuracy checks of the technique and equipment were made at different times.

During the summer each ice-wedge trough in five sites was examined to determine statistically the number that had cracked the previous winter. The amount of ice left in contraction cracks was also recorded to determine the annual increment of ice.

Only some general conclusions resulting from analysis of the many hundreds of readings and observations can be presented here. For example, the data for site 1 show that between early November, 1949, and March, 1950, the high centers of the polygons contracted fairly uniformly, but the low troughs cracked or moved erratically. The active layer of the high centers froze several weeks prior to the active layer of the troughs, because of snow cover and higher water content of the troughs. During freezing, water was drawn to the

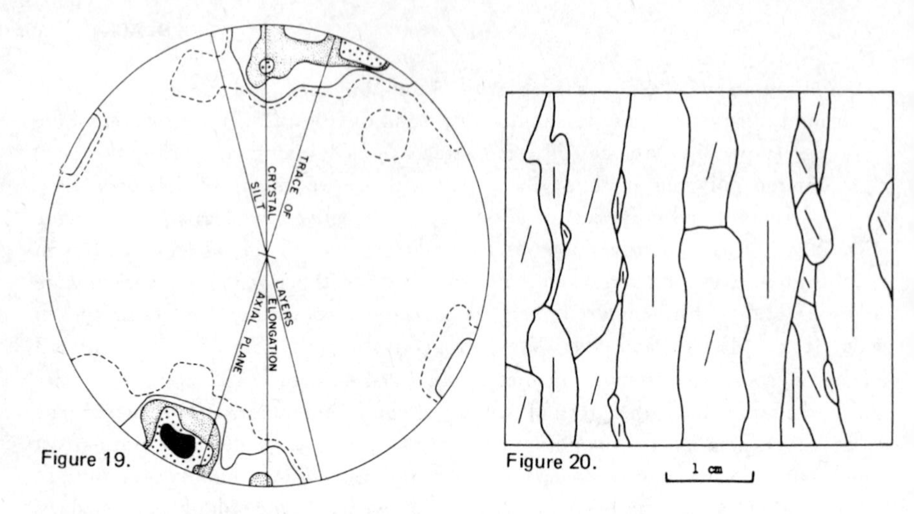

Figure 19.
Fabric diagram of a thin section cut vertical and normal to the horizontal axis of an ice wedge. The maximum containing 18.4-26.7% of 60 grains is 9 degrees from the plane of the thin section and 11 degrees from the vertical.

Figure 20.
Grain plot of part of the same thin section used for Figure 19 shows the alternate plunge of c-axes to left and right which is lost in the fabric diagram.

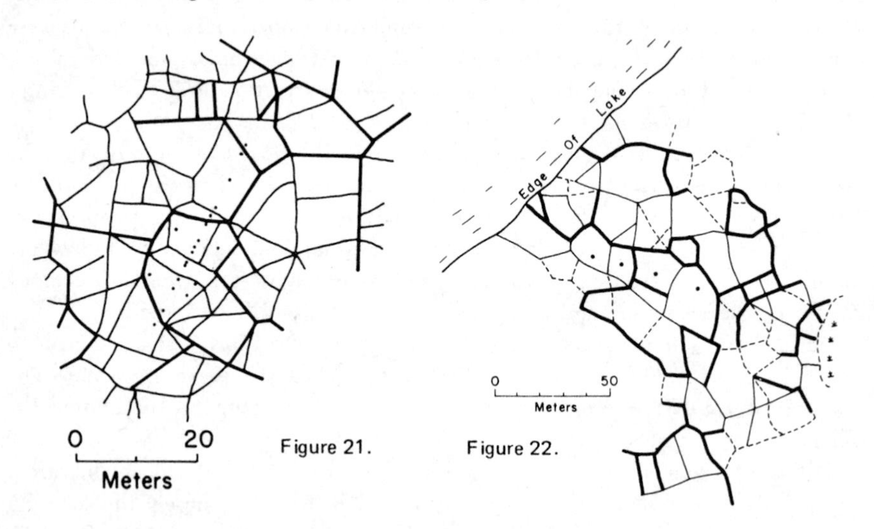

Figure 21.
Sketch of contraction site 1, showing ice wedges that cracked (37% with heavy lines) or did not crack (63% with light lines) and positions of iron rods by dots.

Figure 22.
Sketch of contraction site 2, showing ice wedges that cracked (65% with heavy lines) or did not crack (35% with light lines) and polygons with dots that contain iron control rods. The dashed lines are ice wedges that were not excavated to determine cracking rate.

high centers leaving local parts of the active layer in the troughs undersaturated; other parts of the troughs built up considerable cryostatic pressure before finally freezing solid. Some contraction cracks opened in the mounds before the troughs were entirely frozen. Water and mud from one trough was injected laterally into a horizontal contraction crack that separated one high-centered polygon mound from the ground below.

From March to April, 1950, the ground at site 1 passed through its lowest temperature and began to warm up. However, the contraction measurements showed that the ground was not changing directly with the temperature changes. It responded more slowly; a similar effect was recorded in June. Contraction cracks remained partly open, and other intervals remained smaller to compensate for them.

In spite of somewhat complex results and sometimes contradictory evidence, the general results of the ground-contraction measurements showed conclusively that the ground in all five sites responded to temperature changes more like ice than quartz. The measurements also located contraction cracks under the snow and showed that none of the cracks closed completely the following summer. All cracks added an increment of ice to the permafrost. The measurements suggested further that semipermanent changes took place during contraction which were not reflected during expansion. Coefficients of thermal expansion computed from a drop in temperature could not be duplicated for the same intervals during a rise in temperature. Spring and summer movements were large and irregular.

Growth rate of ice wedges

At contraction sites 1, 2, 4, and 5 respectively, 37, 64, 59, and 46% of the ice wedges examined cracked during the winter of 1949-50. By analogy and theoretical considerations the rate may be expected to be about 70% in site 3. Because of thick loose beach gravel in the active layer, it was not possible to excavate enough wedges in site 3 for a valid statistical sample. In all sites, contraction cracks generally were 3-10 mm wide at the top of permafrost, and decreased in width downward. The widest cracks were in site 1, the area of most relief. They were 7-12 mm wide, and in the active layer, some were as much as 15 mm. The width of a crack seemed to bear no relation to the size of the wedge in which it occurred, nor did the size of a wedge seem to influence the tendency of a wedge to crack. Measurements at the surface showed that cracks were still 2.5-7 mm wide in June, closing somewhat more during July and August.

The amount of hoar frost added to an open contraction crack was about 50% of the void, but it varied widely. The general range was about 25-75%. The amount of ice added from spring melt water was negligible in some cracks and very great in others or in portions of a crack. Hence, the total increment of ice added at any one part of a contraction crack ranged from 100% to almost nothing. It was not possible to determine accurately the annual increments of ice added at depth. At the top of permafrost, the maximum range was 0.4-6.0 mm. On the average the increments seemed to be 1-3 mm, decreasing with depth.

In most parts of the Arctic Coastal Plain the sizes of wedges and polygons vary widely. Generally the largest wedges occur in what may be called the initial or master polygon. Smaller wedges subdivide the large polygons into smaller polygons. A secondary division of polygons is common, and a tertiary division less so. For example, at site 1 primary wedges average 2.5 m in width, secondary wedges 1.5 m, and tertiary wedges about 0.3 m. Apparently as polygons age, subdivision occurs. The number of wedges that crack during the winter decreases from a theoretical maximum of 100%, when the initial polgons first formed, to a minimum during tertiary subdivision. Changes in relief, insulation by snow, distribution of water, vegetation, and other factors cause new contraction cracks in permafrost to open at the expense of former cracks in adjacent ice wedges. Cracking rate is assumed to be a linear function in the early stages, even though infinite complexities exist in nature that could alter the rate. Seemingly, in simple polygons, as in sites 2-4, more than 50% of the ice wedges crack; in compound polygons, as in sites 1 and 5, less than 50% of the ice wedges crack.

To determine the average annual growth rate of an ice wedge it is necessary to know both the average annual increment of ice added and also the annual rate of cracking of that wedge. Increments added to contraction cracks surely vary greatly from year to year, but are expected to approximate the ranges measured for one year. All wedges in a site do not crack at the same annual rate; some may crack each year for an extended period of time, whereas others may not crack at all after an initial growth period. Hence, to determine *average* growth rates of wedges in an area, the average size of large wedges is used, not the largest wedge.

Assuming an average increment of ice of 2 mm is added to each contraction crack, then an annual growth rate of 1 mm per year would occur if a wedge cracked only 50% of the time during its life. Wedges averaging 1 m wide would thus be about 1,000 years old. Using data from the five contraction sites (Fig. 25), percentages of wedges that cracked are plotted against average widths of large wedges. These points fall remarkably close to a line from 100% of cracking (the initiation of wedges in a site) through them toward ice wedges 3.5 m wide. This suggests that large wedges should be about 3.5 m wide when the number of wedges that crack diminishes to zero. However, wedges of 5-7 m in width were seen in two places in the Arctic Coastal Plain, and there is no reason why cracking cannot continue to add ice to the ground even though all mineral matter has been replaced by ice wedges. Thus, the cracking rate may approach, but never become zero, if conditions are conducive to cracking. Wedges probably cannot widen beyond that state wherein their sides are inclined less than 45 degrees, because flow and shear to relieve annual increments of ice will occur entirely within the wedge. Hence,

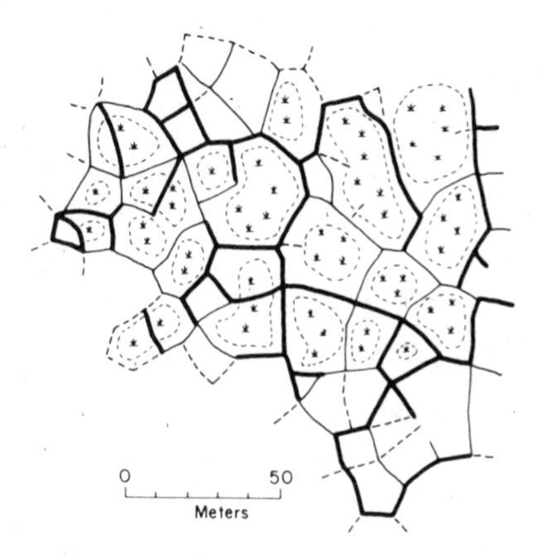

Figure 23.
Sketch of contraction site 3, showing ice wedges that cracked (59% with heavy lines) or did not crack (41% with light lines). Dashed ice wedges were not excavated to determine cracking rate.

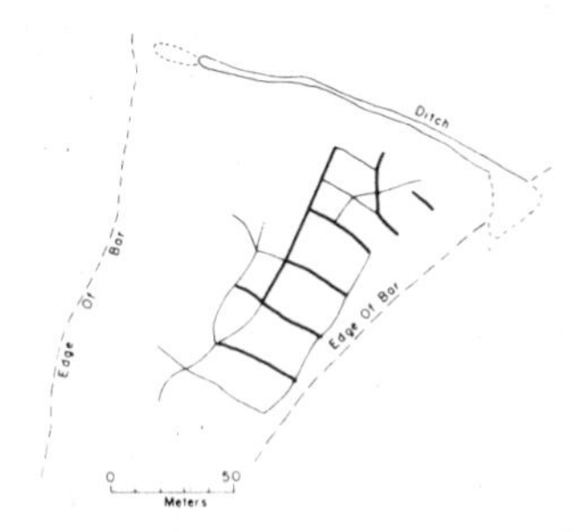

Figure 24
Sketch of contraction site 5, showing ice wedges that cracked (46% with heavy lines) or did not crack (54% with light lines).

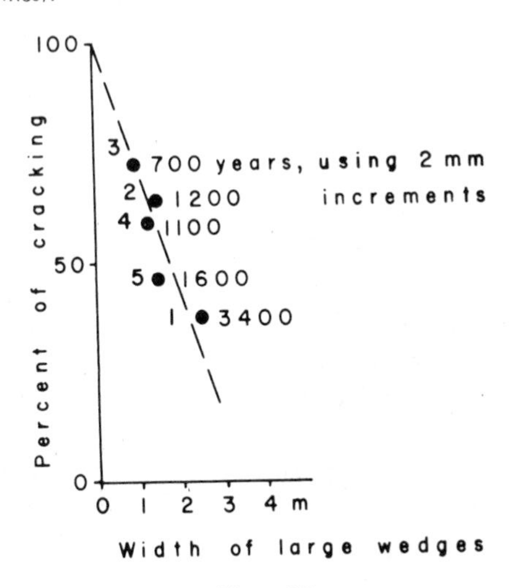

Figure 25
Ages of ice wedges in the five contraction sites as related to percent of cracking and width of large wedges (not the largest).

growth rates of large wedges should diminish to zero with all activity confined to the wedges.

Using widths of wedges and cracking percentages, as indicated in Figure 25, with an average increment of 2 mm, the ages of the five sites are 700- 3,400 years. Such approximations may be modified by knowledge of the thermal coefficients of the ground and the geomorphic expression that controls the amount of ice that may be introduced into the ground each year. For example, site 3 in coarse sand and gravel has a thick active layer and coefficient of thermal expansion slightly less than ice. It should be slightly older than 700 years, as is suggested also by radiocarbon dates (Péwé and Church, 1962). Similarly, sites 1 and 4 should be younger than indicated because of greater than average increments of ice added to contraction cracks in the low troughs. This is substantiated by assuming that the largest wedges in site 1 of 6 m width cracked 100% of the time throughout their lives. At 2 mm per year increase in width they should be 3,000 years old as against 3,400 years old, using the percent of cracking approach. Similarly in the other sites the respective ages of the largest wedges would be 1,000, 500, 1,500 and 1,500 years. Of the three sites on Figure 1, site 5 clearly is oldest and site 4 youngest, more in line with the percent of cracking method. The ages of Figure 25 are reasonably close to those predicted considering that widths of wedges are only approximations from surface inspection. These ages are also in line with radiocarbon dates (Brown and Sellmann, 1973; Carson, 1968; Sellmann and Brown, 1973).

GEOMORPHOLOGY OF ICE-WEDGE POLYGONS

In continuous permafrost ice-wedge polygons occur in practically all unconsolidated materials and locally in bedrock. They are widespread in fine-grained unconsolidated deposits in the discontinuous permafrost zone and are less common in coarse-textured material. They are rare and usually inactive in the sporadic permafrost zone. Casts of former ice wedges have been reported from many areas considered to have had permafrost in the past (Brown and Péwé, 1973).

In the Arctic Coastal Plain of northern Alaska, ice-wedge polygons are essentially ubiquitous and striking landforms. Slopes greater than a few degrees reveal few, if any, complete polygonal patterns on the surface. Nevertheless, excavations and natural exposures revealed ice wedges below the rapidly moving active layer. Modern deltaic and stream floodplain deposits, and recently drained lake beds do not become perennially frozen until at least one winter has passed, and polygons appear later. Thaw lakes that are deeper than 2 m have destroyed the ice wedges that led to the formation of the lakes. Shallower lakes still have apexes of ice wedges beneath their beds. Ice-wedge casts are also seen.

Fabrics of ice wedges, including lineation and foliation of bubbles and other inclusions and of dimensional and lattice orientations of ice grains corroborate our understanding of the origin of ice wedges and of the physical processes and conditions within permafrost. Results of the ground-contraction measurements substantiate that the supersaturated soils of the Arctic Coastal Plain respond like ice to thermal changes. Contraction cracking does occur and increments of ice are added annually from the atmosphere and soil surface to permafrost to form and enlarge ice wedges. The process results from present conditions although many decades or centuries are needed to produce striking polygonal patterns. Temperature measurements and field observations and laboratory measurements of the ice content and lithology of permafrost further substantiate this interpretation of the physical environment and pressures operating within permafrost and the active layer.

The size of polygons varies markedly within a locality and from place to place. Initial polygons are subdivided in centuries to millenia. The larger polygons are found in younger deposits, in coarse-textured deposits, in bedrock, in undersaturated soils, or where temperature changes are not marked. Smaller polygons occur in older deposits, in finer textured material, in supersaturated soils, or where temperature changes are marked. The initial polygons in an area of supersturated fine-grained soils generally are subdivided in time by secondary and tertiary ice wedges into smaller polygons. This results from the abandonment of the cracking process in a wedge, and the release of tension by cracking in a new location. Differential insulation by snow drifts and differences in relief, vegetation, and soil moisture in the active layer bring on the shifts of cracking locations. Because of the wide variability, it is erroneous or misleading to state averages for the sizes of polygons. In the Barrow area, however, primary polygons are commonly 10-100 m in diameter, secondary polygons 4-30 m, and tertiary polygons 1-5 m. Of course, a single polygon in such an area commonly has wedges of two or three generations making up its sides.

Ice-wedge polygons range in shape from squares and elongated parallelograms to irregularly closed or incompletely closed figures. The simplest shape should be a hexagon in an area of infinite dimensions and uniform contraction. However, many irregularities occur. On slopes where gravity is a factor and in small areas where boundary control is effective, such as abandoned stream channels and ponds, contraction cracking forms squares and rectangles.

Primary polygons formed from cracks in a newly exposed surface retain the relief of that surface. Surface slump in the initial crack may produce at most imperceptible relief; thermal expansion in closing the partly filled crack may produce almost imperceptible relief adjacent to the crack. As the polygons age, relief may gradually increase up to 1-2 m in low-centered polygons and to several meters in high-centered ones. Relief is greatest at intersections of wedges, where small ponds commonly form by thaw of the tops of wedges.

These processes produce a double ridge outlining the ice wedge, over which the trough may be at, above, or below the original ground surface. The ridges impede surface runoff, producing ponds in the centers of the polygons, and augmenting erosion and thaw of the subjacent ice in troughs. Relief of low-centered polygons results from upthrusting of material adjacent to the wedges, from preferential growth of vegetation on top of the ridges so produced, and from thawing of the tops of the ice wedges and consequent slumping or settling of the surface Relief of high-centered polygons results primarily from surface thaw of the tops of ice wedges and from erosion of the trough over them. Runoff is generally better integrated than in areas of low-centered polygons. In older subdivided polygons with wide wedges, material is upthrust to the surface adjacent to the wedges in so small a center as to produce a high mound directly. Gullying may remove an entire wedge, leaving the central part of the polygon standing in full relief.

Generally constructive processes vie with destructive thaw and surface erosion in near balance. Many areas of high-centered polygons show on excavation that the centers are the former peat deposits of low-centered polygon ponds.

Figure 26
Air view of patterned ground.

Flat-centered polygons are characteristic of newly-formed polygons, of areas of coarse-grained materials with less susceptibility to plastic flow, and of areas with coefficients of thermal expansion approaching those of rock. The young flat-centered polygons in supersaturated soils may evolve into either low-centered or high-centered polygons depending upon surface slope and runoff, the former in poorly drained and the latter in better drained areas. Low-centered polygons eventually become high centered, but the reverse is not true. In general a few tens of years produce well-defined polygons. Relief of a few centimeters develops within a few centuries at most. Ice-wedge growth with time leads to marked differences in relief, in vegetation, and in the striking polygonal patterns (Fig. 26) so characteristic of the tundra.

During the growth of ice-wedge polygons relief changes at various sites lead to long delays in the freezing of low wet spots in the active layer. Cryostatic pressure locally injects water and mud into horizontal contraction cracks to aid in the formation of various mounds. Or, the early freezing of a bare area permits growth of ice in the active layer there, using the nearby water in the hollows or troughs over ice wedges as a source. A wide variety of mounds associated with the ice-wedge polygons is produced or modified from minor surface irregularities. These form a continuum composed of various percentages of ice, mineral soil, and organic matter (Mackay and Black, 1973).

The addition of ice to the upper part of permafrost by ice- wedge growth over a period estimated to be on the order of at least 5,000-7,000 years (1 mm per year increase in width of wedges 5-7 m wide) makes for an increasingly unstable situation in which much of the upper 2-4 m of permafrost is replaced by ice. Any disruption of the surface vegetation leads to extensive thaw and formation of thaw lakes. No fixed time span for completion of the ice-wedge cycle of growth and thaw can be given, because of the numerous variables that can speed up or delay the growth of wedges or that can initiate thawing or protect the surface locally. However, near Barrow, a time of 3,500-4,000 years since the inception of many oriented lakes has been suggested (Carson, 1968).

Man, of course, can interrupt nature's cycle at any time by his activities. Particularly, he may induce thaw of the massive ice, as in the construction and operation of a hot-oil pipe line, such as that approved for the Prudhoe Bay field.

CONCLUSIONS

Fabrics of ice wedges, ground-contraction measurements, and other field and laboratory observations and measurements on permafrost in the Arctic Coastal Plain of northern Alaska substantiate the contraction theory for origin of ice wedges. Leffingwell's theory needs modification only in slight details. Ice is being added to the upper part of the permafrost, in contraction cracks that occur in polygonal patterns whose size and shape reflect local conditions. On the basis of limited direct measurement and other data an average increase in width of ice wedges is 1 mm per year, suggesting that the larger wedges of the Coastal Plain may grow in 5,000-7,000 years, but much longer time spans may be required in some regions. The buildup of ice through time creates unstable conditions which may lead to the initiation of thermokarst. This process seemingly has been operating for many tens of thousands of years in irregular cycles of different time spans in local areas. Obviously man may initiate a thaw cycle by his activities.

REFERENCES

Bader, H. 1951. Introduction to ice petrofabrics: J. Geol., v. 59, p. 519-536.

Black, R.F. 1951. Eolian deposits of Alaska: Arctic, v. 4, p. 89-111.

Black, R. F. 1952. Polygonal patterns and ground conditions from aerial photographs: Photogrammetric Engineering, v. 18, p. 123-134.

Black, R. F. 1953a. Fabrics of ice wedges: Ph.D. Thesis, Johns Hopkins Univ., 87 p.

Black, R. F. 1953b. Ice wedges and permafrost of the Arctic Coastal Plain of Alaska: U. S. Geol. Survey Open File Report, 788 p. Library copy at the Arctic Research Library, Barrow, Alaska.

Black, R. F. 1954. Permafrost - A review: Geol. Soc. Amer Bull., v. 65, p. 839-856.

Black, R. F. 1957. Some problems in engineering geology caused by permafrost in the Arctic Coastal Plain, northern Alaska: Arctic, v. 10, p. 230-240.

Black, R.F. 1963. Les coins de glace et le gel permanent dans le Nord de l'Alaska: Annales de Geographie, v. 72, p. 257-271.

Black, R. F. 1964. Gubik Formation of Quaternary age in northern Alaska: U.S. Geol. Survey Prof. Paper 302-C, p. 59-91.

Black, R. F. 1969a. Geology, especially geomorphology, of northern Alaska: Arctic, v. 22, p. 283-299.

Black, R. F. 1969b. Thaw depressions and thaw lakes - A review: Biuletyn Peryglacjalny, n. 19, p. 131-150.

Black, R. F. and Barksdale, W. L. 1949. Oriented lakes of northern Alaska: J. Geol., v. 57, p. 105-118.

Britton, M. E. 1957. Vegetation of the arctic tundra: In *Arctic Biology,* H. P. Hansen, ed., Oregon State Univ. Press, p. 26-61.

Brown, J. and Johnson, P. L. 1966. U. S. Army CRREL Topographic Map, Barrow, Alaska: Cold Regions Res. & Eng. Lab Spec. Report 101.

Brown, J. and Sellmann, P. V. 1973. Permafrost and coastal plain history of Arctic Alaska: Arctic Inst. of N. Amer. Tech. Paper No. 25, p. 31-47.

Brown, R. J. E. and Péwé, T. L. 1973. Distribution of permafrost in North America and its relationship to the environment - A review, 1963-1973: In *Permafrost*, Second Intern. Conf., Nat. Acad. Sciences, p. 71-100.

Carson, C. E. 1968. Radiocarbon dating of lacustrine strands in arctic Alaska: Arctic, v. 21, p. 12-26.

Fairbairn, H. W. and Chayes, F. 1949. *Structural petrology of the deformed rocks:* Addision-Wesley Press, 344 p.

Gold, L. W. and Lachenbruch, A. H. 1973. Thermal conditions in permafrost - A review of North American Literature: In *Permafrost,* Second Intern. Conf., Nat. Acad. Sciences, p. 3-25.

Gunn, W. W. 1973. Bibliography of the Naval Arctic Research Laboratory: Arctic Inst. North America Tech. Paper No. 24, 175 p.

Hartwell, A. D. 1973. Classification and relief characteristics of northern Alaska's coastal zone: Arctic, v. 19, p. 162-184.

Hussey, K. M. and Michelson, R. W. 1966. Tundra relief features near Point Barrow, Alaska: Arctic, V. 19, p. 162-184.

Knopf, E. B. and Ingerson, E. 1938. Structural petrology: Geol. Soc. Amer Mem. 6, 270 p.

Lachenbruch, A. H. 1962. Mechanics of thermal contraction cracks and ice-wedge polygons in permafrost: Geol. Soc. Amer. Spec. Paper No. 70, 69 p.

Lachenbruch, A. H. 1966. Contraction theory of ice-wedge polygons - A qualitative discussion: In Proc. Permafrost Intern. Conf., Nat. Acad. Sciences - Nat. Res. Council Pub. 1287, p. 63-71.

Lachenbruch, A. H. 1970. Some estimates of the thermal effects of a heated pipeline in permafrost: U. S. Geol. Survey Circ. 632, 23 p.

Leffingwell, E. de K. 1919. The Canning River region, northern Alaska: U. S. Geol. Survey Prof. Paper 109, 251 p.

Linell, K. A. and Johnston, G. H. 1973. Engineering design and construction in permafrost regions - A review: In *Permafrost,* Second Intern. Conf., Nat. Acad. Sciences, p. 553-575.

Mackay, J. R. and Black, R. F. 1973. Origin, composition, and structure of perennially frozen ground and ground ice - A review: In *Permafrost,* Second Intern. Conf., Nat. Acad. Sciences, p. 185-192.

Nakaya, U. 1956. Properties of single crystals of ice, revealed by internal melting: Snow, Ice and Permafrost Res. Establishment Res. Paper 13, 80 p.

Péwé, T. L. and Church, R. E. 1962. Age of the spit at Barrow, Alaska: Geol. Soc. Amer Bull., v. 73, p. 1287-1292.

Sellmann, P. V. and Brown, J. 1973. Stratigraphy and diagenesis of perennially frozen sediments in the Barrow, Alaska, region: In *Permafrost,* Second Intern. Conf., Nat. Acad. Sciences, p. 171-181.

Shumskii, P. A. 1964. Principles of geocryology, Part 1, General geocryology, Chap. IX, Ground (subsurface) ice: Nat. Res. Council of Canada Tech. Translation 1130, 118 p.

Spetzman, L. A. 1959. Vegetation of the Arctic Slope of Alaska: U. S. Geol. Survey Prof. Paper 302-B, p. 19-58.

Taber, S. 1943. Perennially frozen ground in Alaska - its origin and history: Geol. Soc. Amer Bull., v. 54, p. 1433-1548.

Wahrhaftig, C. 1965. Physiographic divisions of Alaska: U. S. Geol. Survey Prof. Paper 482, 52 p.

Walker, H. J. 1973. Morphology of the North Slope: in M.E. Britton, ed., Arctic Institute of North America Tech. Paper No. 25, p. 49-92.

Washburn, A. L. 1956. Classification of patterned ground and review of suggested origins: Geol. Soc. Amer Bull., v. 67, p. 823-865.

Washburn, A. L. 1973. *Periglacial processes and environments:* St. Martin's Press, 320 p.

Wiggins, I. L. and Thomas, J. H. 1962. A flora of the Alaskan Arctic Slope: Univ, Toronto Press, 425 p.

CHAPTER 10

HISTORY OF GLACIAL LAKE WAWARSING, SOUTHEAST NEW YORK

William B. Heroy *

PREFACE

Although this manuscript has not been previously published other workers of the time were well aware of its existence. Dr. Heroy's work was cited by John Rich in his classic monograph of the Glacial Geology of the Catskills. Furthermore the first half of the work was given to Paul MacClintock for preliminary review. In personal correspondence dated June 25, 1935, Professor MacClintock especially praises the concept and evidence which Dr. Heroy marshalled to document the stagnation theory for landform development in the study region.

The entire original manuscript of more than 100 pages needed to be shortened for this presentation. Unfortunately the numerous photographs that aided to illustrate the text deteriorated so much with age that they are non-reproducable at this time. In spite of these shortcomings I have made every effort to retain as much of the original style as possible, and to do justice to the substantive part of the contribution.

James T. Kirkland
Research Associate
Department of Geological Sciences
State University of New York at Binghamton
Binghamton, New York 13901

* Deceased 1971

INTRODUCTION

One incident during the retreat of the continental ice sheet from southern New York was the formation in Rondout Creek valley, a tributary of the Hudson River, of an extensive glacial lake. The name "Lake Wawarsing" is proposed, from the town of Wawarsing, Ulster County, New York, where the lake was principally situated and its most conspicuous features occur.

THE ICE RETREAT

For a time the ice sheet clung tenaciously to the territory it had captured. Finally the forces that brought the ice sheet into existence became less powerful or the destructive agencies more active. The ice mass was compelled slowly to recede.

Its thickness also diminished and it is possible that higher points of the Catskills may have been exposed as islands before the ice margin had retreated very far northward.

So long as the margin of the ice lay south of the Delaware-Hudson divide the south-flowing streams could remove waters derived from the melting ice. When the retreating ice reached a position that left uncovered, south of the ice margin, north-draining slopes and valleys the opportunity existed for impoundment of small bodies of water between the ice margin and the divide. North-flowing drainages thus dammed by the ice either filled with water until the lowest point in the divide to the south was reached, when they would discharge south over the crest, or they discharged laterally along the ice front. While the ice lingered near the divide numerous small ephemeral lakes were formed in these north-sloping areas. The further northward retreat of the ice permitted these smaller glacial lakes to drain by exposing their normal outlets or permitting them to coalesce into larger marginal glacial lakes.

Such marginal glacial lakes developed in the Delaware drainage basin along the divide between Delaware and Neversink Rivers. Similar lakes formed in the headwaters of Rondout Ck. with outlets either at the ice front or over low cols into the Neversink drainage. At a somewhat later stage waters of the Esopus were diverted into the headwaters of the Rondout and similar glacial lakes in the Esopus drainage were incident to these diversions.

In the Delaware-Rondout valley the retreat of the glacier brought the ice margin to the Summitville divide, freeing the Delaware drainage from ice interference. With the further retreat of the ice north-flowing waters were impounded between the divide and the ice margin having as their only avenue of escape the col at Summitville (Fig. 1). These impounded waters are described as Lake Wawarsing. The glacial lake continued its existence until, through the melting of the barrier ice, eastward drainage channels with elevations lower than the Summitville col were exposed at the north end of Shawangunk Mountain, permitting impounded waters to drain eastward into Wallkill valley thus draining the lake in a comparatively short time.

This series of geologic incidents, with Lake Wawarsing as the most striking feature, is closely interrelated. So that the lake may appear in its proper relationship to other incidents in the late Pleistocene history of this region some events in the retreat of the ice from the Delaware to the Hudson drainage will be described and their resulting physiographic features pointed out.

FEATURES OF THE NEVERSINK-DELAWARE DIVIDE

Across the divide there are a number of low passes, which from their elevation, position and topography, were channels through which glacial waters flowed from the Neversink to the Delaware drainage. From field and topographic map evidence it is believed that most, if not all of these passes, represent spillways over which

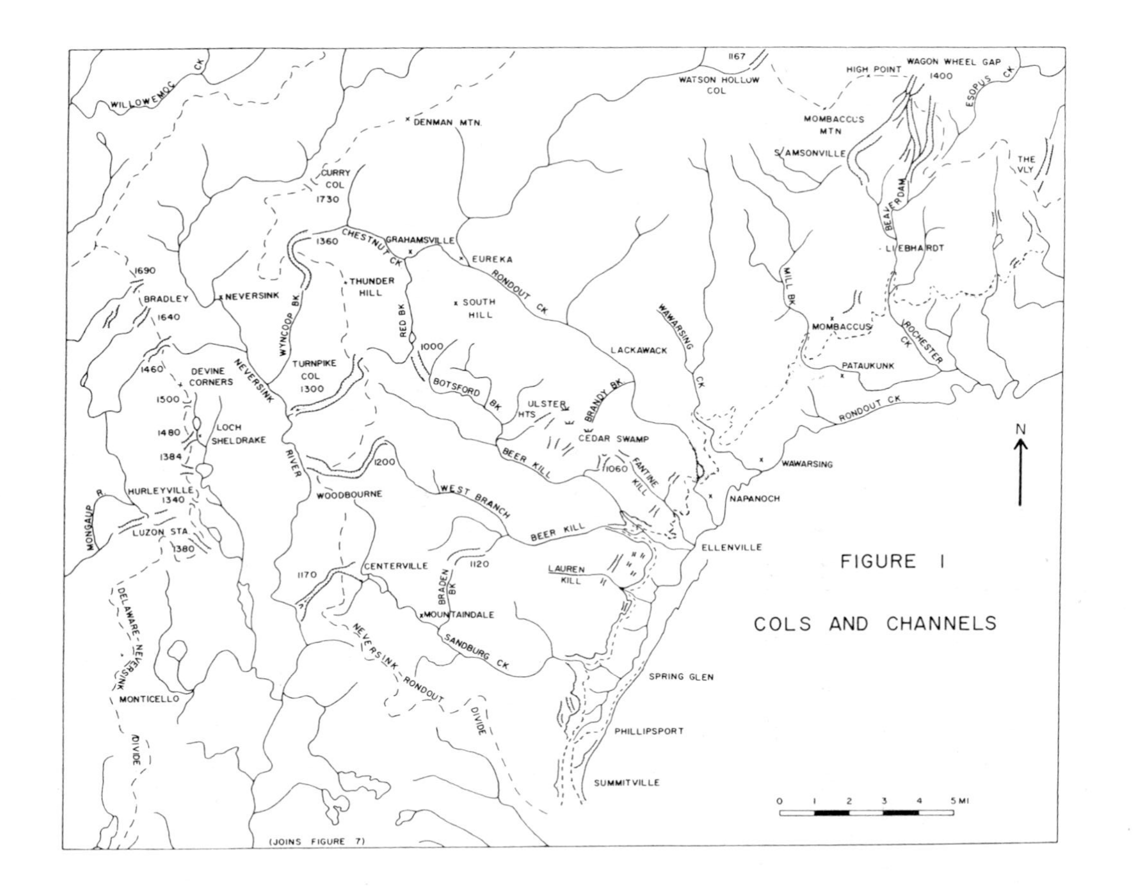

FIGURE I

COLS AND CHANNELS

glacial waters flowed westward into the drainage basins of the Willowemoc and the Mongaup. The following passes or cols are of this character: (1) Bradley col, (2) Devine Corners col, (3) Loch Sheldrake col, (4) Luzon col, and (5) Hartwood col (Fig. 1). The Bradley and Devine Corners channels form a group, the latter of which contained a large stream. The Loch Sheldrake channels are somewhat lower and captured the glacial waters from the Devine Corners col when they became exposed. They could not have been long occupied, for the further melting of the ice would expose the well developed channel at Luzon, thus leading to the abandonment of these higher channels. The Luzon channel was occupied until the retreat or disintegration of the ice exposed a channel near Hartwood, 15 mi south. The size and depth of the latter channel suggest that after it became the course for the escaping waters of the Neversink basin it held them until, through further ice removal, the river was free to occupy its present course to Roses Point.

FEATURES OF THE NEVERSINK-RONDOUT DIVIDE

At their headwaters in the Catskill Mountains the divide between the Rondout and the Neversink is very high. From Balsam Cap to Denman Mountain it is continuously over 2000 ft. The writer knows of no evidence that channels were formed in this part of the divide by glacial waters crossing the divide. To the south the divide is lower, for the most part below 2000 ft, and a number of low cols occur. Some of them are crossed by local drainage channels. These are in succession from north to south: (1) Curry col, (2) Wyncoop Hollow col, (3) Turnpike col, (4) Woodbourne col, (5) Centerville col, and (6) Summitville col (Fig. 1).

LOCALLY IMPOUNDED WATERS IN THE RONDOUT BASIN

It is evident that the withdrawal of the ice front to the north and east of the Delaware-Rondout divide made possible the impounding of water between the divide and the ice margin. As the retirement continued and the ice surrendered larger parts of the Rondout drainage these local bodies of water increased in size, covering the floors of the valleys triburary to the Rondout and discharging south over the divide through the various drainages. The valleys of Chestnut Ck., Red Bk., Sandburg Ck., and various headwater branches of the Beer Kill were temporarily occupied by local glacial waters. As the ice retreated these lakes gradually increased in size until the ice front reached the face of the Catskill escarpment. The confined waters then found exit by passing along the front of the escarpment at the margin of the ice. As these new and lower outlets became available the glacial waters constantly sought the lowest point of discharge, so that complete drainage of the local lakes in the Rondout soon followed.

Sandburg Valley

The level of waters confined locally in Sandburg valley was determined by the elevation of the Centerville channel, about 1170 ft. This provided opportunity for the deposit of alluvial materials in the valley north of Woods lake and probably elsewhere in the upper valley. The time during which the Centerville outlet functioned may be termed the "Centerville Stage" (Fig. 2). Withthe withdrawal of the ice front to the mouth of Sandburg valley lower outlets than the Centerville channel became accessible. The first point at which such discharge was possible was along the face of the hill 2 mi northwest of Phillipsport. Close examination of this hillside reveals evidence of ice-contact channels. These channels functioned as

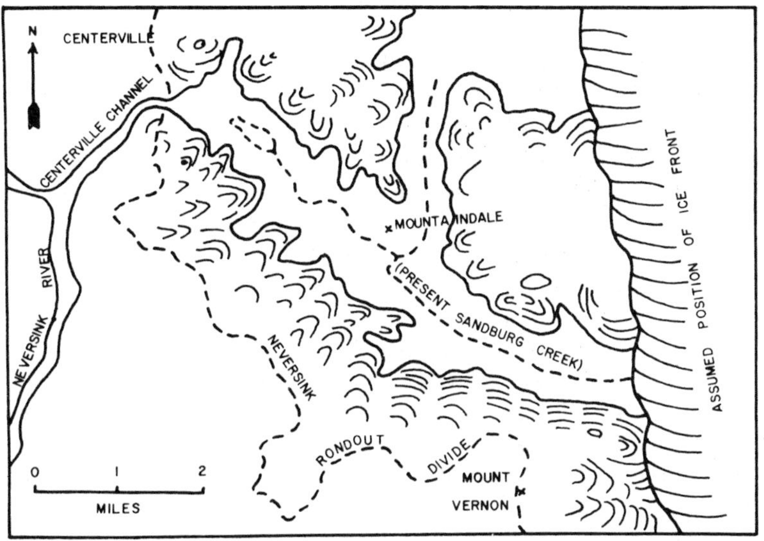

Figure 2. Maximum extension of local glacial waters in the Sandburg valley, "Centerville Stage".

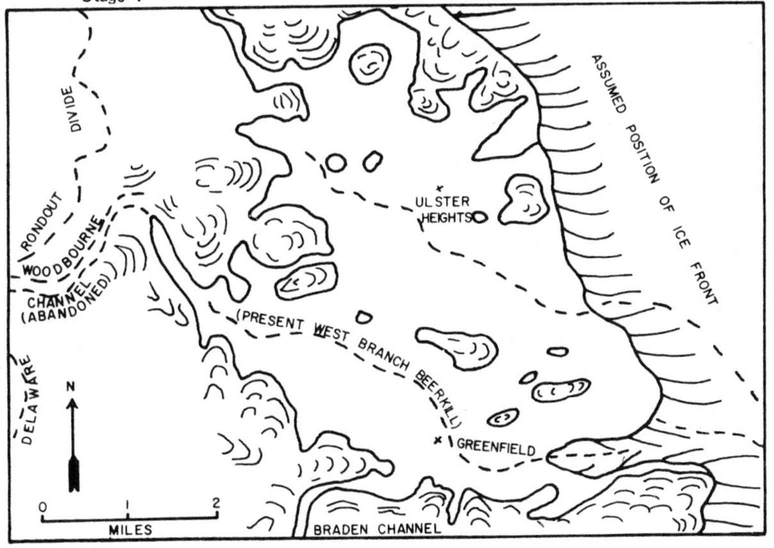

Figure 3. Maxiumum extension of local glacial waters in the Beerkill valley, "Braden Stage".

spillways and thus controlled the successive stages in the lowering of the waters in Sandburg valley. They carried waters for considerable periods of time as indicated by the presence in Sandburg valley of well-marked lakeshore deposits which took a number of years to accumulate.

Beer Kill Valley

The lowest point along that part of the Delaware–Rondout divide which borders on the present drainage basin of the Beer Kill is at the Woodbourne col. The recession of the ice margin in the upper Beer Kill valley led to the impounding of glacial waters between the divide and the ice margin; the elevation was controlled by the height of the 1200 ft col. During this stage, the "Woodbourne Stage", the ice front was probably not very far east of the Woodbourne channel and there was, consequently, no very extensive body of water in the upper Beer Kill basin. Eventually Braden Bk. outlet was uncovered - "Braden Stage" (Fig. 3). When this area was freed of ice the level of the lake in the upper Beer Kill basin was lowered to that of the waters held in upper Sandburg valley. The Woodbourne channel was abandoned and the Centerville channel became the route for west-flowing glacial waters. Thus, during the Centerville stage the waters held in the upper part of the Beer Kill and Sandburg basins were united in a single glacial lake. This lake ended with the opening of lower outlets along the Catskill front west of Phillipsport. When this occurred the waters in Sandburg valley were lowered some 300 ft. The waters in Beer Kill valley did not have a similar subsidence because the elevation of the divide at the head of Braden Bk. is about 1120 ft and the col acted as a spillway to retain the waters in Beer Kill basin at a higher level.

It is clear that when the level of the Sandburg glacial lake was lowered below the level of the Centerville outlet the lower end of Beer Kill valley was still dammed by the ice. The waters impounded in it were held to an elevation of about 1120 ft by Braden col, which acted as a spillway. The outflow thus formed the channel and continued down the present Braden valley to the Sandburg. The existence of this lake was terminated by the opening of channels along the ice front about 2 mi west of Ellenville, which permitted waters to drain south along the margin of the ice in Rondout valley.

Middle and Upper Rondout Valley

The retreat of ice from Rondout valley above Napanoch gave opportunity for the development of local bodies of water between the ice front and the divide in the same manner as in Sandburg and Beer Kill drainages. Because of the topography of the upper Rondout basin the post-glacial history is somewhat more complicated than in the two parallel valleys to the southwest. the upper Rondout, above Eureka, has a southwest course, the stream turning a right angle near that village. At Eureka, Chestnut Ck. flows into it and a mile above Eureka, Red Bk. flows into Chestnut Ck. from the south.

With the withdrawal of the ice front east of the divide the first event was the formation of a small glacial lake in Chestnut Ck. valley. This lake, controlled by Wyncoop channel, was short lived, for the uncovering of the Turnpike channel at the head of Red Bk. must have followed very quickly—"Turnpike Stage" (Fig. 4). The Turnpike channel is much more strongly marked and in comparison functioned for a longer period.

Further retreat of the ice exposed the divide between Beer Kill and Rondout drainages in the vicinity of the Sullivan-Ulster county line, 3 mi northeast of Ulster Heights. On the south slope of South Hill the divide between the head of Red Bk. and that of Botsford Bk. is quite low, slightly over 1000 ft. When ice exposed the hillslope west of this pass at an elevation of 1300 ft the waters held in Red Bk. valley could flow south between the hillslope and the margin of the ice discharging through the Woodbourne channel into the Neversink. The opening of this route caused the level of impounded waters to fall below that of the Turnpike channel, which was accordingly abandoned. Along the course of the south-flowing waters two minor channels were formed where the ice front rested against hill-spurs and the waters were forced across cols behind them. One of these minor channels just north of the east end of the Woodbourne channel has an elevation of about 1240 ft. This channel came into use about the time glacial waters first had access to the Woodbourne channel. Its elevation is about the same as the original col across the divide in the location of the Woodbourne channel, before it was cut to its present depth. If this assumption is correct, then the total amount of cutting done in the formation of the Woodbourne channel is fully 120 ft.

Figure 5 shows the drainage relations in this locality just as the Woodbourne channel came into use and the route followed by glacial waters flowing out of Red Bk. valley. The exact position of the ice margin is open to discussion but clearly must have been aligned north-south and must have approximated the position shown to make possible the formation of the indicated drainage channels. A further recession cleared the col at the head of Botsford Bk. uniting the lake in the Red Bk.-Chestnut Bk. area with that in the upper Beer Kill valley. The level of the water was determined by the elevation of the Woodbourne channel.

The release, as previously noted, of waters held in the Sandburg valley made the Braden channel the control point for waters in the Beer Kill valley. The divide at the head of Red Bk. has approximately the same elevation (1110 ft) as the Braden channel, so that during the Braden Stage a continuous body of water occupied the upper Beer Kill and Red Bk. valleys. With the opening of outlets in the lower Beer Kill valley the part of this water lying south of Red Bk. col was drained. The ice had not yet melted from the middle Rondout valley and a dam still existed between Lackawack and Eureka, holding the water in Red Bk. and Chestnut Bk. valleys. The lowest outlet for this water was the col at the head of Red Bk. valley. A channel leading across the col into the head of Botsford Bk. now marks the route by which at that stage the water discharged. The channel is not deeply cut and the period during which the waters followed this route was probably short. This event, the "Botsford Stage" (Fig. 6), was terminated by the retreat of ice in the middle Rondout valley sufficiently far to permit the waters held in the upper Rondout drainage area to find an outlet into the valley of lower Beer Kill.

Just south of Lackawack a small stream, Brandy Bk., flows into the Rondout from the southwest. This stream heads in two extensive areas of swamp land

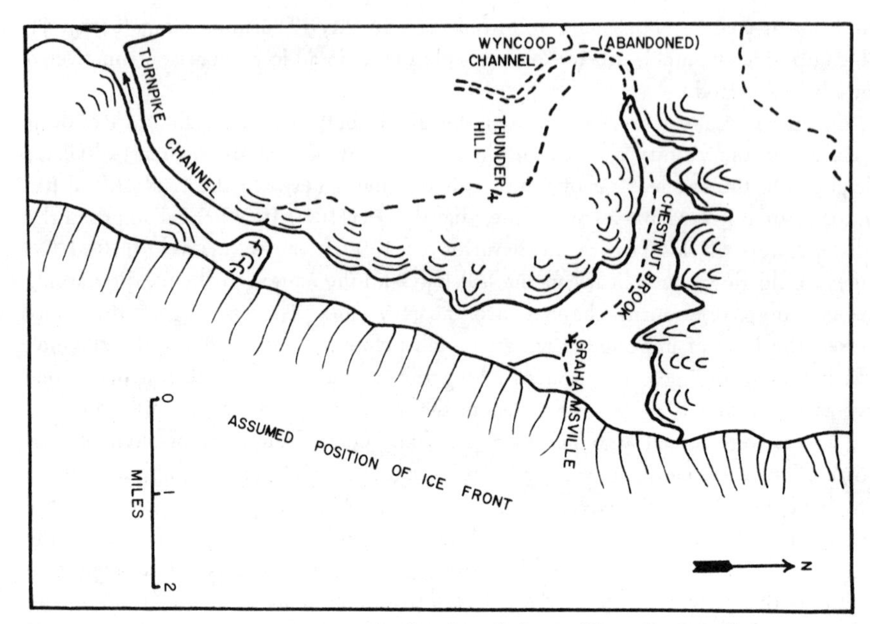

Figure 4. Local glacial waters in the upper Rondout drainage, "Turnpike Stage".

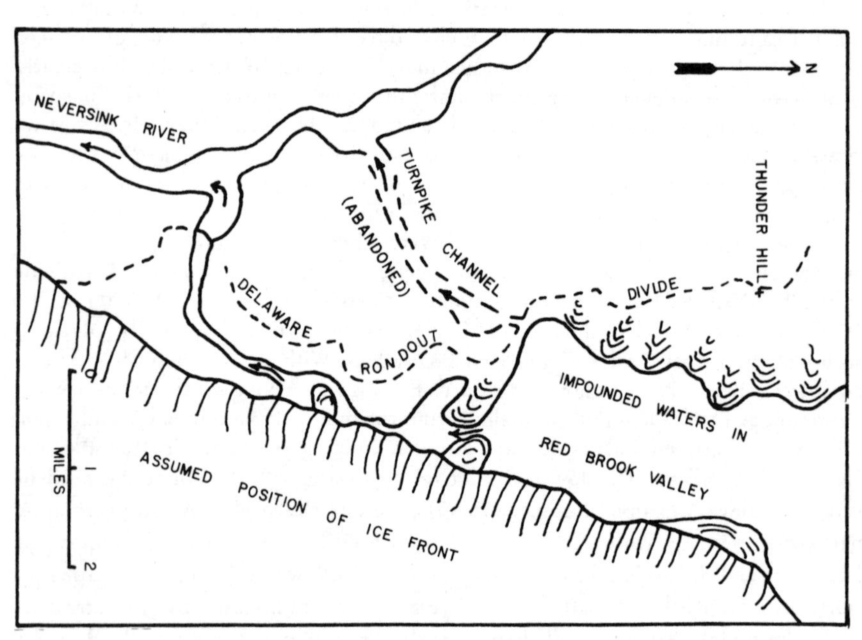

Figure 5. Route of discharge of glacial waters from Red Bk. valley through Woodbourne channel to Neversink River, "Woodbourne Stage".

between Beer Kill and Rondout valleys. The west edge of the more westerly of these swamps appears to have been the route which first became available for discharge of waters held in Rondout valley across the Beer Kill-Rondout divide. The outflow was first discharged to the southwest along the small valley which passes through the village of Ulster Heights. As ice continued its retreat channels somewhat farther east were opened. At the southeast end of the eastern swamp a sharply defined channel heads in two forks and extends nearly 1 mi southwest toward the Beer Kill. This channel marks the final stage in the diversion of glacial waters from Rondout to Beer Kill drainage. The level of water in Beer Kill valley by this time was lower than 1000 ft and the ice had retreated eastward practically to the Catskill front.

At the east end of the swamp a scourway leads toward the head of Fantine Kill valley. This scourway represents a later stage in the post-glacial drainage history, the "Fantine Kill Stage". The water thus diverted from the Rondout to the Fantine Kill valley was evidently held in the latter valley by the ice and found its way to the southwest along the ice margin. The hill-spur between the Beer Kill and Fantine Kill is crossed by a high swampy col which clearly marks the position of a glacial stream. Its elevation is slightly over 800 ft. A similar channel behind the spur 1 mi west of Napanoch was formed by discharge of waters form Rondout valley along the ice front when its margin stood just west of Napanoch. This channel represents the next stage after abandonment of the scourway at the head of Fantine Kill and is the lowest outlet that carried glacial waters locally impounded in Roundout valley.

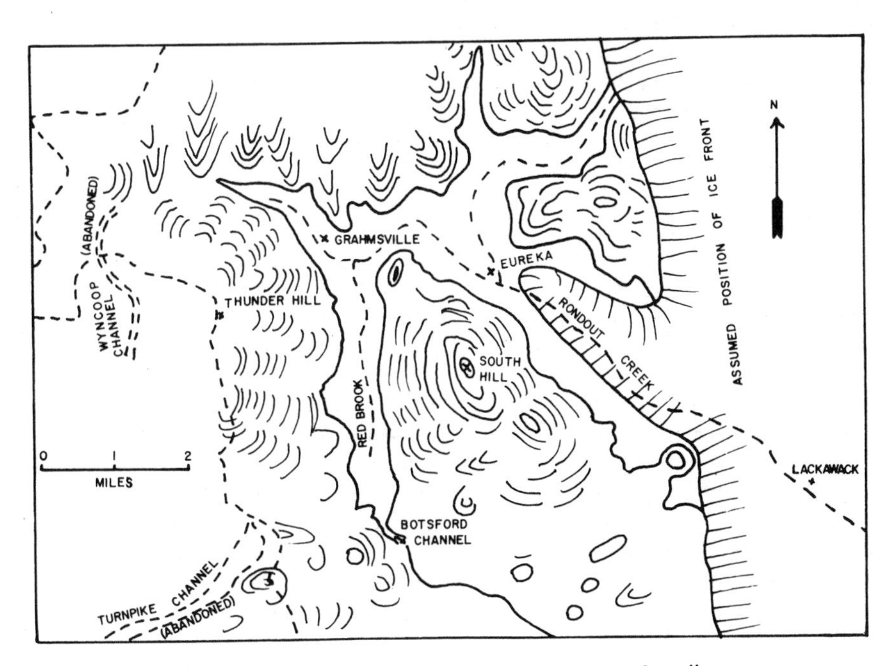

Figure 6. Drainage relations in the upper Rondout Valley, "Botsford Stage".

THE BEGINNINGS OF LAKE WAWARSING

With the withdrawal of the ice margin, at first to the eastern edge of the Catskills and then into Rondout valley, the local lakes in the tributary valleys of the region gradually became extinct. The position of the ice front was such that northern tributary valleys held local lakes to a later period than the Sandburg valley. During this stage of ice retreat a change occurred in the character of the ice occupying this region. The form of the main continental ice sheet was apparently modified from that of a lobe or salient until it approached that of a tongue. The ice shape was more in the nature of a triangular projection extending outward from the main ice sheet rather than the valley glacier type.

It was against the western side of this ice mass that the channels and scourways along the Catskill front were formed.. At one time this tongue-like extension must have extended down the Neversink valley toward Port Jervis. Later it melted back to the Summitville divide and further dissolution and withdrawal to the northeast side of the divide made possible the first accumulation of glacial waters in the valley proper. At this stage the history of Lake Wawarsing began.

The Summitville and Phillipsport Cols

The divide between the Delaware (Neversink) and Rondout drainages crossed the Delaware-Rondout structural valley at Summitville, the lowest point having an elevation of 540 ft. This col was the highest point on the old Delaware and Hudson canal and for this reason the village was named Summitville, although topographically it is in the bottom of the valley.

The Summitville col was the control which established the elevation of the water surface of Lake Wawarsing (Fig. 1). Unlike previous cols, the Summitville col was not crossed by a single glacial channel but had a more complex history.

An important factor is the deposition on the valley floor of a large mass of glacial debris. The most prominent feature of this glacial filling is a hill, "central hill", 0.5 mi north of the village of Summitville and extending northward beyond Phillipsport. It is an ice-margin deposit and has two large depressions on its surface as well as many smaller pits. One of the large depressions is about 50 ft deep, funnel-shaped, with many boulders and formed by the melting of an ice block. Good exposures along the sides of the hill show it to be composed of sand and gravel with many large boulders. Such features indicate this hill is a morainic deposit that choked the structural valley at its narrowest point and obscured preexisting topography.

A broad terrace extends from the southern end of central hill southward about 1 mi. The terrace is continuous with the floor of the eastern valley. The eastern margin of the terrace is along the foot of the mountain, and its western margin forms an escarpment on the east side of the western valley. The terrace area is indicated on Figure 1 by stippling. It has an elevation of about 620 ft, about 80 ft higher than the floor of the valley to the west. Along the west side of the valley opposite Summitville there is a much narrower terrace at the same elevation, also shown on Figure 1 by stippling. These remnants indicate an outwash plain formerly existed in the valley south of central hill. Into this outwash plain the southern end of the western valley has been eroded to a depth of 80 ft. Both valleys on either side of central hill are channels through which glacial waters discharged from Rondout drainage into Delaware drainage. The eastern and higher channel will be called the "Summitville channel" and the western and lower as the "Phillipsport channel". The stage in the history of glacial lake Wawarsing when its waters discharged through the Summitville channel will be referred to as the "Summitville Stage" and the stage of discharge through the Phillipsport channel as the "Phillipsport Stage". The Summitville channel and the shoreline of the Summitville Stage are indicated on Figure 1 by a solid line, while a dashed line is used to indicate the Phillipsport channel and the shoreline of the Phillipsport Stage.

As the Summitville channel has a spillway elevation of 620 ft, with a corresponding shoreline elevation of about 640 ft, it is clear that during the Summitville Stage the waters of Lake Wawarsing stood at an elevation considerably higher than during the Phillipsport Stage. Because these channels formed during retreat of the ice sheet, the tendency of the impounded waters was to find a lower outlet, thus, the higher channel corresponds to an earlier event. But in order that the Summitville channel could be occupied by glacial waters the Phillipsport channel must have been blocked to such a depth that the outflow was compelled to seek the higher outlet. This is indicated by the fact that the waters discharging through the Summitville channel were able to grade the main valley to the southwest and to form the previously described outwash apron.

With the retreat of the ice northward and with the deposition of the central morainic hill, there was left in the western valley a block or tongue of ice filling it to a higher level than any ice filling in the valley east of the hill. Thus the outflow of glacial waters was first established in the eastern or Summitville channel. That the Summitville channel was not entirely ice-free at this time is suggested by the presence of a kettle hole in the middle of the channel. As the ice barrier in the western valley melted an outlet was formed through which the discharging waters found a lower outlet than the Summitville channel. Once opened, the erosional action of the large volume of water which was being discharged from Lake Wawarsing rapidly cleared a broad passage giving the Phillipsport channel its present size and form. The diversion of the waters from the Summitville channel to the Phillipsport channel must have been abrupt, for a gradual transition would have resulted in the channeling of the outwash apron at the south end of the Summitville channel, whereas this apron is in fact undissected.

The complete downcutting of the Phillipsport channel must have resulted in a 80 ft lowering of the lake level in the valley to the north. The Phillipsport channel is much broader than the Summitville channel suggesting that it either carried a larger volume of water or functioned for a longer time or both. As it exists today, the Phillipsport channel is a fine example of a glacial channel. Its banks are smooth and steep. Its western wall is composed largely of shale while its eastern bank is composed of glacial debris. Its floor has been graded and shows no evidence of ice block topography. It was evidently excavated in morainic material and no bedrock shows in its floor. The elevation of the present floor of the channel is about 540 ft and the depth of water in the channel must have been fully 20 ft, for the corresponding shore features have an elevation of about 560 ft.

SHORE FEATURES OF LAKE WAWARSING

Lake Wawarsing came into existence after the clearing of glacial ice from the Summitville col. As the ice receded the Lake gradually extended northward, its northeastern shore being determined by the ice front. Ultimately the recession of the ice freed an outlet across the northern end of Shawangunk Mountain lower than the Phillipsport channel and closed the Phillipsport Stage of the lake's history. The highest outlet thus opened is located 1.5 mi southeast of High Falls village.

The southeast shoreline of the lake was along the slope of Shawangunk Mountain. This slope is steep and its streams have small catchment basins. The drainage area of Lake Wawarsing on the southeast side was small. The volume of sediment which entered the lake from this side was comparatively small and shore features correspondingly inconspicuous. On the other hand, streams entering the lake from the northwest had the advantage of much larger drainage areas. Some of them even carrying waters from outside the Rondout basin, and thereby acquired much larger loads. The deposition by these streams, combined with the action of the waters of the lake itself, led to the development of large and conspicuous shore features on the northwest shore. The following are the most prominent.

Sandburg Delta

A series of shore features comprising the Sandburg delta occur at the northern end of the Phillipsort channel near the village of Phillipsport and extend 3 mi along the northwest side of the valley. They are built from materials transported into the lake by the glacial Sandburg Ck. Two terrace levels at 600 and 700 ft on the delta correspond to the Phillipsport and Summitville Stages.

The volume of material in the Sandburg delta is of interest. Its area is approximately 1200 acres and its average thickness exceeds 100 ft. Its volume would then be at least 600,000,000 cu yd. The area and volume of this delta are disproportionate to the size of the present drainage basin of Sandburg Ck, supporting the interpretation that glacial discharge in the Sandburg area was augmented by waters from outside its present basin.

Ellenville Delta

Some minor ice margin and shore deposits can be observed along the northwest side of the Sandburg valley in the vicinity of the Lauren Kill. The next prominent shoreline feature is the delta which extends along the base of the Catskill front from a point directly west of Ellenville northward beyond the Fantine Kill, a distance of almost 2 mi. So thick is this deposit that there is practically no break in its continuity between Ellenville and Napanoch. Just north of the Fantine Kill the band of material is thinner. This point is arbitrarily used as a northward limit for the Ellenville delta, although shoreline deposits extend along the edge of the valley north to Napanoch without a break (Fig. 1).

South of the Beer Kill the delta deposits are restricted to a narrow terrace which forms a part of the hill directly west of Ellenville. This deposit seems to correspond in elevation with the Phillipsport Stage. The amount of alluvial material deposited southwest of the Beer Kill is surprisingly small when compared with deposits north of the stream. In contrast to this meager accumulation a broad terrace is developed north of the Beer Kill. The inner margin of this terrace approximates the 600 ft contour. The outer margin of the terrace extends east to the road giving the delta a width over 0.5 mi. Near the Fantine Kill the terrace has been eroded, and northeast of the Fantine Kill the spur capped by the 600 ft contour is also part of the delta. The terraces extend along both sides of the Fantine Kill valley (Fig. 1). All of these deposits north of the Beer Kill are capped by a single terrace with an elevation of about 600 ft, rather than two terraces as in the Sandburg delta.

Along the face of the spur between the two forks of the Beer Kill there is a series of delta deposits, the highest of these at an elevation of about 760 ft extends about 1 mi westward along the north side of the valley of the west fork of the Beer Kill (Fig. 1). The deposit does not extend northward into the valley of the north fork suggesting that deposits formed at this elevation in the gorge of the north fork were removed by the torrents which poured down this gorge a little later in the glacial history. The principal terrace in this region corresponds to a shoreline elevation of about 640 ft. There is a beautiful little bench of tilled land on the spur between the forks of the of the Beer Kill at this level and just across the north fork another terrace at the same level can be traced eastward nearly 1 mi. On the face of the spur between the forks of the Beer Kill there is still a lower shoreline with an elevation of 560 ft. This feature is less conspicuous than the terrace above it and does not form a broad terrace. The lower level does correspond to the upper surface of the majority of the Ellenville delta. In the succession of shorelines and corresponding terraces found on the slope between the forks of the Beer Kill is preserved the record of the lowering of the level of Lake Wawarsing from the Summitville to the Phillipsport Stage.

Napanoch Delta

Rondout Ck. flowed into Lake Wawarsing at Napanoch. Between the villages of Napanoch and Wawarsing it formed a large delta, perhaps the largest and most conspicuous shore feature of the lake, stippled pattern (Fig. 1). From its inner margin where it is in contact with the hard rocks of the Catskills to its outer margin along Rondout Ck. it extends 1.5 mi and about the same length from Rondout to Wawarsing Ck. Its inner margin is slightly above 600 ft and it is capped by a level terrace which has a maximum width of about 1 mi.

As originally deposited the delta was doubtless continuous from the northern margin of the Ellenville delta eastward beyond Wawarsing. It probably included all of the isolated hill occupying the valley floor east of the village. Erosion since the glacial period has dissected it and removed much of the original material. A ridge which extends from 1 mi west of Wawarsing northeast toward Mombaccus has near its summit several swampy cols which were channels for glacial waters flowing at higher levels than Lake Wawarsing. The shoreline of the lake followed the 600 ft contour along the southern slope of this ridge. Much of the slope is wooded but the shoreline is clearly marked along most of the ridge by a narrow bar of alluvial material as well as evidence of wave action on some of the bedrock exposed along the ridge.

Pataukunk and Rochester Deltas

A mile northwest of the village of Pataukunk, Mill Bk. passes from the Catskills to the Rondout valley through a gorge-like valley. On both sides of the mouth of the gorge extensive delta deposits have been developed along the Lake Wawarsing shoreline (Fig. 1).

INFLOW FROM THE ESOPUS BASIN

During the period of glacial retreat from the Rondout valley the volume of water occupying the Rondout Basin was greatly augmented by waters of similar origin which passed over the divide from the basin of Esopus Ck. The divide between the Esopus and the Neversink is only about 10 mi in length and is high. The Winnisook gap is the only comparatively low point between the Neversink and Esopus basins and is about 600 ft lower than the divide on either side of it.

Eastward between the head of the Rondout and the head of Bush Kill on the Esopus side there is a gap, Watson Hollow (Fig. 1). The divide again rises and passes at higher elevations eastward across Mombaccus Mountain and High Point, and is the southeast salient of the Catskill Mountains. It then drops abruptly and continues eastward at elevations of 700-900 ft for a distance of 7 mi into the Delaware-Rondout structural valley.

As a general statement it is probably true that at the time of its recession, the ice remained longer in the Esopus basin than in the more southerly basins of the Neversink and Rondout. While these latter valleys were being cleared the ice must have remained high in the valleys and gorges of the Esopus basin. The evidence shows, however, that before the ice had completely left Rondout valley it had also cleared much of the Esopus-Rondout divide, and, lying across its normal outlet into the Hudson valley, also impounded waters, not being able to follow their normal channels, were forced across divides into neighboring basins.

For a short time at least, glacial waters of the Esopus basin flowed across the divide at Winnisook into the headwaters of the Neversink. As soon as the col at the head of Bush Kill was freed of ice the waters left the Winnisook col and used the Watson Hollow outlet, which is about 1000 ft lower than Winnisook. Passing over thw Watson Hollow col waters would reach the head of the Rondout and at that stage would be added to the glacial waters of the Rondout.

The Watson Hollow channel continued to carry the glacial waters so long as the ice front lay high against the east slopes of South Mountain and High Point. With further ice retreat, the glacial waters found their way along the east flank of High Point to its eastern spur. The enormous volume of glacial waters held in the Esopus valley began cutting across the spur at an elevation of about 1650 ft creating an immense cleft, 250 ft deep, 650 ft wide at the top, and extending 0.5 mi across the eastern spur of the mountain. Thus was created Wagon Wheel gap, a feature so conspicuous that it can be seen for many miles north and south. It is an outstanding example of a glacial spillway and is *one of the geological marvels of the Empire State.*

The bottom channel elevation where it is cut across the crest of the mountain is 1400 ft. The north slope of High Point is somewhat more abrupt than the south slope in the vicinity of the gap so that the narrowest and steepest part of the gap is toward the north. Here the steep and water- scoured walls excavated in the hard Catskill flagstones form a wonderfully impressive example of the tremendous force of moving waters. Talus blocks, fallen from the steep wallls have somewhat filled the bottom of the channel which was therefore deeper when the waters were discharging through it.

A series of successive channels were exploited by waters from the Esopus basin. These include Samsonville, Mombaccus, Beaverdam and Krumbille channels.

Still further retreat of the ice exposed the Vly channel (Fig. 1), the lowest channel through which Esopus waters discharged into Lake Wawarsing. With further retreat, waters were discharged from the Esopus basin through a channel which passes through the village of Atwood but by this time Lake Wawarsing had ceased to exist.

CLOSING STAGES OF LAKE WAWARSING

When the ice sheet had receded so that its front held a line across Rondout valley from a point near Kripplebush to a point on the crest of Shawangunk Mountain 1.5 mi east of High Falls, Lake Wawarsing reached its maximum extent. At this stage the ice acted as a dam 5.5 mi long, sufficiently impervious to impound water in the valley to a depth of 350 ft. From this dam the lake extended up valley to a distance of over 20 mi. It had a shoreline about 70 mi and covered an area about 60 mi. Including the Esopus drainage which fed into it, the total drainage must have been at least 500 sq mi.

When the recession of the ice northward along the ridge of Shawangunk Mountain exposed the 600 ft contour waters of Lake Wawarsing found an outlet eastward into Wallkill Valley. With further retreat a series of channels were exploited with elevations between 600 and 400 ft. As the ice receded northward and entirely freed the northern spur of Shawangank Mountain glacial waters in the Wallkill, Rondout and Esopus vallleys coalesced at about the 400 ft level to form another glacial lake, probably even larger in extent than Lake Wawarsing.

With this episode the history of Lake Wawarsing was brought to a close. Its remarkable assemblage of shore features and inlet and outlet channels make it a most interesting example of the "fossil" lakes associated with the glacial period.

CHAPTER 11

THE RELATION OF RIVER TERRACE FORMATION TO GLACIATION IN THE SHOSHONE RIVER BASIN, WESTERN WYOMING

John H. Moss

ABSTRACT

A strong difference of opinion exists regarding the relation of river terraces to glaciation in the Rocky Mountains. The fundamental questions are: did the ice override previously formed terraces or is there a genetic relationship between glaciation and the terraces, many of which extend downstream for many miles from glaciated mountains to adjoining basins? If the terrace formation is climatically controlled, an associated question is: when in the glacial cycle did the river aggrade and when did it degrade?

Because of the impossibility of tracing terraces through most mountain gorges, field evidence to determine the relationship between mountain valley moraines and terraces in adjoining basins is commonly inconclusive. However, in the valley of the South Fork of the Shoshone River in the volcanic Absaroka Mountains, Wyoming, lie two morainal complexes to which the Cody and Powell terrace systems can be traced far out in the adjacent Bighorn Basin. Approaching the better preserved later moraine, the gravel of the Cody terrace—with weathering characteristics similar to the morainic deposits—thickens, rises in height above present stream grade, and merges with outwash materials extending downstream from drainage channels in the moraine. Since the outwash gravel sheet of the Cody terrace is traceable almost continuously from the mountains far into the basin, aggradation or lateral spreading of volcanic gravel was associated with glaciation. Downcutting, which seems to have continued to the present, was associated with deglaciation and interglacial climatic conditions, although it was more pronounced in or near the mountains than in the center of the basin.

INTRODUCTION

When Dr. Coates invited me to speak at this symposium on the relation of terraces to glaciation, I first considered a presentation divided into 3 parts, each focused on a geographic area where different Pleistocene terrace problems could be well demonstrated. The first area would have been the eastern United States, where in many areas stagnation was an important process in ice retreat and the relation of local kame terraces and more extensive glaciofluvial terraces to moraines and ice margins presents special problems. In the second area, the Middle West, the complex stratigraphy of Pleistocene river terraces presents many problems centering around interpreting the relationship of glaciofluvial materials, loess, soils, peat, and lacrustrine deposits. The third area would logically be the Rocky Mountains with their record of mountain glaciation and a conspicuous array of river terraces, the origin of some of which is related to the changing climate of the Pleistocene.

It is obvious that a comprehensive discussion of terrace-glaciation problems in all these areas would require a series of monographs. Accordingly, I have decided to limit this paper to the Rocky Mountain area, in fact to the area I know best, the western Bighorn Basin and Absaroka Mountains (Fig. 1). I do this because terraces are unusually well preserved in this area, because they occur in conjunction with moraines, and because a controversy has long raged on how to interpret their relationship.

Establishing the relation of glaciation to river terraces extending far downstream from moraines in the Rocky Mountains has engaged the attention of geologists for many decades. Determination of the relationship is complicated in some areas by the difficulty of distinguishing river terraces from pediments and other erosion surfaces, all of which extend in spectacular fashion from the mountain massifs into adjoining basins. Another complicating factor is the impossibility in many cases of tracing river terraces through mountain gorges to the moraines (Bryan and Ray, 1940; Ray, 1940).

In the western Bighorn Basin of Wyoming and Montana, distinguishable remnants of pediments and river terraces are exceptionally well preserved, with the highest dated as Plio-Pleistocene (Rohrer and Leopold, 1963). Since the adjacent mountain ranges were glaciated and in some valleys moraines remain, the area is an especially propitious one in which to study the relationship of terraces to glacial deposits. Furthermore, in the Shoshone River basin, it is possible, with the assistance of deep cuts provided by road construction, to trace the terraces from the basin through the one intervening canyon to the moraines.

The critical questions with respect to river terraces in this area are: do the terraces pre-date the moraines they adjoin or was their formation related to the climatic oscillation which caused ice advance and moraine building? Also, by what mechanism might changing climate have led to terrace formation?

In some areas, such as the Michigan River basin (Eschman, 1955) and Frying Pan valley in Colorado (Nelson, 1954), interpretation of terrace formations associated with glaciation and moraine building has not been challenged. Agreement about a genetic relationship between terraces and moraines is lacking particularly in the case of terraces traceable far into adjacent basins where their origin has been differently explained. This paper will deal largely with interpreting the controversial terrace-moraine relationships in the valley of the South Fork of the Shoshone River in the Absaroka Mountains of Wyoming, which are an example of terraces with remarkable continuity extending from the mountains across an adjoining basin.

In the South Fork valley and Bighorn Basin different investigators viewing much

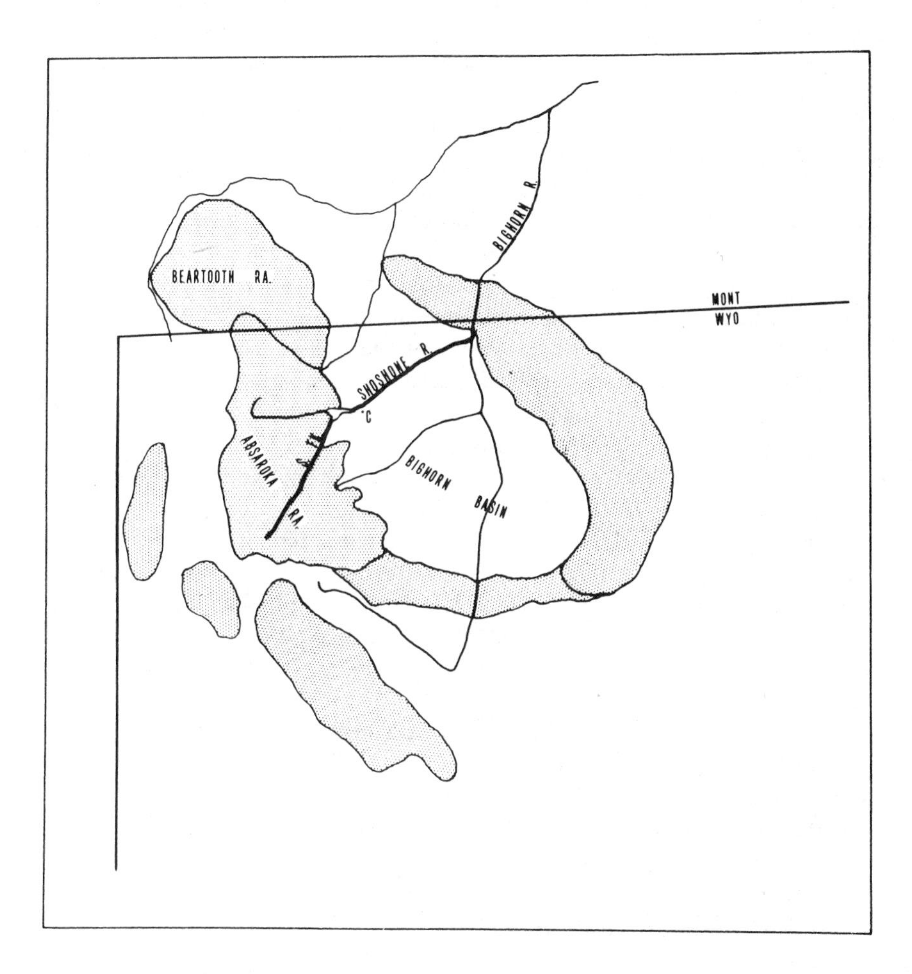

Figure I. Index map showing the course of the Shoshone River, Wyoming, flowing from the Absaroka Mountains across the Bighorn Basin to thejunction withthe Bighorn River. The letter "C" indicates the location of the town of Cody. West of Cody are a gorge and dam forming the Buffalo Bill reservoir covering the junction of the North and South Forks of the Shoshone.

the same field evidence reached diametrically opposite conclusions regarding the mode of origin of the river terraces and their relation to the moraines. In his classic paper on the erosional history of the Bighorn Basin, Hoover Mackin (1937) supported the view of Alden (1932) and Rouse (1934) that the terraces in the South Fork valley were not "genetically associated with the valley moraines" (Mackin, 1937, p. 868). They believed that the formation of the major terraces preceded glaciation and that the moraines were deposited on previously formed river terraces. The terraces, Mackin believed, were formed by laterally swinging graded streams which progressively deposited a thin sheet of channel deposits on a relatively smooth bevelled bedrock surface. Pierce and Andrews (1941) and Pierce (1968) also concluded that the highest Cody terrace predated glaciation. Other workers, with access to data which indicate that the terrace gravel is in places over 100 ft thick and the relief of the subgravel bedrock surface in excess of 60 ft, related the terrace formation to the glacial cycle, pointing out that the gravel component of the terraces can be traced into the outwash deposits fronting the moraines (Moss and Bonini, 1961; Moss and Whitney, 1971).

Close scrutiny of this controversy, however, reveals that part of the difference of opinion stems from a difficulty in terminology—a difference in significance assigned to the bedrock and sediment components of the term "terrace."

It is the purpose of this paper to try to analyze the confusion in terrace terminology, to investigate the relationship of terraces to moraines in the complex Shoshone valley, and to present evidence on the mechanisms of formation in this area of terraces related to Pleistocene climatic changes.

GEOLOGIC SETTING

The South Fork of the Shoshone River rises at an altitude of approximately 11,000 ft amid the high peaks of the Absaroka range (Fig. 1). These mountains are composed of a thick sequence of generally flat-lying Tertiary volcanic rocks consisting of breccias, tuffs and basaltic lava flows cut by scattered dikes and stocks. These rocks are underlain by less resistant Paleocene rocks overlying older Mesozoic and Paleozoic strata. The mountains are deeply dissected, with relief up to 5,000 ft. Where it has cut through the volcanics into the softer Tertiary sedimentary rocks, the South Fork has opened up a broad valley which contrasts strikingly with its narrower valley in the volcanics.

From its source the South Fork flows northeastward to the Buffalo Bill reservoir west of Cody, where it is joined by the North Fork and swings eastward through a canyon onto the softer rocks of the Bighorn Basin on its way to its junction with the Bighorn River at an altitude of 3,600 ft.

There is a marked range in precipitation between the mountains and basin. Both precipitation and vegetation are related to altitude, the sparse sagebrush of the basin giving way with increasing altitude to conifers in the mountains. Mean annual rainfall at Lovell near the junction of the Shoshone and the Bighorn is 6 in, in contrast to more than 20 in in Yellowstone Park in the mountains.

The main valley of the South Fork, together with its upper tributaries, has been intensely glaciated. The ice extended 40 mi down the South Fork to the vicinity of Ishawooa, altitude 6,000 ft, latitude 44° 20', where it deposited massive moraines throughout a 5 mi stretch of the valley. These moraines today rise more than 500 ft above the river. Extending downvalley from the moraines are flights of river terraces which can be grouped into two complexes traceable to the well-known Cody and Powell terraces—first described by Mackin (1937)—in the Bighorn Basin (Figs. 2 and 3). These terraces are absent upvalley from the moraines.

The present South Fork, downcutting mostly in the fine-grained friable Mesozoic and Tertiary sedimentary rocks between the moraines and the Buffalo Bill reservoir, has a gradient of approximately 40 ft/mi in contrast to the gravel and boulder terraces which increase in gradient from 60-120 ft/mi in the same reach.

It is in the vicinity of Ishawooa, where the terraces meet the moraines, that disagreement exists regarding how the terraces and moraines are related. This report will concentrate on this area of controversy.

RIVER TERRACE TERMINOLOGY

River terrace terminology is in a state of considerable disrepair through proliferation of terms and failure to establish common usage. The literature is filled with a myriad of descriptive terms such as rock-cut, strath, lateral, fill, fill-strath, fill-top, fan, erosional, cut-in-fill, cut, depositional, valley plain, alluvial, slip-off slope, rock defended, gorge, delta, and others used in different papers over the

Figure 2. View of the Powell (upper) and Cody (lower) terrace complexes in the South Fork valley. South Fork River flows through the trees in the lowest part of the valley.

Figure 3. Four mi east of Cody in Bighorn Basin, Powell terrace on skyline; two treads in Cody complex (cut in same fill) below. Shoshone River can be seen cutting into tilted Cretaceous bedrock underlying Cody gravel. (Photo by Jack Richards)

years (Miller, 1883; Davis, 1902; Bucher, 1932, Cotton,1940; Quinn,1958, Howard, 1958; Leopold, Wolman and Miller,1964). Howard, Fairbridge and Quinn in Fairbridge (1968) present a summary of most of the terms, but writers in this field would do well to define the exact meaning they are attributing to a particular term. Some of this proliferation in river-terrace terminology results from using for the same type of terrace different terms which stress either its geographical location, its mechanism of formation, or the material composing it (see also Leopold and Miller, 1954).

In this paper, the term "river terrace" is used to describe the steplike topographic embankments lining river valleys and marking former higher elevations at which the river once flowed. As in a flight of stairs, the term "tread" or "top" will be used to designate the relatively flat upper surface of the terrace, the term "riser" or "scarp" to denote the slope–usually steep in the Rocky Mountains–separating terrace treads. To form a terrace requires two different stream processes: lateral cutting or aggradation to produce the tread; downcutting to form the riser.

After stream deposition ceases, terrace treads are commonly mantled by colluvium or fan deposits from valley sides or higher terraces, with the result that the height of the original terrace top may be difficult to determine. As pointed out by Johnson (1944) many problems are involved in correctly correlating terraces downstream. In the South Fork, Cody and Powell remnants are continuous for long distances and tracing can be done with considerable confidence. In addition, construction of a new road through the canyon of the Shoshone west of Cody has produced giant roadcuts enabling the terraces in the basin to be correlated with those in the South Fork valley.

TERRACE TYPES

The vast array of terrace levels present in the Bighorn Basin and surrounding mountains is part of a series grading from two end members. The end members can best be defined in terms of the characteristics of the riser. At one end is the "fill terrace," in which the riser, and hence the terrace, is composed entirely of sediment laid down during a period of aggradation (Fig. 4A). The terrace is formed when aggradation is followed by downcutting which produces the scarp, leaving the top of the fill as the terrace tread. If, as is true in the Rocky Mountains, progressive downcutting was interrupted more than once by periods of aggradation, more than one fill terrace may be present in the same valley (Fig. 4B). Fill terraces are called "paired terraces" because remnants of the tops of the fills commonly stand at the same elevation above the present stream (Fig. 4D).

The other end member of the terrace series is the "strath" or "rock-cut terrace," in which the riser consists virtually entirely of bedrock capped by only a thin veneer of gravel (Fig. 4C). The origin of this type of terrace in the Bighorn Basin has been eloquently described by Mackin (1937). Laterally migrating streams cutting on the outside of meanders bevel the bedrock across which they shift, concurrently depositing channel gravel equal in thickness to the depth of flood scour. Overbank silts deposited during floods commonly cover the widening sheet of gravel which mantles the eroded bedrock surface. The laterally cutting streams are also slowly downcutting so that as the meander belt sweeps back and forth across the valley a series of unpaired terraces is formed on opposite sides of the stream.

The essential difference between these two end members is that to form a fill terrace, a period of aggradation is essential. For the rock-cut terraces, lateral cutting is required. Another significant difference is that in the latter, the bedrock surface beneath the gravel parallels the terrace top, whereas in a fill terrace the bedrock surface on which the gravel rests may be uneven.

Although both end members of the series of terrace types described above are present in the Shoshone drainage basin, it is interesting to note that terraces with characteristics intermediate between typical rock-cut and typical fill terraces also are present. The most striking variable is the thickness of the alluvial cover over bedrock in the terraces, which ranges from an average of 15-25 ft in typical strath terraces (Mackin, 1937, 1948) to over 120 ft in some of the fill terraces. As examples of an intermediate stage, several remnants in the South Fork valley standing 120 ft above the present stream consist of 60 ft of gravel overlying 60 ft of

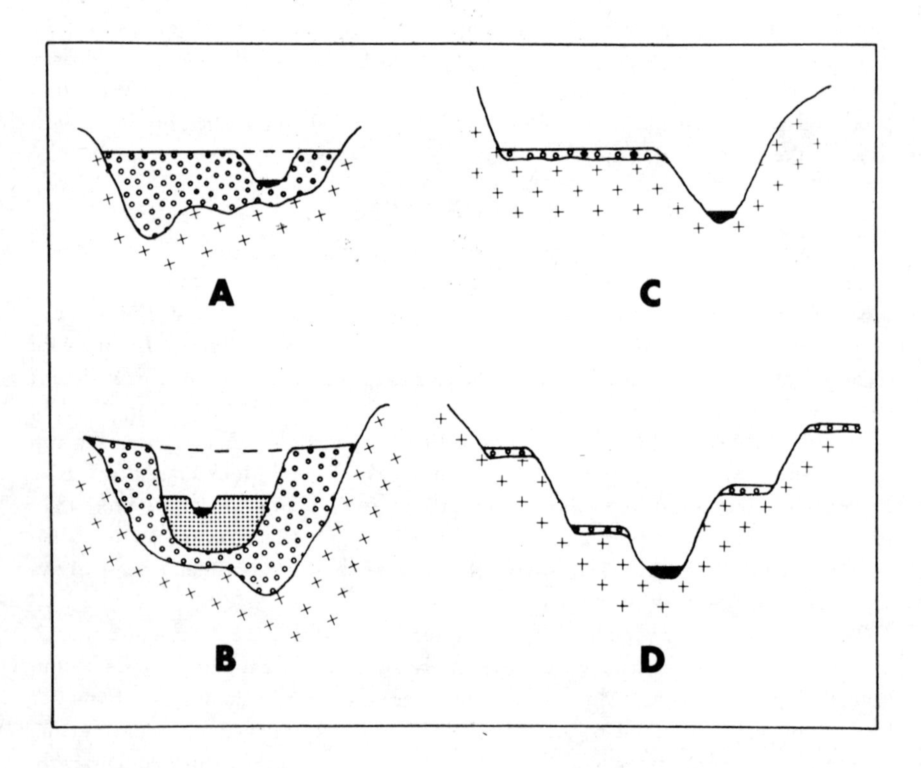

Figure 4. End members of terrace series. A and B, fill terraces; C and D, strath or rock-cut terraces.

bedrock (Fig. 5). In downcutting following deposition of the 60 ft of fill, the South Fork cut through the gravel and deeply into the soft bedrock beneath. Another variant occurs at the Horner Early Man site 3 mi east of Cody, where 5 prominent terrace treads are cut into the same fill (Fig. 6) with no parallelism between the terrace treads and the underlying bedrock (Moss and Bonini, 1961).

ISHAWOOA MORAINES

Two moraine complexes have been mapped by the author and John Whitney (1971) throughout a 5-mi stretch of the South Fork valley south of the abandoned Ishawooa post office (Fig. 7). Remnants of an older moraine (Mbl) occur on both sides of the river near the downstream end of the morainal complex. The largest mass is a subdued hummocky area 400 ft above the present river on the east side of the valley near the north end of the morainal mass (Fig. 8). Pierce (1968) has identified these deposits as older landslide deposits. Unfortunately, no cuts occur in them and they are extremely difficult to excavate with hand tools. The top 2 ft of the deposit are poorly sorted, compact, deeply iron-stained, containing both fresh and rotted cobbles. Some of the fresher cobbles are notably cracked. Petrologically, most of the deposit is related to the Absaroka volcanics outcropping in the nearby valley walls and further upvalley. However, the presence of dacite cobbles recovered from the deposit indicates that at least some of the material was transported downvalley and is not locally derived, since no intrusives of this type have been reported in the adjacent valley walls. Grading into this hummocky area are the last upvalley remnants of the Powell terrace (Tpo).

On the basis of geographic location, topographic expression and weathering characteristics of the sediments, this older moraine was tentatively identified as Bull Lake or early Wisconsin in ranges of the middle and southern Rockies as defined by Blackwelder (1915), Moss (1951), Holmes and Moss (1955), and Richmond (1960).

The younger moraine is a massive hummocky pile of till containing many boulders and cobbles covering the southernmost blocks of the South Fork detachment thrust and other bedrock upstream. Although till occurs up to an elevation of 400 ft above the present river bed, bedrock outcrops in the morainal area indicate that the till deposits are much thinner and the bedrock surface beneath it uneven. Most of the morainic material lies on the east side of the valley and east of the present course of the South Fork. Its highest elevation is to the east, from which it descends in a series of lateral moraines separated by drainage ways extending toward the river. At river's edge the moraine-bedrock contact is 90-120 ft above present river level. Either the ice tongues scoured deepest and remained longest near the high southwestern valley wall, or the deepest part of the pre-Pinedale valley was on the western side of the present valley.

Numerous roadcuts expose the till comprising the lateral moraines. In many of its characteristics this till resembles Pinedale till elsewhere in the Rocky Mountains. It is composed of a loose, easily excavated cobble-boulder till with a sand-silt matrix. The upper 24 in are stained with iron oxide but exposures generally present

Figure 5. Riser of terrace opposite mouth of Rock Creek in South Fork valley. Terrace top stands 120 ft above river; gravel is approximately 60 ft thick.

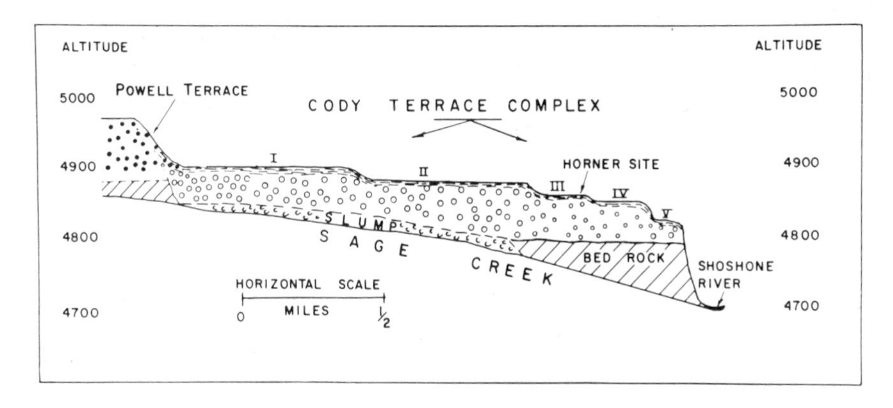

Figure 6. Cross-section along Sage Creek near junction with Shoshone River showing relationship of Cody and Powell terrace gravels. Note terrace steps cut in Cody gravel and lack of parallelism between bedrock beneath gravel and terrace treads.

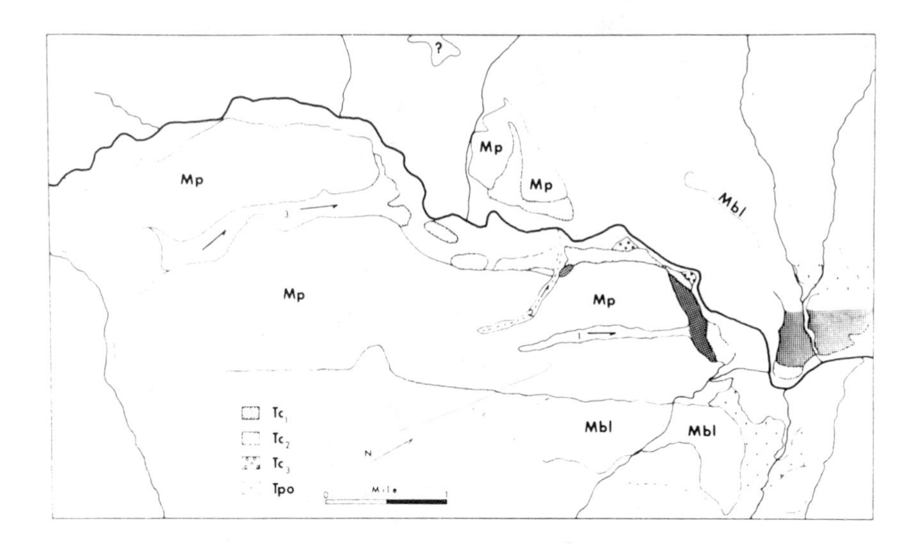

Figure 7. Map of moraines and terraces near the abandoned Ishawooa post office in the valley of the South Fork of the Shoshone River at an altitude of 6,000 ft. Mbl—Bull Lake (early Wisconsin?) moraine; Mp—Pinedale (late Wisconsin?) moraine; Tpo—Powell terrace; Tcl, Tc2, Tc3—successively younger terraces cut in Cody fill.

Figure 8. Subdued hummocks of Bull Lake moraine on east side of the South Fork valley approximately 400 ft above present river. Between the hummocks can be seen valley train deposits slanting downstream to right forming Powell terrace.

a light-colored appearance. Development of a Cca horizon is weak, although the underside of some cobbles is coated with caliche. The relatively high percentage of rotted cobbles, 25% at some localities, is more typical of a Bull Lake age. However, since the rotted cobbles are mostly relatively easily weathered mafic volcanic rocks, a Pinedale age is tentatively favored by the author.

Several gravel-floored drainage ways, labelled 1, 2 and 3 on Figure 7, extend from the adjacent gravel terraces upward into the moraine. When the ice extended into the area, outwash streams flowed down across the moraine to form valley trains now represented by the highest Cody terrace (Tc_1 on Fig. 7). No age difference is discernible in the weathering characteristics of the gravels on the terraces and in the drainage ways leading down from the moraine.

The largest drainage way, labelled 3, was cut later by powerful streams flowing from the ice at a time when it was in contact with the upstream end of the moraine and from Carter Mountain to the east. Angling through the moraine, this outwash channel is a flat-floored valley 100-200 ft wide and 50-75 ft deep. It grades into a prominent terrace 30 ft below the highest Cody (Tc_2 on Fig. 7). It apparently represents a period of cutting of the original Cody fill as the ice activity in the area began to wane. Channel 2 is still occupied by an intermittent stream; channel 1 is not.

As pointed out above, downvalley from the moraine the South Fork valley is filled with flights of Cody and Powell terraces. Above the moraine the valley changes strikingly, the ice having eroded a broad U-shaped trough without terraces into which there has been some landsliding, and tributaries from the valley sides are building postglacial fans on the present valley floor.

Between the moraines and the Buffalo Bill reservoir covering the junction of the two forks of the Shoshone west of Cody, the Powell and Cody terrace complexes have multiple treads, with the difference in elevation between the highest Cody and the lowest Powell commonly 90-150 ft. Upstream from the junction of the rivers both terraces emerge from beneath the surface of the reservoir. The sediment component of the terraces consists of volcanic gravel cobbles and boulders which become coarser upstream, reaching intermediate diameters of over 40 in at the moraine front. No differences in weathering and soil profile development between the moraines and the Cody terrace alluvium in the South Fork valley are apparent. This material is largely alluvium with admixtures of tributary fan deposits and colluvium from the steep valley sides. The height of the terraces above the present stream increases upvalley, with the highest tread of the Cody reaching a height 200 ft above present river level, the Powell approximately 400 ft.

Although for the first few miles above the reservoir large areas of the older Powell terrace are well preserved, dissection has been more intense further upstream and determination of the terrace height is made difficult by a cover of fan and colluvial deposits. By contrast, the younger Cody terraces, characterized by numerous treads, can be traced virtually continuously from the reservoir to the moraine front. In addition to increasing in height above present stream level and becoming coarser upvalley, the gravel of the Cody terrace thickens notably upstream and, as pointed out above, finally merges with the glaciofluvial deposits carried by outwash streams during the time of moraine building. In short, the author and his co-workers believe that both the Cody and Powell gravels are glaciofluvial, although field evidence in support of the Cody is stronger.

RELATION OF TERRACES TO MORAINES

As stated before, earlier workers, apparently not entertaining the possibility that the thick gravels downvalley from the moraines might be glaciofluvial, failed to see any genetic relationship between the terraces and the moraines. Rouse (1934), following Alden's lead, postulated that the Powell and Cody terraces were the latter's Number 2 and 3 terraces. According to this interpretation, the terraces are older than the moraine, the ice having advanced over previously planated, gravel-strewn surfaces to lay down the morainal deposits. Figure 9A is a copy of Rouse's diagram showing the relationship between the moraine and the benches in front of it as he interpreted it. The difference in elevation between Terraces 3 and 3' is 30 ft. Unfortunately, no indication is given as to Rouse's thinking on the nature or thickness of materials forming either the moraine or the benches. The extension of Number 3 bench under the moraine, indicated by the dotted line, is purely hypothetical.

Figure 9B shows the relationship of the moraine and terraces as interpreted by the author and John Whitney. Bedrock outcrops found at different altitudes within the moraine indicate that the glacial deposits must have been laid down on a highly irregular bedrock surface, not on a horizontal bench. Also, because lithologic characteristics of the gravel comprising the terraces are similar, it appears more likely that the gravel is part of the same fill, with the lower tread carved out by the river as it cut down.

A terminology difference, namely a difference in the use of the term "terrace," may also be involved in these two different interpretations of the field evidence. Alden and Rouse may have thought the gravel to be inconsequential and in stating that the terrace passed under thy moraine may have been referring to the bedrock surface on which the gravel lies. The bedrock does indeed pass under the moraine, although probably not with the planed off configuration shown in Figure 9. In today's studies, in addition to the general terrace configuration, the characteristics and distribution of the sediment component are carefully noted and included in interpreting the origin of terraces. The distribution and thickness of the gravel at the margin of the principal Ishawooa moraine indicate that whereas the bedrock of the terrace passes beneath the moraine, the gravel extends up outwash channels to the top of the moraine.

Mackin also failed to see or realize the significance of the large gravel thicknesses in the vicinity of Cody and in the South Fork valley. He interpreted all the Powell and Cody terraces as strath terraces formed by lateral planation of bedrock capped by a maximum of 25 ft of channel gravel and fine-grained overbank flood and fan deposits. In addition to proposing that periods of filling were part of the late Pleistocene history, the author believes that the field evidence supports the conclusion that the gravel on some of the lower treads may have been at one time considerably thicker, with erosion having removed part of the original thickness.

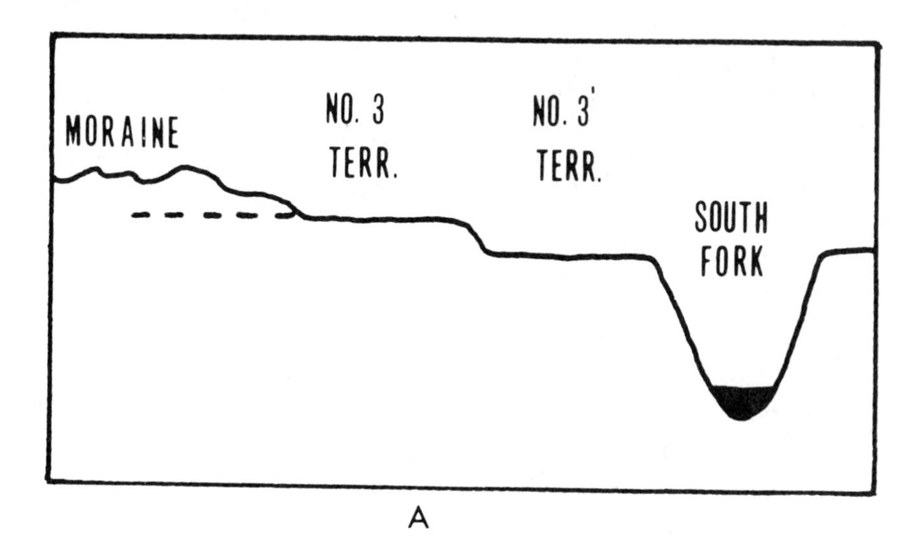

A

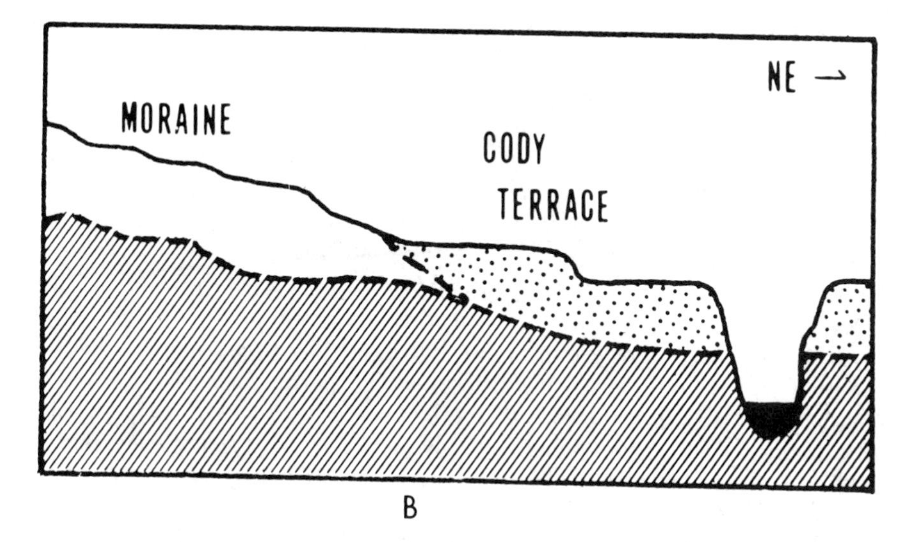

B

Figure 9. Interpretation of relationship of Cody terrace to moraine in South Fork valley according to Rouse (1934) above and Moss and Whitney (1971) below.

This is clearly evident where more than one terrace has been developed in the same fill. East of Cody at the Horner site, as shown in Figure 6, Tread V contains only about 20 ft of gravel, but that is because the overlying 40-50 ft have been eroded away.

Mackin also states that "the great breadth of the old planation surfaces indicates a remarkable stability in the delicate balance of load and discharge of the streams which is not to be expected during glacial stages" (Mackin, 1937, p. 868). Although logically sound, he failed to investigate thoroughly the possibility that the broad terrace treads at Cody and in the South Fork valley might be the tops of fills, which during a glacial period could have been broadly spread over an uneven bedrock surface in a previously widened valley. He envisaged the moraines at Ishawooa as sitting on a laterally planated bedrock surface which continued downstream as the Cody terrace, which was formed in the same manner. He did not explain where the glaciofluvial gravels, which would have inevitably been produced by glaciation, went, nor did he account for the 60-120 ft thickness of gravel in the terraces between the moraines and Cody.

Pierce and Andrews (1941) reported gravel thicknesses of 65 ft for the Cody terrace ½ mile west of the Cody railroad station and 85 ft for the Powell terrace in the South Fork valley. They concluded, however, that the main Ishawooa moraine overlay the Cody terrace and stated that the moraine must be pre-Wisconsin. They too failed to account for the large thickness of the gravel comprising the upper part of the terraces in the South Fork valley.

After publication of the Moss and Bonini paper (1961) containing the data providing additional proof of gravel thicknesses up to 120 ft on the Cody terrace and suggesting that the gravel of the terrace was glaciofluvial, Pierce (1968) reexamined the terrace-moraine relationships at Ishawooa and made a new distinction between the highest terrace, which had been grouped under the name Lower Cody. He reaffirmed his agreement with Rouse that the moraine rests on the gravel of the Cody terrace (uppermost level) and therefore was older than the terrace. The lower terraces of the complex, beginning with the one 30 ft below the Cody (C_2 on Fig.7), he included in the Lower Cody group, all of which he stated to be post-moraine in age. In dating the Cody terrace as older than the moraine, he cited as evidence seeps which begin "a short distance below the contact of the moraine and the terrace" (Pierce, 1968, p. 241). Examination of these scattered wet spots by the author and his colleagues failed to reveal any difference in the characteristics of the material of the terrace and the moraine that would account for the seepage. In the South Fork Valley, springs more commonly issue near the contact between bedrock and overlying sediment, which may be masked by slumped morainal material at the seep localities in question.

It is the thesis of this paper that the Cody terrace gravel is glaciofluvial in origin and that from Cody to the moraines it was deposited as a fill over bedrock of whatever configuration it had acquired in pre-Cody time. The best proof of its relation to glaciation is that its gravels can be traced into the outwash channels originating in the moraines at Ishawooa. The floors of these water ways are graded to the highest Cody level (C1 on Fig. 7). The lower Cody levels, such as C2 and C3 on the map (Lower Cody of Pierce and Andrews, 1941; and Pierce, 1968), are terrace treads cut in the original Cody fill during deglaciation. Lateral planation was important during downcutting, as indicated by the large number of broad terrace treads developed in the South Fork valley and around Cody. As downcutting continued, the river cut into the underlying bedrock, where its lateral cutting was restricted and a series of narrower treads, named the Inner Terrace flight by Mackin (1937), was formed. These are strath terraces with a thin layer of gravel overlying bedrock. They extend downward in steplike fashion to the floodplain. The notable difference in the width of the terrace treads between those cut in gravel and those cut in bedrock is due either to difference in discharge or difference in resistance to lateral cutting of the two materials.

DOWNVALLEY EXTENSION OF TERRACES

The remarkable persistence of the terraces for 80 mi down the Shoshone from the moraines in the mountains across the basin to the junction with the Bighorn provides an unusual opportunity to observe how terraces change throughout a long stretch of river valley and to relate these changes to the mechanisms of formation. In general, with increasing distance from the mountain front, there are notable changes in both the Powell and the Cody terrace complexes. In the case of the Cody, the number of terrace treads becomes fewer, the gravel component thins, and gravel size diminishes (Fig. 10). In addition, the elevation of the highest Cody terrace treads above river level becomes less, the Inner Terrace flight is no longer distinguishable, and the Cody terrace merges with the floodplain 15 mi from the junction with the Bighorn. The Powell undergoes similar changes except that it persists as a terrace to the junction, where it stands 80 ft above the river and merges with a terrace on the Bighorn. Within the mountains, as pointed out above, the Powell and Cody terraces rise in elevation above river level from the Shoshone valley to the moraines, and their gravel components increase in thickness.

In the basin between the canyon and the junction with the Bighorn, the terraces change gradually as gravel thickness diminishes from the fill-type end member of the series to the strath or rock-cut end member. The changes in the terraces seem to have been principally influenced by differences in climate and altitude between mountain and basin that affect river hydrology, differences in resistance of rock types between the two areas, and distance from the mountains.

Schumm (1965) summarized the current state of knowledge about changes in paleohydrology related to the climatic swings of the Pleistocene. The relationship between temperature, precipitation, runoff, sediment yield and sediment

Figure 10. Farthest downstream remnant of Cody terrace on Shoshone River near Byron in Bighorn Basin. Note thinness of terrace deposits, low height of terrace tread above river and relatively small size of gravel.

concentration is highly complex. In the Absaroka Mountains, which rise to over 11,000 ft and are comprised largely of resistant volcanic rocks, glaciation—as in other parts of the West—was presumably accompanied by a decrease in temperature, greater precipitation, more intense mass wastage due to frost action, greater erosion due to glaciation, and an increase in the load-discharge ratio. The fills, consisting of glaciofluvial and fan deposition together with landslide and other colluvial material, are particularly well displayed in roadcuts in the Shoshone canyon west of Cody and attest to a period of high sediment yield causing aggradation during glacial periods. Scouring of the upper mountain valleys by glaciers and more intensive frost action in periglacial regions provided a coarse load of resistant boulders, cobbles and gravel which was later dissected as the glacial climate waned, load dwindled, and the streams returned to degradation. Evidence in the South Fork valley does not support the conclusion of Scott (1968) in the Colorado piedmont that periods of alluviation correlate with interglacial periods.

The volcanic detritus transported from the mountains to the basin forms most of the coarse gravels comprising the thick fills extending downstream from the mouth of the Shoshone canyon west of Cody and also the thinner terrace veneers further downvalley. The basin, although presumably also affected by the climatic swing to glacial conditions, is underlain largely by poorly consolidated Tertiary rocks which were easily pulverized to sand, silt and clay in the mill of volcanic load derived from the mountains and thus contributed little gravel. As a result the terrace gravels on the Shoshone are almost entirely composed of volcanic detritus carried down from the mountains. Some tributaries originating in the basin have terraces composed of locally derived sediments corresponding to the Powell and Cody of the Shoshone; in others, terraces are poorly preserved due to erosion. That sediment yields were higher in the basin in glacial times is indicated by the thick fan and colluvial deposits which interfinger with and overlie the terrace gravels. Thick aprons of resistant limestone-rich colluvial deposits radiate out in all directions from Heart Mountain north of Cody. Prominent fans extend outward onto the Powell and Cody terraces from McCullough Peaks and other dissected high areas in the basin. The 100 ft thick gravels reported by Swenson (1957) from drill holes and seismic studies northeast of Heart Mountain are probably in large part limestone colluvium, since the volcanic gravels on the Shoshone have diminished in thickness to less than 20 ft on the river to the east.

The decrease in quantity of volcanic materials downstream explains the change from fill terraces to rock-cut terraces in the mid-basin area. The more limited quantity of gravel which reached this area was either spread as thin sheets on broad, previously formed valley floors or provided the cutting tools to develop broad floodplains which remain as terraces today. In any case, the preservation of terraces along the Shoshone, in contrast to the poor preservation of terraces in some of the streams originating within the basin, results from the protection afforded by the resistant volcanic gravels forming the caps of the Shoshone terraces.

In the mountains the change to glacial climatic conditions had the effect of causing aggradation in the Shoshone valley, whereas in the basin downcutting was arrested or slowed, and lateral cutting, during which broad, thin gravel sheets were spread across the valley bottoms, occurred. In this model, following deposition in late Pliocene or early Pleistocene time of the Fenton Pass formation on Tatman Mountain 1,400 ft above present river level (Rohrer and Leopold, 1963), the Shoshone entered into a long cycle of downcutting which was interrupted during glacial episodes by intervals of increased gravel production and deposition. These periods halted downcutting, causing filling in the mountains and lateral spreading of the gravel in the basin.

Because the number of terrace treads is large and varies from place to place in the drainage basin, uplift and rejuvenation can hardly be invoked to explain the flights of terraces. As pointed out by Rouse (1934), there is no evidence for regional tilting to account for either the downcutting between successive treads of the Cody and Powell complexes or the more distinctive period of downcutting between the Cody and Powell. The continuity of terrace gravels between the basin and mountains on the Shoshone and across border faults along the Beartooth front (Ritter, 1967) argues against any uplift of the mountains relative to the basin. In addition, recently completed studies by Merrill in the Greybull drainage basin, which also heads in the Absaroka range 20 mi south of the Shoshone, has led him to conclude: "The main stream alluvium of the major terrace remnants are [sic] composed of aggradational deposits that are for the most part related to the glacially-induced increased sediment load transported by the Greybull River. The glaciofluvial contribution increases downstream . . . . . Thus it seems that the major terraces represent former floodplains that formed during glacial times in response to both glaciofluvial and fluvial processes" (Merrill, 1973, p. 278). Ritter reaches much the same conclusion for streams issuing from the Beartooth range. "Much of the gravel capping the benches in the Beartooth region may have been deposited during periods of mountain glaciation, at which times large amounts of coarse detritus were released from the wasting ice. Meltwater floods and floods caused by the destruction of ice-damned lakes transported the coarse-grained load to the fringing plains. The effects of these floods, however, could not be sustained, and gradually the streams became incompetent to move their loads. Thus aggradation occurred and persisted until the sudden influx of coarse debris was arrested and the melting ice was no longer able to provide flows adequate to move all the channel material" (Ritter, 1967, p. 481).

It should not be inferred that all terraces in the Bighorn Basin have been caused solely by climatic change. Ritter (1967, 1972) has demonstrated convincingly that stream piracy, along with glaciation, played an important role in terrace development on the flanks of the Beartooths. Capture of main streams from the mountains by low-gradient tributaries originating in the basin diverted the coarse load of the former into the latter, depositing thick fill. Subsequent downcutting left this fill as a terrace. The lower reaches of the Shoshone and Greybull were also diverted to new courses by piracy (Mackin, 1936, 1937).

Although conclusions presented here fit the particular climatic, geographic and geologic conditions present in the Shoshone River basin, it would be unwise–as Flint (1957) has pointed out–to generalize or correlate from this area to every glaciated basin and range area in the western United States. From studies in the Wind River Mountains (Holmes and Moss, 1955) and in the valley of the Boulder River flowing northward from the Beartooths to the Yellowstone River, it is apparent that although there may be general similarities, detailed patterns of glaciation, hydrologic conditions and terrace formation vary not only between mountain ranges but also between nearby valleys in the same range. Two essential needs for the future are detailed field data on the characteristics of terraces and of terrace-moraine relationships in more valleys and, as stated Schumm (1965), a better understanding of the hydrologic changes resulting from climatic change.

ACKNOWLEDGEMENTS

The Shoshone River project has been a joint effort in which many individuals have participated throughout the last decade.

I would particularly like to express my appreciation to Dale F. Ritter for reading this manuscript and for making helpful suggestions. I am also indebted to a number of Franklin and Marshall students, especially John Whitney, John Guswa, Peter Patton and George Ballard, for valuable field work on the Shoshone River terrace problem. Professors W. Roy Phillips of Franklin and Marshall, and Donald U. Wise, now of the University of Massachusetts, aided in the early part of the work at the Horner Early Man site.

This work could not have been carried out without financial support from the National Science Foundation Undergraduate Research Participation program, the National Geographic Society and my niece, Miss Anna Washburn. Dr. Harold McCracken, Director of the Whitney Gallery of Western Art, and Robert Edgar of Cody gave generously of their time in discussions regarding geological problems in connection with the Mummy Cave Early Man site on the North Fork of the Shoshone. Professional photographer Jack Richards of Cody took some of the photographs used in the project, and Ralph Lewis, a Franklin and Marshall student, did the drafting for this paper. Maria Cattell rendered valuable editorial assistance.

REFERENCES

Blackwelder, E. 1915. Post-Cretaceous history of the mountains of central western Wyoming: Jour. Geol., v. 23, p. 97-117, 193-217, 307-340.

Bryan, K., and Ray, L. L. 1940. Geologic antiquity of the Lindenmeier site in Colorado: Smithson. Instn. Misc. Coll. Publ. 3554, 76 p.

Bucher, W. H. 1932. "Strath" as a geomorphic term: Science, v. 75, p. 130-131.

Cotton, C. A. 1940. Classification and correlation of river terraces: Jour. Geomorphology, v. 3, p. 26-37.

Davis, William Morris. 1902. River terraces in New England: Mus. Comp. Zoology Bull., v. 38, p. 77-111.

Eschman, D. F. 1955. Glaciation of the Michigan River Basin, North Park, Colo: Jour. Geol., v. 63, p. 197-213.

Fairbridge, R. W., ed. 1968. *Encyclopedia of Geomorphology:* Reinhold Book Corporation, 1295 p.

Flint, R. F. 1957. *Glacial geology and the Pleistocene epoch:* John Wiley & Sons, 553 p.

Holmes, G.W., and Moss, J.H. 1955 Pleistocene geology of the Southwestern Wind River Mountains, Wyoming: Geol. Soc. Amer. Bull., v. 66, p. 629-653.

Howard, A.D. 1959. Numerical systems of terrace nomenclature: a critique: Jour. Geol., v. 67, p. 239-243.

Johnson, D. 1944. Problems of terrace correlation: Geol. Soc. Amer. Bull., v. 55, p. 793-818.

Leopold, L., and Miller, J.P. 1954. A post-glacial chronology for some alluvial valleys in Wyo: U.S. Geol. Survey W.S.P. 1261, 90p.

Leopold, L.B., Wolman, M.G., and Miller, J.P. 1964. *Fluvial processes in geomorphology:* W.H. Freeman & Co., 522 p.

Mackin, J.H. 1937. Erosional history of the Bighorn Basin, Wyoming: Geol. Soc. Amer. Bull., v. 48, p. 813-893.

Mackin, J.H. 1948. Concept of the graded river: Geol. Soc. Amer. Bull., v. 59, p. 463-511.

Merrill, R.D. 1973. Geomorphology of terrace remnants of the Greybull River, Northwestern Wyoming: Unpublished Ph.D. dissertation, University of Texas at Austin.

Miller, Hugh. 1883. Methods and Results of River Terracing: Roy. Phys. Soc. Edinburgh, v. 7.

Moss, J.H. and Bonini, Wm. 1961. Seismic evidence supporting a new interpretation of the Cody terrace near Cody. Wyo: Geol. Soc. Amer. Bull., V.72, p. 547-556.

Moss, J.H., and Whitney, J.H. 1971. Diversity of origin of the Cody and Powell terraces along the Shoshone River, Bighorn Basin, Wyoming: Abstracts with Programs, Geol. Soc. Amer., v. 3, p. 652-653.

Nelson, R.L. 1954. Glacial geology of the Frying Pan River drainage, Colorado: Jour. Geol., v. 62, p. 325-343.

Pierce, W.G., and Andrews, D.A. 1941. Geology and coal resources of the region south of Cody, Park County, Wyoming: U. S. Geol. Survey Bull. 921-B, p. 99-180.

Pierce, W.G. 1968. The Carter Mountain landslide area, northwest Wyoming in Geological Survey research 1968, chap. D: U. S. Geol. Survey Prof. Paper 600-D, p. 235-241.

Quinn, J.H. 1959. Paired river terraces and Pleistocene glaciation: Jour. Geol., v. 65, p. 149-166.

Ray, L.L. 1940. Glacial chronology of the southern Rocky Mountains: Geol. Soc. Amer. Bull., v. 51, p. 1851-1917.

Richmond, G.M. 1960. Glaciation of the east slope of Rocky Mountain National Park, Colo: Geol. Soc. Amer. Bull., v. 71, p. 1371-1382.

Ritter, D.F. 1967. Terrace development along the front of the Beartooth Mountains, Southern Montana: Geol. Soc. Amer. Bull., v. 78, p. 467-484.

Ritter, D.F. 1972. The significance of stream capture in the evolution of a piedmont region, southern Montana: Zeitschrift fur Geomorph. N.F., v. 16, no. 1, p. 83-92.

Rohrer, W.L., and Leopold, E.B. 1963. Fenton Pass Formation (Pleistocene?), Bighorn Basin, Wyoming: U.S. Geol. Survey Prof. Paper 475-C, p. 45-48.

Rouse, J.T. 1934. The physiography and glacial geology of the valley region, Park County, Wyoming: Jour. Geol., v. 42, p. 738-752.

Schumm, S.A. 1965. Quaternary paleohydrology: in *The Quaternary of the United States:* Princeton Univ. Press, p. 783-794.

Scott, Glenn R. 1965. Nonglacial quaternary geology of the Southern and Middle Rocky Mountains: in *The Quaternary of the United States:* Princeton Univ. Press, p. 243-254.

Swenson, F.A. 1957. Geology and ground water, Heart Mountain and Chapman Bench divisions, Shoshone irrigation project, Wyoming: U.S. Geol. Survey W.S.P. 1418, 55 pp.

CHAPTER 12

GLACIAL GROOVES AND PERIGLACIAL FEATURES IN THE SAHARAN ORDOVICIAN

Rhodes W. Fairbridge

ABSTRACT

The history and proof of the continental glaciation of West Africa during the late Ordovician is reviewed in its broad lines for the benefit of English-language audiences. (A comprehensive, detailed report is available only in French.) Some special problems are discussed in more detail.

Remarkable deep and almost rectilinear glacial grooving can be followed across almost flat terrain for tens of kilometers. Scattered evidences confirming the glaciation are now known that embrace an area of at least 8 million km^2(the size of the U.S.). Many of the grooves and associated ripplemark, are scoured in continental shelf and outwash-type sands, which are believed to have been partly in the form of permafrosted sandurs. Extraordinarily well-preserved sand volcanoes, sandstone dikes and plugs, kettles, eskers, parallel beach ridges, possible pingos and permafrost fracture fillings are described and illustrated.

INTRODUCTION

In view of the fact that the uncovering of the Sahara Ordovician glaciation represents one of the least expected and remarkable field discoveries of the mid-twentieth century, and inasmuch as most of the descriptive literature is in French, it is thought to be appropriate to present here a generalized account of the phenomena for the sake of the non-specialist readers. This extended introduction will be followed by more detailed discussions of some of the specific features that have attracted the writer's attention. It should be stressed that his visit to some parts of the immense glaciated region had been made possible only by the generous cooperation of the Republic of Algeria, the Institut Francais du Petrole and the Institut Algerien du Petrole. Needless to say, the logistics for mounting an extended trip into the central Sahara require considerable equipment and local cooperation. In January 1970, an international group led by B. Bijou-Duval and P. Rognon (France), and A. Bennacef, A. Benamar and O. Merabet (Algeria), was invited to review the discoveries in Algeria and this was followed by a two-day conference and review of results. Members of the group, besides the writer, included Percival Allen (U.K.), J.J. Bigarella (Brazil), Alexis Bogdanov (U.S.S.R.), Jean Dresch (France), Yvonne Gubler (France), Paul Potter (U.S.A.), J.P. Mangrin (France), Anders Rapp (Sweden), Stefan Rozycki (Poland), Adolf Seilacher (West Germany), Nils Spjeldnaes (Denmark) and L.M.J.U. Van Straaten (Netherlands), and Andre Vatan

(France). While the opinions expressed in this chapter are obviously the responsibility of the writer, the value of the ideas and contributions made by the above-mentioned field companions in discussion should not be discounted. To all concerned, the writer is eternally indebted.

REVIEW OF THE SAHARA GLACIATION

One of the great attributes of geological science is its ability to reconstruct the paleogeography of past landscapes. It is about 100 years since Ami Boue (1875) coined the term "paleogeography." In the course of years, the first faltering attempts at drawing word pictures and maps of ancient land features–which certainly contained many errors– have been replaced by the multiple and cumulative experience and methods of modern geology that permit the use and testing by the technique of multiple hypotheses. The conclusions so reached in turn bring about a convergence of evidence from such disparate lines of reasoning that no further doubt remains. Pragmatic in his working procedures, the geologist, nonetheless, emerges as one blessed with imaginative vistas that at times almost seem able to vie with the inspirations of the great musicians. This is the theme–the watchword–of this review.

Some twelve years ago, there was initiated a systematic series of geological survey expeditions sent out by the Institut Francais du Petrole (of Paris) to explore and map the large areas of the Sahara of southern Algeria (Fig. 1) with the object of extending their knowledge of the formations that constitute the reservoir rocks of the extensive North African oil fields. Comparable surveys have been made in Morocco and Mauritania and on a reduced scale in Mali, Niger, Chad, and other West African countries.

Figure 1. Scene in the Central Sahara, west of Djanet, Algeria. The rugged plateau in the distance is a glaciated surface of Upper Ordovician age. The blocky Ordovician sandstones are essentially flat- lying and rest on a peneplaned surface of Precambrian, into which the foreground pediments are cut.. (Photo: R.W.F.)

In one of the most extraordinary geological surprises of the present century a landscape of former continental glaciation was discovered in rocks of the Ordovician Period, approximately 450 million years old. Initially, there was some skepticism, indeed incredulity, for the contrast in environments was as great as could be imagined from that of the present day. What is today one of the hottest places on the surface of the globe, with shade temperatures up to 135° F was believed to have been then under subzero conditions. Many of the details were worked out in extensive and careful reconnaissance surveys of French and Algerian scientists during the years 1962-1970; The first suggestion was hinted at by Zimmermann (1960) in a report to the International Geological Congress in Copenhagen, and the first formal published note was by Sougy (1963), followed by several papers by Beuf and Bijou-Duval (1966) and others. Finally, a magnificently illustrated volume was published in 1971 by S. Beuf, B. Bijou-Duval, J. Debyser, O. de Charpal, P. Rognon, O. Gariel and A. Bennacef. Bennacef (and the same associates) presented a summary in English in 1971 (Fig. 2). The writer first drew attention to these discoveries in the English language (Fairbridge 1969, 1970a, 1970b) and answered criticisms posed by Schermerhorn (1971).

Several critical questions have to be asked. First, do the long rectilinear grooves, gouges and scratches that cross the landscape truly represent glacial phenomena? Careful studies over many years in all continents now confirm that, although landslides and other phenomena produce local scratching of rock surfaces, only the slow motion of glacier ice, armed with "teeth" of rocks and boulders, makes these distinctive striations (Fig. 3).

Secondly, if we are sure about the answer to question one, is it not possible that the striations were produced by mountain glaciers such as may exist in

Figure 2. Deep glacial grooves in the Upper Ordovician of the Mouydir region, Algeria. Due to repeated advances and retreats of the continental ice many of the glacial tongues advanced into paleovalleys partly cut during isostatic-eustatic interplay near the former land-ocean boundaries. (Photo: courtesy A. Bennacef, Institut Algerien du Petrole.)

Figure 3. Glacial striations in the Upper Ordovician of the Wadi Tafassasset, Algeria. Note how the grooves are closely overlapped by ripple-marked and cross-bedded sands, suggesting that the ice front was advancing onto the continental shelf. In the distance modern giant dunes advance. (Photo: R.W.F.)

equatorial latitudes at very high altitude? This question has been closely studied in the mid-Sahara. The crystalline basement rocks (of Precambrian age–see Fig. 1) were planed off by erosion over many hundreds of millions of years after their general consolidation. They constitute what geologists call a continental shield, a long–stabilized nucleus many thousands of kilometers in diameter, comparable to the shields of Canada, Brazil, Siberia, Australia...The contact between the old rocks and cover of Ordovician was carefully examined. In many places all across North Africa the top of the older rocks is almost horizontal, showing traces of deep weathering with kaolinitic clays and patches or layers of hematite, a red iron oxide, commonly found only in tropical latitudes today. This is a very important paleogeographic deduction. It means that the North African shield was planed during pre-Ordovician time in an environment of warm, humid nature.

Question three naturally asks: was this warm-humid environment suddenly overwhelmed by glaciers? Again, a close study of the contact formations disclosed that there were pebbles and sand grains at the base of the Ordovician sequences that showed the distinctive "frosting" of grains subjected to strong wind action, as in the desert sands today. Also, some of the "ventifact" pebbles have triangular facets (known, from the German as "dreikanters") which only form by eolian erosion. So, we have to conclude that after the hot-wet equatorial erosional history of the basement surface was complete, a phase of arid desert erosion occurred. By late Precambrian and early Cambrian time, about 700-550 million years ago, glaciers were forming in the mountainous region of Morocco and Mauritania (and in many other parts of the world). Windswept tundra, steppes and deserts were to be expected over the interior plains.

The first Ordovician sediments (of about 450 million years) to be preserved were sands, closely cross-bedded, caused by shifting currents of a shallow advancing sea, coming in over what is now the mid-Sahara from the north. How can we tell such sands are marine–and not fluvial or eolian? Here we have point four: the sandstones contain tracks of creeping animals (Fig. 4), trilobites, tubes of worms, and impressions of seaweeds. Also, the geometry of crossbedding found in shallow-water marine bars and sand banks is quite distinctive.

Into this shallow sea came tongues of advancing glaciers. Traces of them are found now all across North Africa, from Mauritania and Morocco, to Algeria, Niger, Libya and Chad. Question five: how do we know that the ice did not come down from the North Polar region? By comparison with recent glacial striations, we know that crescentic or lunate fractures or "chatter-marks" are produced in the underlying formations by the friction of the ice and they are mostly concave to the downstream side. There is a slightly different structure, produced by subglacial water scouring, a high pressure melt-water phenomenon, that produces deeper crescentic scours, convex to the downstream side (Fig. 5). A further indicator is found in the individual scratches or grooves. These start off with a sharp gouge, followed by a gradual diminution downstream. A fourth clue is provided by the shape of the glacial valley junction; the dendritic Y-shaped patterns show convergence downstream. All types of features positively prove the direction of the

Figure 4. Distinctive tracks of trilobites recorded on the bedding planes of the Upper Ordovician sandstone in the glaciated region of the Algerian Sahara (Tassili N'Ajjer). In places the whole fossils are found, together with brachiopods and other marine species, confirming the Upper Ordovician age. (Photo: R.W.F.)

Figure 5. Water scour marks in the Ordovician of southern Algeria (Wadi Tafassasset) comparable to those produced by high-pressure flow beneath glaciers. These are closely associated with the striated and grooved pavements, and help to prove that the Ordovician ice moved from south to north (right to left in this illustration). (Photo: R.W.F.)

ice motion. Observation: the ice moved from south to north in all cases, all across North Africa. A North Polar source is out of the question (Fig. 6).

The moving ice scraped off fragments of the underlying formations as it moved northwards. As the glaciers melted, they left behind thousands of km^2 covered with the rock debris they had carried. This debris is called "tillite", the rock materials of glacial moraines. Repeated advance and melting took place so that on some hillsides now one may observe three or more successive layers of tillite. In the same way in the Alpine foothills today, one may often see traces of three or four moraines, reflecting advances and retreats of the ice during the present ice age. Is the lithology of a tillite always diagnostic for glaciation? Here there can be some doubt. A rather similar disordered jumble of boulders, sand and clay ("diamictite") can be caused by landslides. But, we have explained, the topographic relief is gentle–the general landscape is (and was) a flat plateau, major landslides would not be expected. Furthermore, the boulders in known glacial situations are often found to be faceted, with fine, parallel striations on the facets. This type of boulder, called an "erratic" block, is never found in any situation other than glacial. The type of rock constituting the erratics is characteristically quartzite, granite and other crystalline types found *in situ* today only great distances to the south. In some of the tillites there is a mixture of "exotic" crystalline boulders (rounded) with angular local blocks of the subjacent sandstone in a bedding suggestive of local sea-ice erosion and transport (Fig. 7).

Are we sure that the age given for these phenomena is correct? The seas into which the glacier heads encroached were rich in marine life: worms *(Slokithos)*,

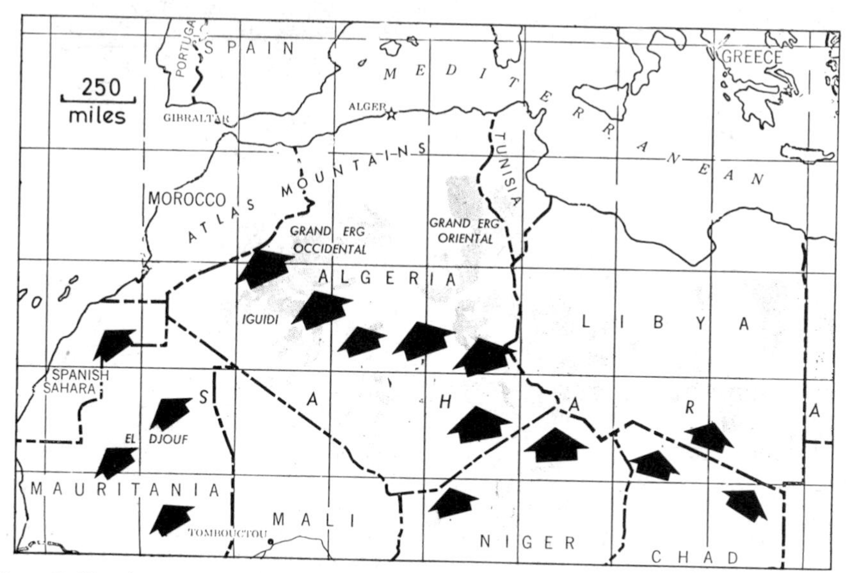

Figure 6. Sketch map of North Africa showing generalized distribution of traces of Ordovician glaciation. Not shown are more distant evidences (some of possibly drift ice nature) ranging from Mali to Sierra Leone, to South America and South Africa.

brachiopids, trilobites. The latter, and particularly their trails (known technically as *Cruziana)* , are very widespread (Fig. 8). The species are typical of Upper Ordovician (Caradocian stage). Some earlier reports of "Silurian" glaciation are misleading; these relate to the old-fashioned use of "Silurian" to include Ordovician. After the Ordovician glaciation in the mid-Sahara there was a general transgression of the ocean and the Silurian sea came right across the continent, with no further evidence of polar conditions.

When the glaciers melted, gigantic quantities of meltwater were liberated, carrying out the finer, sandy sediment, which was then deposited in broad plains or on the continental shelf. Deep scours and ravines were filled by washed sands (Fig. 9). Today, flying across this country, typically in the Tassili Plateaus of southern Algeria, we can still discern the meandering patterns of these proglacial streams, extending over thousands of km^2. Former beachlines can be made out, and these

Figure 7. Mixture of "exotic" blocks (erratics of granite and quartzite that have travelled long distances) associated with sharp, angular debris derived from underlying sandstones. Interbedded with washed outwash sands, this type of bedding suggests sea-ice erosion and transport. Upper Ordovician at Monts Gautier, NE of Erg Killian in southern Algeria. (Photo: courtesy A. Bennacef, Institut Algerien du Petrole).

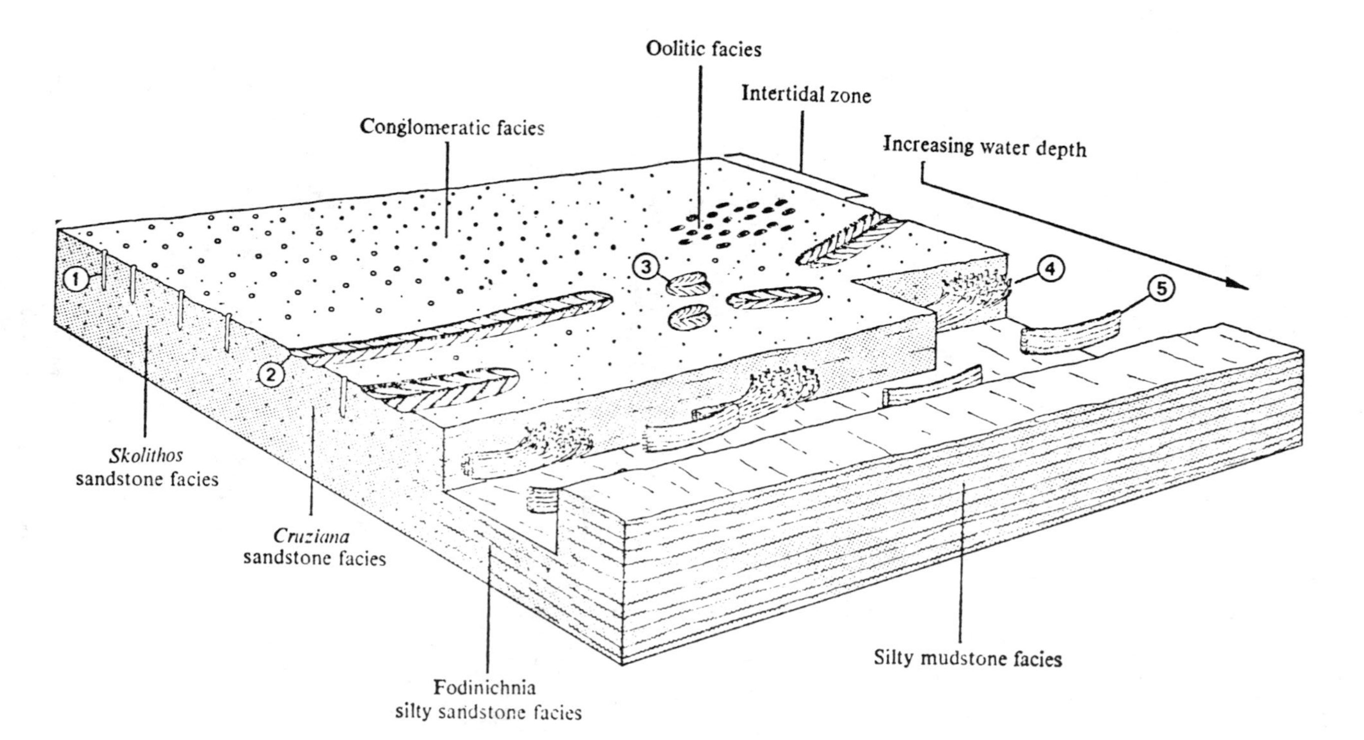

Figure 8. Synthetic diagram illustrating evidence of the type of organic life preserved in the continental shelf sands immediately below and in front of the glacial gacies of the Upper Ordovician in the Central Sahara. The diagram was prepared, however, from the remains found in western Britain; the ecologic conditions were evidently almost identical (after Crimes, 1970).

are comparable to those of the Holocene postglacial beaches of Scandinavia, Spitsbergen, Greenland and Arctic Canada.

A great load is added to the earth's crust by any major ice sheet, which usually builds up to about three kilometers in thickness. Ice sheets generally do not grow higher, because the principal moisture–bearing clouds in high latitudes are usually below 3000m. The density of water and ice is approximately 1, whereas the mean density of rock is about 3. Accordingly, since the earth is essentially in a sluggish hydrostatic equilibrium, the crust responds, admittedly slowly, subsiding under the load of ice, in the ratio of about 1:3, i.e. subsiding 1 km for every 3 km of ice. This isostatic response is reversed during deglaciation and the crust rises once more. Recovery of the crustal attitude of Scandinavia and Arctic Canada after the last glacial maximum was about two-thirds completed during the first 10,000 years of deglaciation. In the mid-Sahara there is evidence that valleys were deeply eroded (to depths of 300 m) near the ice margin which would have been an impossible dissection for rivers in a normally flat shield terrain. Isostatic response was accompanied by secondary fluvial dissection during post-glacial episodes. The same valleys were evidently re-occupied during subsequent glacial readvances, for the walls were striated and smoothed by the newer glacier tongues leaving distinctive roches moutonnées (streamlined and asymmetrically plucked hillocks).

Some remarkable small-scale details are to be seen on these ancient proglacial sand plains. For example, in many places there are closely spaced cracks in the

Figure 9. Evidence of gigantic flooding, causing water scour, with filling of homogeneous washed sands without any fine-grained components. Note the curious "elephant skin" jointing, characteristic of uniformly (silica) cemented sandstone. This proglacial formation is capped by a tillite, product of another glacial re-advance. Locality: western edge of Tassili du Tafassasset, 250 km. S.W. of Djanet, Southern Algeria. (Photo: R.W.F.)

sandstones, fractures today filled with silica-cemented sand, but strongly reminiscent of frost cracks and ice- wedges such as may be found in polar regions (Fig. 10).

Or again, there are small "sand volcanoes", which are spring deposits that are commonly caused by a heavy load forcing the subsurface melting ice-water (with sand) up through cylindrically bored escape channels marginal to the area that is loaded. Somewhat similar features are found today in the tundra regions called "pingos", "suffosion cones" and "tundra craters". Those are subaerial, but the Tassili examples must clearly have been submarine, because they were perfectly preserved by being covered over by younger sediments, but not disturbed by weathering or wave action.

The critic may wonder how it is possible that such a major phenomenon could have remained unnoticed for so long. It is not easy to reply convincingly to this point. It is a fact, regrettable indeed, that vast areas of the earth still lack geological maps, even on a "reconnaissance scale" of say 1:500,000. Many of the new developing nations still lack topographic and geologic survey departments. But North Africa, especially Algeria, was relatively well off in this respect.The French left behind a magnificent legacy of fine maps and reports. In the arid landscape of the central Sahara, the geological formations stand out clearly and desert sands only

Figure 10. Belts of curious silica-filled fissures traversing the Ordovician sandstones more or less vertically. These are believed to represent frost cracks that may have been partially or wholly filled by ice wedges before the silica-fill that probably occurred very much later during diagenesis. (Photo: R.W.F.)

cover parts of them. But still the glaciation had remained unrecognized. Perhaps the explanation is partly to be found in human nature. The geological explorers *simply did not expect to find ancient glacial evidence in the Sahara!*

INDEPENDENT EVIDENCE OF THE ORDOVICIAN SOUTH POLE SITE

By an extraordinary set of coincidences, at the same time as the geologists of the Institut Francais du Petrole were beginning to explore the mid-Sahara, a number of paleontologists were studying the marine fossils of the Ordovician Period in western Europe. A Norwegian, Prof. N. Spjeldnaes, in 1960 worked out a series of ancient climatic zones for that period, on the basis of the tropical-temperate-cool zonation suggested by the various biological remains. Independently of Spjeldnaes, in 1967 two American workers, W.B.N. Berry and A.J. Boucot, also related the fossils and sediment types of the period to a paleoclimatic pattern. In *both cases,* the climatic reconstruction pointed to a temperature gradient with an ancient equator somewhere north of Europe and a polar region in West Africa.

Yet a third group of workers are the paleomagnetists. These scientists are in the field of geophysics, dealing with the measurement of magnetism of ancient rocks. The phenomenon works like this. When an iron-bearing mineral is formed today, as it cools or crystallizes, it preserves in itself a magnetic alignment corresponding to the present magnetic field of the earth at the particular place where the mineral in question solidified. Measurement with delicate instruments of the magnetic orientation of an ancient rock provides bearings on the direction of the ancient pole of that time. If this operation is repeated at several localities, well separated, a geometic "fix" can be obtained where several bearings intersect. This establishes an ancient magnetic pole. The geographic pole is never far away, and is best determined by getting a large number of magnetic pole points, drawing a circle around them, and designating the middle of the circle as the best approximation.

In the case of the southern hemisphere ("Gondwana") pole, some workers active in Australia, South Africa and South America, assuming a pre-continental drift re-assembly of continents, found that the South Pole position has been gradually displaced through time. About 100 million years ago it lay near Australia, 200 million years ago near South Africa, 300 million years ago in central Africa and 450 million years ago in northwest Africa. Thus we reached the astonishing conclusion that it was the former South Pole that lay north of the present equator, in the Sahara of Northwest Africa in the Ordovician Period (Creer, 1968; McElhinny, 1973).

Thus our scientific watchword has been achieved. Not only by every detail of reconstruction of the paleogeographic landscape of the Ordovician Period in the mid-Sahara, but by several totally distinct lines of reasoning, we are led to an identical conclusion: *the South Pole of that period lay in Northwest Africa.*

REFERENCES

Bennacef, A. et al. 1971. Example of cratonic sedimentation: Lower Paleozoic of Algerian Sahara: Amer. Ass. Pet. Geol. Bull., v. 55, (12), 2225-2245.

Beuf, S., Biju-Duval, B., Stevaux, J. and Kulbicki, G. 1966. Ampleur des glaciations 'siluriennes' au Sahara: leurs influences et leurs consequences sur la sedimentation: Rev. Inst. Franc. Petrole, v. 21, (3), 363-381.

Beuf, S. et al. 1971. *Les Gres Du Paleozoique Inferieur au Sahara:* Institut Francais du Petrole Publications, Paris, 464 p.

Boue, A. 1875. Einiges zur paleogeographischen Geographie: Sitzb. Akad. Wiss. Wien, Math.-nat. K1., v. 71, no. 1, 305-425.

Creer, K.M. 1968. Arrangement of the continents during the Paleozoic era: Nature, v. 219, 41-44, 246-250.

Crimes, T.P. 1970. A facies analysis of the Arenig of western LLeyn, North Wales: Proc. Geol. Assoc. London, v. 81, no. 2, 221-239.

Dionne, Jean-Claude. 1974. Mud cracks and polygons on ice push ridges, in tidal flats of the St. Lawrence estuary: Canadian Jour. of Earth Sci., v. 11, no. 3, 489-494.

Dylik, J. and Maarleveld, G.C. 1967. Frost cracks, frost fissures and related polygons: A summary of the literature of the past decade: Meded. Geol. Sticht., N.S., no. 18, 7-21.

Fairbridge, R.W. 1969. Early Paleozoic South Pole in Northwest Africa: Geol. Soc. Amer. Bull., v. 80, 113-114.

——————, 1970a. South Pole reaches the Sahara (Meeting): Science, v. 168, 878-881.

——————, 1970b. An ice-age in the Sahara: Geotimes, v. 15, 18-20.

——————, 1971. Upper Ordovician Glaciation in Northwest Africa? Reply: Geol. Soc. Amer. Bull., v. 82, 269-274.

Macar, Paul. 1969. A peculiar type of fossil ice fissure: in Pewe, Troy L. (ed.) *The Periglacial Environment,* McGill-Queen's University Press, p. 337-346.

McElhinny, M.W. 1973. *Palaeomagnetism and Plate Tectonics:* Cambridge University Press, 368 p.

Schermerhorn, L.J.G. 1971. Upper Ordovician glaciation in Northwest Sahara? Discussion: Geol. Soc. Amer. Bull., v. 82, 265-268.

Sougy, J. 1962. West African fold belt: Geol. Soc. Amer. Bull. v. 73, 871-876.

Zimmermann, M. 1960. Nouvelles subdivisions des series antegothlandiennes de L'Afrique occidentale (Mauritania, Soudan, Senegal): in 21st. Int. Geol. Congr., Copenhagen, v. 8, 26-36.

PART 4

PRACTICAL APPLICATIONS

In these days of environmental concern and relevance for today's problems, it is fitting that this volume close by showing some of the contributions glacial geomorphology can provide in practical matters. The science of geomorphology is that discipline which specializes in knowledge of the earth's surface. Because man lives, works, and plays on this locale nearly all his activities transforms the ground in some manner. This is the geomorphologist's domain, so he is the obvious candidate for knowing the best procedures that will minimize man's metamorphism of the land and for providing input into land-related decision-making plans. Since ice covered 1 million sq mi of coterminous United States, most of Alaska, and even parts of Hawaii (and of course most of Canada and much of Great Britain), it should be the glacial geomorphologist who can provide the best information on land use in these regions. Indeed this topic is really *environmental geomorphology* with a glacial twist.

The geomorphologist's knowledge of soils and the glacial substrate is necessary for helping select the most suitable sites for subdivision development and rural home location. For example approval of a site is now often dependent upon the ability of the earth materials to exceed the percolation requirements for effluent. In the third symposium *Quantitative Geomorphology* I pointed out the differences in the glacial valley fill sediments; streamflow and aquifer yields are several times stronger in south-trending valleys than in north- trending valleys. Even knowing the differentials in till thickness has practical application. Because till is not an aquifer, hillside and upland wells in the southern Tier of New York generally must penetrate at least 100 ft of bedrock to obtain sufficient water for household use. Cost of wells when located on south-facing slopes is often double that of wells drilled elsewhere because of the extraordinary till thickness on these shadow zones.

A recent application of glacial geology has emerged in respect to new quarry regulations by the New York State Department of Transportation. They require reports by a certified geologic specialist for all sand and gravel operations whose materials are used on state-approved contracts for roads. This important service now being rendered by glacial geologists is aiding New York State in determining the quality, quantity, and origin of sand and gravel resources. Many other examples of practical use could also be cited, such as those agriculturalists on Long Island who were knowledgeable that the best potato yields occur on that part of the outwash plain that contain the thickest loess deposits.

Chapter 13 by White shows the importance to planners in recognizing the character of buried landforms. This topic has not previously been reviewed so the author has performed a real service in showing the nature of vertical and lateral extent of such features and their usefulness in making decisions on water supply and foundations.

Chapter 14 by Legget documents the influence of glacial landforms, from both historical and modern viewpoints, in civil engineering projects in Canada. He makes it abundantly clear that recognition of such features can be a vital component to the success of the construction and building trades.

Chapter 15 by LaFleur provides insight into the direction our profession should go in making a larger impact in the planning process. He provides a case history study and outlines procedures that show the importance of glacial studies in zoning and land-use analysis.

CHAPTER 13

BURIED GLACIAL GEOMORPHOLOGY

George W. White

ABSTRACT

Most glacial deposits are made of different materials deposited at different times. The surface of the latest deposit is the present land surface -- the present geomorphology. The surface of each buried deposit is a surface of paleogeomorphology, either entirely depositional, entirely erosional, or partly of each. The buried surfaces may range from relatively smooth to very irregular. One buried surface may truncate another. Buried examples are known of all the glacial geomorphic forms -- ground moraine (till plain), end moraine, drumlins, kames, kame terraces, eskers, outwash plains, valley trains, sand dunes, and loess plains. Buried erosional forms are known -- valleys may be especially prominent. At each buried surface the material may change, at places markedly, and influence the economic, environmental, and engineering parameters. The interpretation of the buried surfaces -- unconformities -- is essential to determine the history of that part of the Pleistocene represented not only by the deposits, but also by the buried geomorphology of each.

INTRODUCTION

The surface geomorphology of glacial deposits is both depositional and erosional. The surface of a buried glacial deposit may be the original depositional surface, or it may have been more of less modified by erosion before or during burial. Unlike buried bedrock surfaces, little attention has been paid to buried glacial geomorphology and very little literature bears exclusively on it. Bedrock surfaces buried beneath drift have been studied in more or less detail, especially to depict buried bedrock valleys, which are so important for water supply. The buried bedrock surface of Illinois has been mapped (Horberg,1950; Piskin and Bergstrom, 1967), as has that of the glaciated part of Ohio (Cummins, 1959). Reports on the glacial geology of counties and other parts of states now routinely include maps of top of bedrock and this may also be shown in cross sections (Winslow and White, 1966; Christiansen,1971 and references; 1973; Bluemle, 1967).

With the realization that repeated episodes of glaciation have produced a whole series of deposits, one upon the other, glacial studies have become more and more stratigraphic and include not only the character of the surface drift and its morphology, but also the whole sequence of deposits upon the bedrock. These studies have resulted in elaborate analyses and descriptions of till sheets (see references in White, 1973). Detailed printed sections, sometimes illustrated by columnar sections, show the stratigraphy in detail, but these give little information about the shape of the geomorphic surfaces which are the contacts between the units. Longitudinal sections, constructed from observations of outcrops along superhighway excavations, or along quarry or strip mine faces, provide the best opportunity to trace stratigraphic units continuously over some distance, which even closely spaced bore holes cannot equal. Examples of such diagrams may be seen in reports by White, Totten, and Gross (1969), Johnson et al (1971), Johnson (1971), and Kempton et al (1971 and references therein). Continuous sections along lake or ocean bluffs may also provide such information (Dreimanis, 1970, and references therein; Carter, 1974).

In the stratigraphic reports the contacts -- many of which are buried geomorphic surfaces -- are discussed only slightly, or not at all; the thrust of the reports is on the material itself and its lateral and vertical variation. That buried valleys can occur in drift, as well as in bedrock has been noted by Norris and White (1961) and by Bluemle (1967, p. 26, pl. 2). Buried moraines have been discussed and illustrated by White (1962) and by Totten (1969, with references).

Earlier glacial surfaces may be buried by till, by outwash, by lacustrine deposits, by eolian deposits, or by mass wastage. Most of the surfaces have been buried by advancing ice, which may, and probably has, more or less eroded the underlying drift. The minimum extent of the areas of no erosion are shown by preservation of paleosols, and the uncommon occurrence of paleosol over large parts of the United States and Canada is a general indication that at least some erosion has resulted from the ice sheet that deposited the overlying drift. Paleosols buried by tills are generally restricted to the few tens or scores of miles adjacent to the outer margins of the overlying till sheet (for a fine example see Horberg, 1953, Fig. 8) because glacial erosion predominates upice while glacial deposition predominates near the margin of the ice sheet (Gross and Moran, 1971). The removal of interglacial or interstadial soil and the leached and oxidized material below the soil records the minimum amout of erosion. In such cases fresh unoxidized till overlies fresh till which is very much older, and only tracing the contact along a longer or shorter distance will disclose not completely removed "pockets" of weathered material on top of the lower unit. If gray till overlies gray till, the maximum amount of ice erosion is very difficult to determine.

Of possibly greater importance is the fluvial erosion that has taken place in a unit after deposition and before its burial by later material. The amount of erosion of the present surface of the drift in the United States ranges from very little on some flat areas to very much, or even removal, in areas of greater relief. If the northern United States were again glaciated and drift deposited upon the present surface, the now buried surface would range from buried plains, to buried kames and moraines, to buried valleys, and ravines cut into drift or through drift to bedrock. Interglacial episodes were all far longer than the time since the disappearance of the ice from the northern part of the glaciated area of the United States and some interstadial episodes may have been as long, or longer. It will be realized, therefore, that the interfaces, the contacts between drift sheets, are important paleogeomorphic elements for elucidation of Pleistocene history. Figure 1 from Frye (1973) shows the position in time and the duration of episodes of "valley incision" throughout the Pleistocene. Frye's important paper deals with the depositional and soil forming records of these episodes as aids in correlating continental and marine history. The erosional unconformities – the paleo-surfaces -- merit equally penetrating analysis. This chapter can only call attention to the presence and their importance, with the hope that areal studies will lead to mapping and analysis of the paleosurfaces in drift sequences, just as buried bedrock surfaces have already been mapped in varying detail in many areas. (Examples of very detailed mapping of Pennsylvanian paleosurfaces of deposition and erosion are the classic studies of Wanless, 1970, and references therein).

DEPOSITIONAL SURFACES

In order to discuss paleogeomorphology of buried depositional surfaces, it is necessary first to discuss the geomorphology of those glacial deposits now at the surface in order to compare them with any buried surfaces. An extensive modern description and discussion of geomorphology of surface deposits is that of Flint (1971, p. 198-226) and of Price (1973), so it will not be necessary to go into an extended description here.

The morphology of ice-laid deposits consists of ground moraine (till plain), end moraine, and drumlins. The surface features of glaciofluvial deposits are eskers, kames, kame terraces, outwash plains, and valley trains. Those of glaciolacustrine deposits are lacustrine plains and beaches. Glacioeolian deposits may be loess "blankets" and sand dunes.

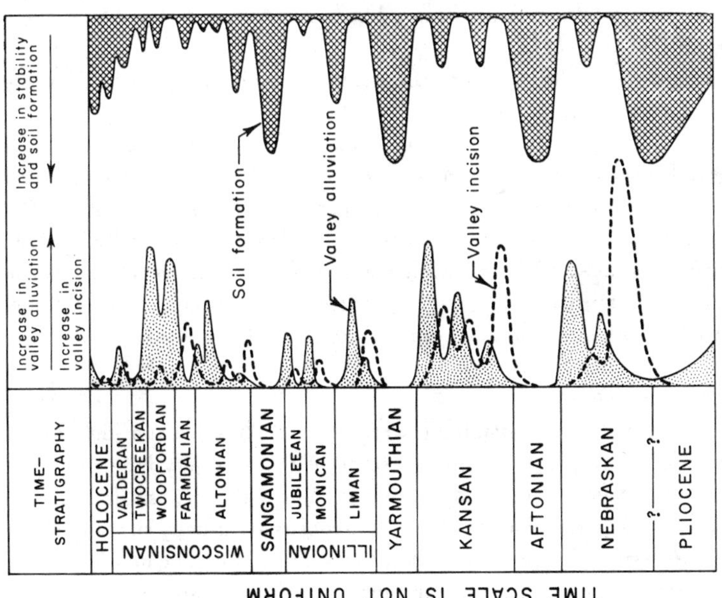

Figure 1. Generalized curves show average conditions in the continental interior for intensities of (1) valley incision, (2) valley alluviation, and (3) soil formation accompanying times of relative stability of much of the surface area. Time scale is not uniform (From Frye, 1973, Fig. 2).

Surface of ice laid deposits

Ground moraine is used here in not only the sense of a deposit with reasonably level, or only slighly undulating surface, but also of a till sheet that covers the underlying irregular bedrock without producing a topographic expression of its own. (If a till sheet covers an underlying irregular surface not bedrock, such as an end moraine, kame terrace, or a drumlin, shall we call it "till sheet over end moraine?"Happily, we do not need to resolve this question in this chapter!)

End moraine may be a ridge-like, or a series of ridge-like deposits -- usually till, but not seldom with gravel included. Some end moraines have kames associated with them or may even be kame moraines. The surface is hummocky, sometimes with kettle holes between the hummocks. An end moraine tract (hummocky topography)may be irregular in outline and without definite lineation. These commonly occur along valley sides in hilly country. They grade into tracts, sometimes very hummocky and very large, which have been called "ice-disintegration features," or "dead ice moraine."

Drumlins, elongated inverted spoon-bowl or cigar-shaped hills are included here because they are almost entirely till, although it is recognized they are "ice molded," and therefore regarded by some as technically erosional features. (See papers by Muller and by Clayton and Moran, this volume.)

Surface of glaciofluvial features

Kames are more or less isolated irregular hills of gravel, which may occur in groups with kettle holes between the kames.

Kame terraces are terraces banked against the sides of a valley. These surfaces may be very irregular and composed of massed kames and kettle holes, or they may be quite flat, with few or no kettle holes.

Eskers are ridges, generally sinuous, composed of sand and gravel. They are generally, but not always, in valleys.

Outwash plains and valley trains are similar in that the former is composed of sand and/or gravel spread out to form a level, or very gently sloping surface over a considerable area, while the latter is of similar material confined to a valley. The surface may be very smooth, or may retain the linear depressions, usually shallow, of ancient stream channels. Some outwash plains and valley trains contain kettle holes, sometimes so extensive that the term "pitted plain" is appropriate. Some kettle holes may be miles in extent and contain lakes.

Surface of glaciolacustrine features

Lakes may be formed by ice damming water between the ice and higher land beyond. Others are formed in kettle holes – some of them of very large size. The surface of lacustrine deposits is usually very flat, unless the deposit is thin and draped over a pre-existing irregular surface. Lake deposits are usually silts and clays, but may be coarser, especially near the shores, either those along the melting ice, or at points where streams flowed into the lake and formed deltas, such as the famous deltas of the New York Finger Lakes region, or the great delta of the Assinaboine in Manitoba. Along the landward shores of large lakes, beaches may be formed, expecially if the shores are gently sloping. The beaches of the glacial Great Lakes and of glacial Lake Agassiz are examples.

Glacioeolion surface features

Plentiful deposits of sand and silt in outwash deposits from retreating ice were available for wind transportation and deposition.The sand was sometimes deposited as a sheet deposit, but more often as dunes of various shapes.

Fine material, generally silt, but including some clay and some fine sand, was blown further from the glacial front and from outwash plains, and especially valley trains, often far from the melting ice. This fine material -- loess -- was deposited in a blanket, decreasing in thickness from the source. The blanket covered the surface, generally mirroring that surface, but sometimes of a sufficient thickness to form topographic features of its own.

Other glacially related surface features

Adjacent to the ice, and for a greater or less distance from it, permafrost features may be formed. These take the form of ice wedges, but the major geomorphic effect is to smooth sloping surfaces by solifluction, with associated congeliturbate on the surface. Such a buried surface with preserved frost formed wedges has been described by Totten (1973, Fig. 19).

Another type of paleogeomorphology is that of surfaces buried by postglacial action, such as alluviation, mass wastage (Lessig, 1959), and the work of man, either in the distant or more recent past. The work of man brings us to the field of archeological geology, which will not be dealt with here. However, such buried surfaces, and especially buried watercourses, are of great practical importance, particularly in long-established cities such as London (Head, 1907).

PALEOSURFACES

Only the marginal part of the glaciated area of the United States, ranging in width from less than a mile to several tens of miles, has been invaded by just one ice sheet. Elsewhere ice has invaded the area two to more than ten times. Each advance was over a surface that had the geomorphology of the preceding glaciation, either as it was deposited, or as it was modified in interglacial or interstadial intervals or in the shorter interval between advances within a stage. The drift of most of the country consists not of a homogeneous deposit of a single ice advance (it may under certain conditions), but of a packet of two or more drifts, each pair separated by an "interface," which is an unconformity. The recognition of this is important for interpretation of Pleistocene history and of great significance for engineering projects, hydrogeology, solid and liquid waste disposal, tunneling, and similar operations.

Paleo-ground moraine. The most common buried glacial surface is that of earlier ground moraine, as illustrated in all but lower part of Figure 2. (For other illustrations, see White, Totten and Gross, 1969, Figs. 17, 22, 23, 25, 40, 41, 42.) The interface between the upper and lower material is usually an almost level surface, which may be unrecognized, especially in borings, if the upper unit is thin, as it often is (White, 1971; Clayton and Moran, this volume). If the unconformity is between two unoxidized (gray) tills, it is even more difficult to recognize. And yet, the difference between the two tills may be very significant for engineering projects. These differences may be in composition, structure, permeability and density. The paleosurface (the interface) may provide a pathway for fluid flow.

Boulder pavements. A boulder pavement (White, Totten and Gross, 1969, Fig. 23) may be present at the interface between two tills, and in excavations may provide a spectacular line to show the surface of the lower unit, which may be level or undulating. In the subsurface a boulder pavement, discovered in tunneling, may provide the first clue to a paleosurface. If the possibility of a boulder pavement at a paleosurface is not realized, the discovery can be shocking indeed when the tunneling machine encounters it! The most extensive boulder pavement now known extends over an area of many tens of thousands of square miles from Manitoba across Saskatchewan to Alberta. It is at the base of the Battleford Till, which overlies the Floral Till at depths of less than one ft to more than 20 ft.

Paleo-end moraine. Almost all end moraines in the Allegheny Plateau investigated by the present author and his associates are composite (White, 1962; Winslow and White, 1966, Fig. 11; Totten, 1969). They are composed of two or more drift units (Fig. 3). The upper till is usually only a few feet in thickness; it may be only 1-2 ft. This upper till is therefore really a till sheet draped over underlying hummocky topography. The implication for elucidation of Pleistocene history is frightening! An end moraine may not, and in the Allegheny Plateau usually does not, record the marginal position of the last ice sheet, as the youngest till, and even the next youngest till, may be traced beyond it. It may not record the marginal position of the penultimate, or even antipenultimate ice sheet – it is hazardous to assume the validity of any Pleistocene history of the last ice advance based on end moraine patterns. This history must be based on till sheets, not end moraines. (See paper by Clayton and Moran, this volume).

In the Allegheny Plateau the bulk of the great Kent Moraine around the Grand River Lobe is Titusville Till (Altonian) and its accompanying gravel, is overlain by thin Woodfordian Kent Till and in part also by thin and discontinuous Lavery Till (White, Totten and Gross, 1969). The moraines (at any rate, the outer ones) of the Killbuck Lobe are similarly composed of tills of Altonian age (Millbrook Till), with later tills draped over it.

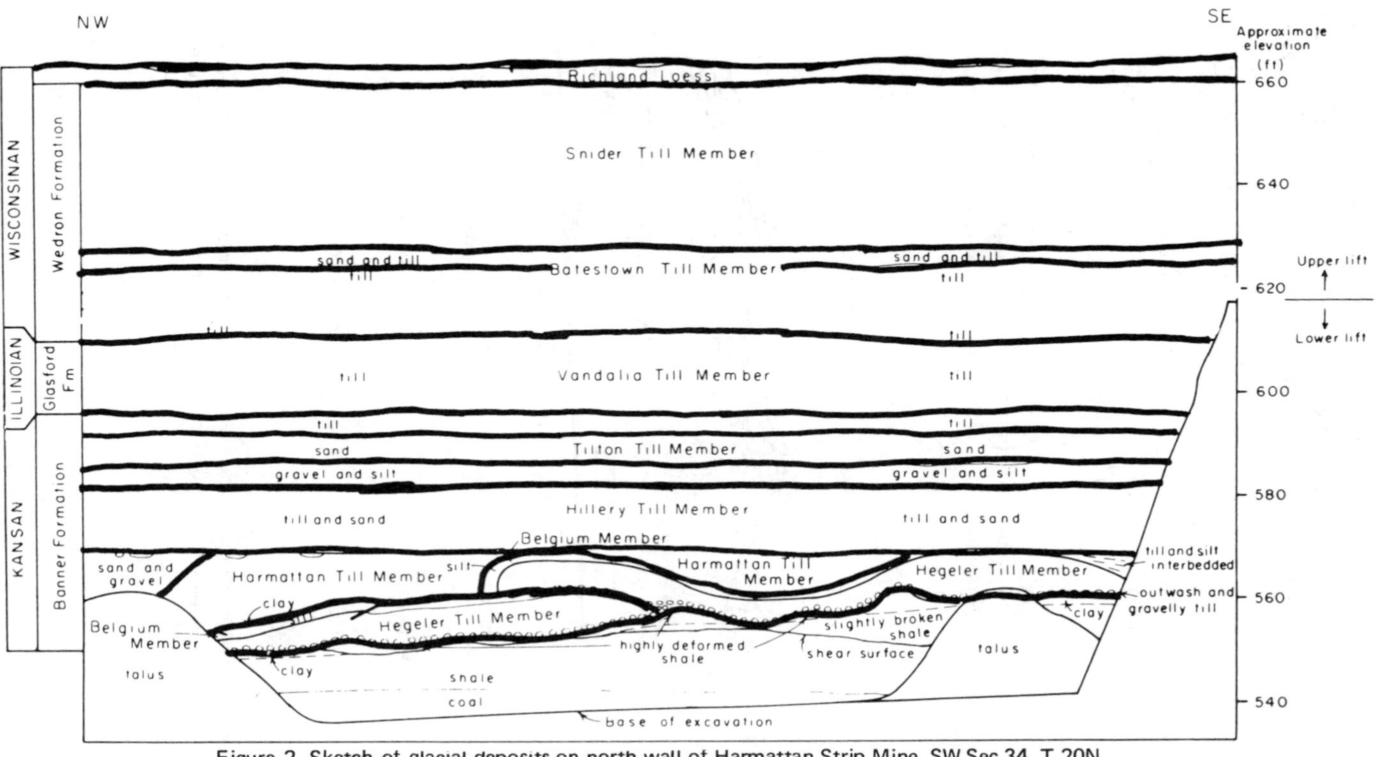

Figure 2. Sketch of glacial deposits on north wall of Harmattan Strip Mine, SW Sec.34, T.20N., R. 12 W, Vermillion County, Illinois, 3 miles west of Danville. Length of section about 100 yards. Lowest till, Hegeler, may be Nebraskan (?), and the unconformities -- buried surfaces--range in age from Nebraskan (?) to Wisconsinan. Buried surfaces -- unconformities -- shown by heavy lines: at least 12 buried surfaces range in age from Nebraskan (?) thru Kansan, Illinoian, to Wisconsinan. Significance of contorted drift units below Hillery Till discussed in text. Diagram modified from Johnson (1971), Fig.2.

Figure 2 provides an interesting speculation: the several drift units below the Hillery Till are folded and thrust as is common in some moraines (Moran, 1971). It is suggested that these disturbed units are the remaining parts of an end moraine, which was reduced to a level surface by fluvial and/or glacial erosion before the overlying Hillery Till was deposited. At any rate, this paleosurface is one of very considerable erosion.

An end moraine's surface may lie below lacustrine deposits as shown by recent investigations in Lake Michigan. Figure 4 shows the very front of an uneroded moraine lying beneath a series of lacustrine deposits.

Paleodrumlins. Abundant records show that many drumlins have a covering of later till and are therefore buried surfaces. In some cases a later drumlin pattern is superimposed upon an earlier one with a different direction (see also other papers this volume).

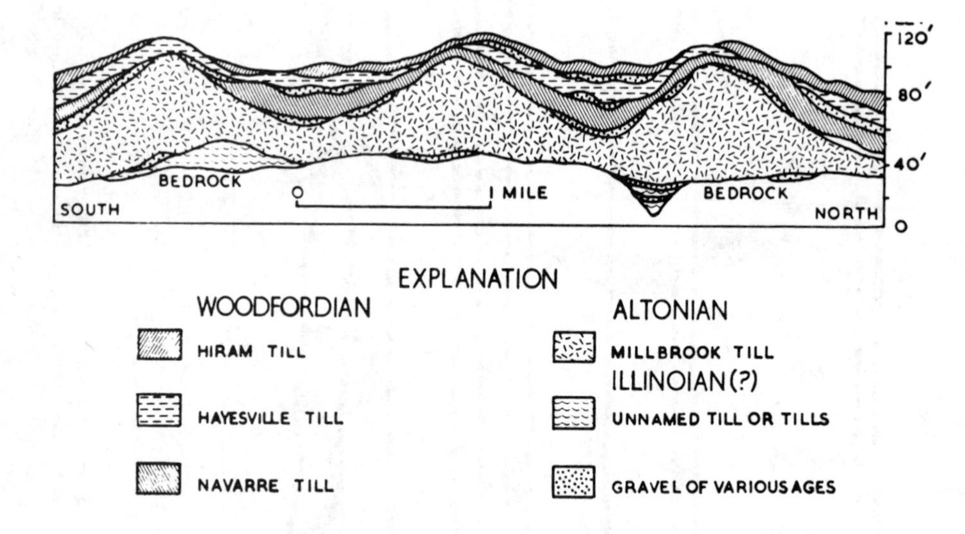

Figure 3. Idealized cross section through three elements of the Mississinewa moraine in the Killbuck basin. In part hypothetical but based on stratigraphic sections. (From Totten, 1969, Fig. 4).

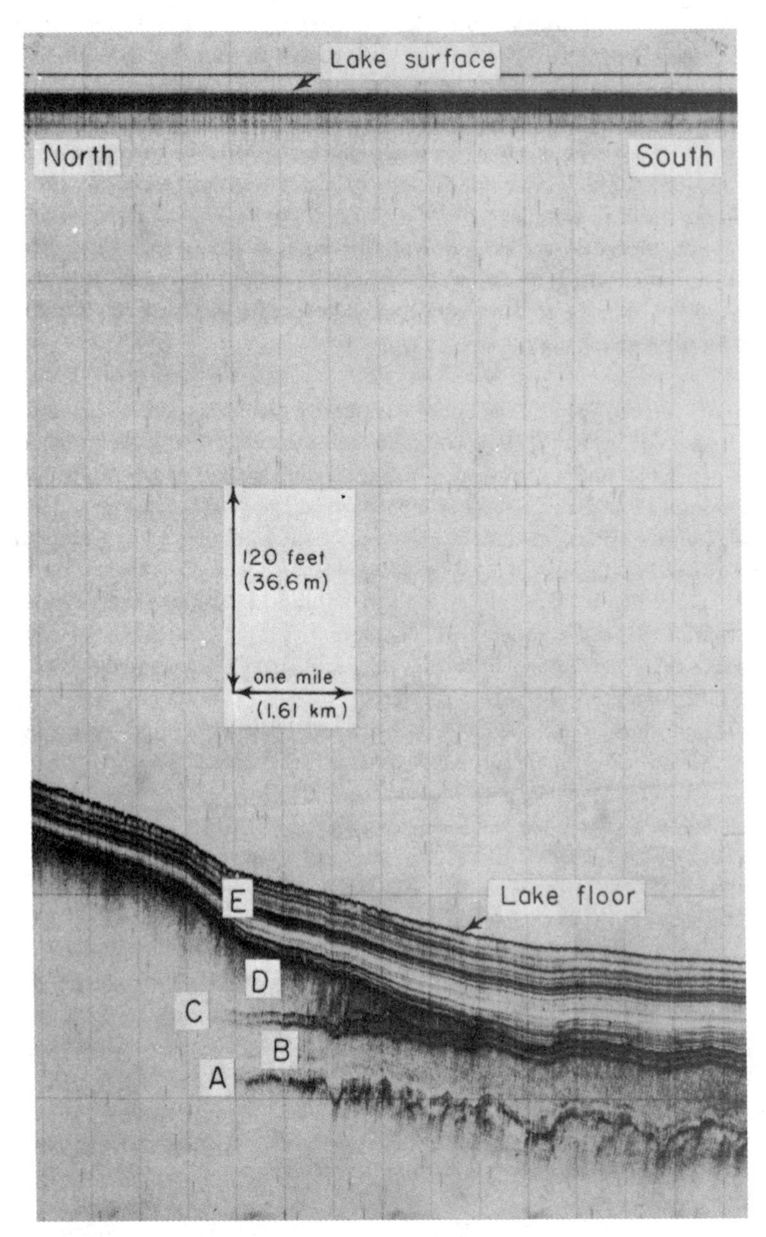

Figure 4. A high-resolution reflection seismic profile from an area east of Milwaukee showing Pleistocene materials beneath the water of Lake Michigan. The dark line (A) is the top of Paleozoic bedrock, the thick gray zones (B and D are late Woodfordian (Wisconsinan) tills, the two dark lines (C) are beds of outwash, and the many finely spaced lines (E) represent a thick sequence of thinly bedded lacustrine silts and clays. The pinchout of till unit (D) is a buried submarine moraine and the margin of the upper till. The moraine is buried by 30-60 feet of lacustrine sediment and is overlain by 500 feet of water. (Modified from Lineback, Gross, and Meyer, in preparation).

Paleokames. Almost all kames of the western Allegheny Plateau are earlier in age than the last ice advance, and later till covers the earlier gravel. The covering may be so thin that gravel is present at the surface and the paleosurface coincides with the present surface. The covering may be many feet in thickness, but not thick enough to do more than modify the underlying kame form, or the covering may be so thick that the underlying kame is well concealed and its presence found only in deep excavations or in drilling.

The recognition of the presence of this kind of paleosurface is particularly important in engineering works, water supply, and waste disposal. If not too far from the surface, the better underdrainage in the kame gravel affects the soil type developed on the overlying till.

Paleo-kame terraces. Just as kames are covered by later deposits, so kame terraces may be covered. All of the great kame terraces of the Allegheny Plateau in northeastern Ohio and northwestern Pennsylvania are of Altonian age, and are covered by one or more till sheets of Woodfordian age. Most of the great gravel pits in the plateau require removal of overburden of later till, which may range from a few feet to tens of feet in thickness (Fig. 5). The paleosurface is that of the typical kame terrace; it may be fairly level, or it may have kames and kettles, making for great variation in the amount of overburden to be removed. An irregular paleosurface may be reflected in the modern surface, or may be completely masked, as in Figure 5.

Many deeply buried paleo-kame terraces remain to be discovered. As the demand for aggregate in urban areas increases (it is already critical around some cities),

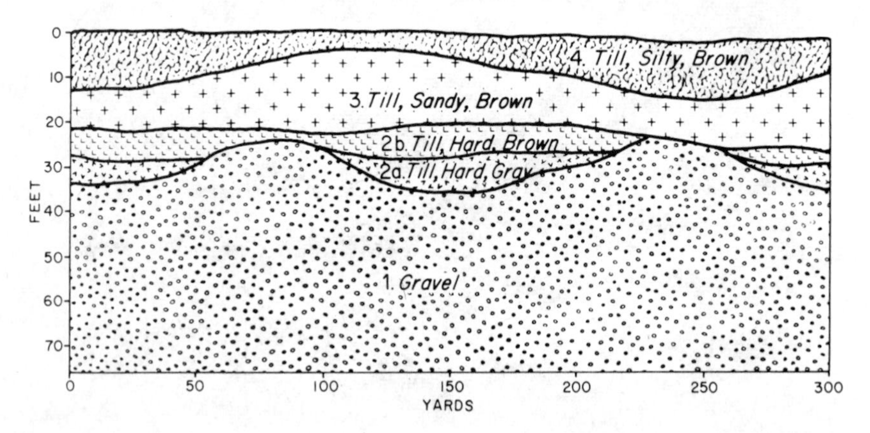

Figure 5. Sketch of glacial deposits exposed in gravel pit of Rupp Construction Company, 3 miles east of Marshallville, Wayne County, Ohio. The gravel, I, is part of a large kame terrace, probably early Altonian in age. Units 2a and 2b are Millbrook Till (Altonian); unit 3 is Kent Till (early Woodfordian); unit 4 is Hayesville Till (late Woodfordian). Erosion during at least 10,000 years modified the surface of till 2; a somewhat shorter period of erosion modified the surface of till 3. Note that the hummocky surface of the buried kame terrace is completely masked.

gravel must be hauled from greater distances and greater thicknesses of overburden must removed. At a location near Cleveland, as much as 60 ft of overburden consisting of two tills, two loesses, and more than 10 ft of lacustrine silt has been removed to secure the underlying gravel (White, 1968, Fig. 1).

Paleoeskers The remarks about paleokames apply also to eskers and no extended treatment is necessary.

Paleo-outwash plains and valley trains. Just as kames and kame terraces may be covered by later depostis, so outwash plains and valley trains may be covered (White, 1960, p. A-7,8). The paleosurface of a buried outwash plain is usually level and quite obvious in excavations or in drilling, as shown in Figure 2 by the "sand and till" below the Snider Till and the "gravel and silt" beneath the Tilton Till. The interfaces at top and bottom of a buried outwash plain may be discovered by seismic exploration, as shown by unit C of Figure 4.

Paleo-valley trains are very prevalent in buried ("preglacial") valleys. Even if the buried valley is truly preglacial, the filling is glacial and its upper surface is a paleo-glacial surface. The valley form may still be evident, or the valley may be so completely covered by later till deposits that its presence is not obvious from the surface (Selkregg and Kempton, 1958, Fig.8; Christiansen, 1973). The paleo-valley train in the Mahomet Bedrock Valley in east-central Illinois is the source of a tremendous supply of groundwater (Stephenson, 1967, Fig. 9).

It is well known from drilling that gravels in buried valleys may vary vertically. The variation may consist of layers of gravel of different composition, which are probably valley trains of different ages. Paleosurfaces may be betrayed by organic deposits upon them. Till sheets may lie between buried valley trains of different ages; these are frequently encountered in exploration for water supplies in buried valleys (Schaefer, White, and Van Tuyl, 1946). The gravel surface below the till is, of course, a paleosurface and the surface of the overlying till, upon which later gravel is deposited, is a later paleosurface. Little work has been done in mapping these important paleosurfaces, but their importance is obvious (Kazmann, 1949; White, 1967, p. 31).

Paleo-valley trains contain gravel of economic importance, but as they are generally below water level, and as the cover may be alluvium over a till sheet, the recovery of the gravel requires removal of the material over the gravel and the excavation of the gravel by dredging, dragline, or wire cable scraper.

Paleolacustrine plains. The buried lake deposits are so far little known, except in North Dakota and a few other places. The history of the late glacial and post-glacial Great Lakes is now well understood, but there must have been similar conditions at many times in the Pleistocene. Pre-last glacial lake deposits are only very meagerly known at various places around the Great Lakes. The Chicago paleolacustrine deposits are the best studied, because the paleosurface of some of the lake silts and clays presents special engineering problems (Peck, 1969, and references).

Drilling (Farrand, 1969), gravity coring (Gross and others, 1970; Dell, 1972), and seismic studies (Lineback and others, 1971 and 1972; and Lineback and Gross, 1972) in the basins of the Great Lakes and other lakes (Chase and Hunt, 1972) is just now being developed, and earlier lake deposits, outwash, and tills are being discovered beneath the sediments on the lake floors (Fig. 4). As these deposits have not suffered any subaerial erosion, their paleosurfaces are entirely depositional and they provide some of the best continuous records of the Pleistocene. Figure 4 shows the amount of detail of paleogeomorphology of lake sediments and of glacial material beneath them that is now being obtained by seismic methods. When seismic methods are supplemented by drilling and gravity coring a magnificent record of Pleistocene history is revealed. Buried beaches are interesting paleogeomorphic features; they may have important effects on engineering structures if not recognized. Building collapse and other disasterous effects due to pore water pressure within the buried sand ridges may be possible.

Paleoglacial eolian features, Buried sand dunes are rarely reported in the literature, but such features are known (Winslow and White, 1966, p. 34, Fig.16). As buried loess sheets are of common occurrence, it may be that paleodunes are more common than realized.

Paleoloess sheets. Loess sheets lying below later till are widespread. In Illinois the Roxanna Loess ("Farmdale" of earlier reports) is widely distributed beneath overlying till of Woodfordian age (Horberg, 1953; Willman and Frye, 1970). In the Allegheny Plateau, loess deposits near Cleveland are overlain by lacustrine silt, which is in turn overlain by till (White, 1968). Paleoloess, with or without associated organic material, underlies till at many places in central Illinois. The paleosurface is from 5-25 ft below the land surface. This paleosurface must be carefully mapped in any area on which large structures, such as a power plant, are to be constructed, because the paleoloess beneath that surface has much different bearing properties from the till above. The paleo-surface is also an interface that influences groundwater movement.

Beyond the glacial boundary, several loess sheets may be superimposed. Those below the uppermost one are paleoloesses, and characteristically have different composition and structure from the uppermost one. The difference may be so marked that the paleosurface may be mapped over large areas by drilling or augering (Follmer, 1970; Mason, 1973a, 1973b).

Buried erosion surfaces on drift. The paleosurface of a drift may be that of the original deposit, now preserved by the overlying material. However, the earlier drift surface may have been modified by erosion of the later ice which covered it, or by stream erosionafter deposition, but before the following glaciation.

Glacial erosion may remove some, or all, of an earlier drift (Fig. 2, below Hillery Till). In the first case some or all of the weathered material on the earlier drift may be removed so that fresh till lies upon earlier fresh till. Such an unconformity between unweathered tills (gray upon gray) is difficult to identify, but the interface may be water bearing, or if traced laterally it may pass into a tract where some of the weathered lower till is preserved.

If all the earlier drift is removed, the later till may rest upon a still earlier till or upon bedrock, and all record of the first unconformity will be lost. It must be realized, therefore, that a paleo-surface may actually truncate another paleosurface, just as in bedrock one unconformity may truncate another, as shown in Figure 6. (For other examples see Johnson and others, 1971, Figs. 1, 6, and 7).

In some places glacial erosion may be very selective, as along valleys parallel to the direction of ice motion. Here "finger lake type valleys" may be eroded in glacial drift, as well as those better known examples eroded in bedrock.

Stream eroded paleosurfaces may exhibit every variation of stream erosion on present day surfaces. Gullies, ravines, and valleys of varying depth cut into the drift have been buried by later glacial deposits (Figs. 7 and 8). These inter-till valleys may be filled with silt, sand or gravel. In superhighway excavations, most cuts are deep enough to penetrate the upper till and even more than one upper till. The interfaces are pathways for water movement. When the cut transects even the smallest paleovalley (inter-till valley), water seeps out in springs along the cut and may cause more or less severe slope failure. Some paleovalleys are the source of quite large springs and cause problems in construction and in maintenance (Norris and White, 1961). Disaster may occur if such paleovalleys are intersected in deep excavations for building foundations or in tunneling.

Another type of fluvial erosion on drift surfaces is that of pedimentation. The surface is then covered by later deposits, wind blown or other. The elaborate studies in Iowa by Ruhe (1969) demonstrate the extent and importance of such paleopediment surfaces.

Another type of erosion of drift is by subglacial erosion when a later ice sheet covers the earlier drift. Carruthers (1953, p. 14) speaks of fluvial erosion beneath ice and recently Wright (1973) has dealt at length with great valleys (tunnel valleys) excavated beneath an ice sheet. These predate the last ice advance, for he describes these tunnel valleys as although "partially obscured ... by younger drift, as many as 12 can be recognized across a breadth of about 100 Km" (p. 253). Similar features described by Farrand (1969) in Lake Superior may be of the same origin. It may be that similar so far unrecognized tunnel valleys may exist elsewhere, completely buried by later drift.

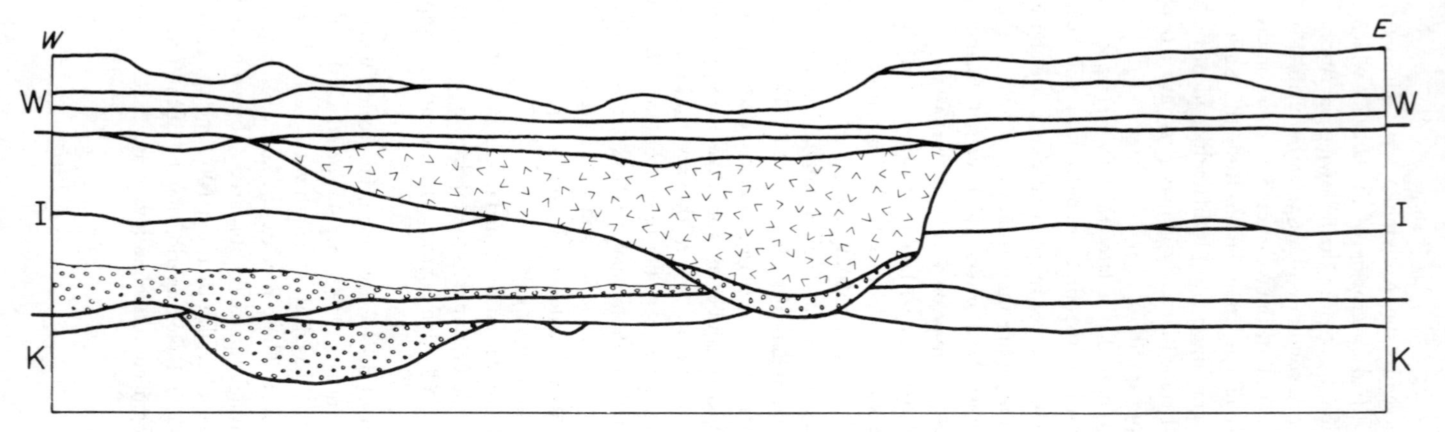

Figure 6. Generalized diagram of glacial deposits in north wall of Harmattan Strip Mine, 3½ miles W of Danville, Illinois. Length of section 600 feet, height 60 feet. Units are K --Kansan, I--Illinoian, and W--Wisconsinan. They are mainly tills. with outwash associates with some of the tills. Bedrock 10 feet below base of this exposure. Major outwash, sand and gravel, shown by circle pattern. Large buried valley in central part of diagram contained much more sand and gravel along much of its course and is conspicuously water-bearing. Thin loess and soil at top of section not shown. Adapted from Figure 16, Johnson and others (1972), which see for further details and formation names.

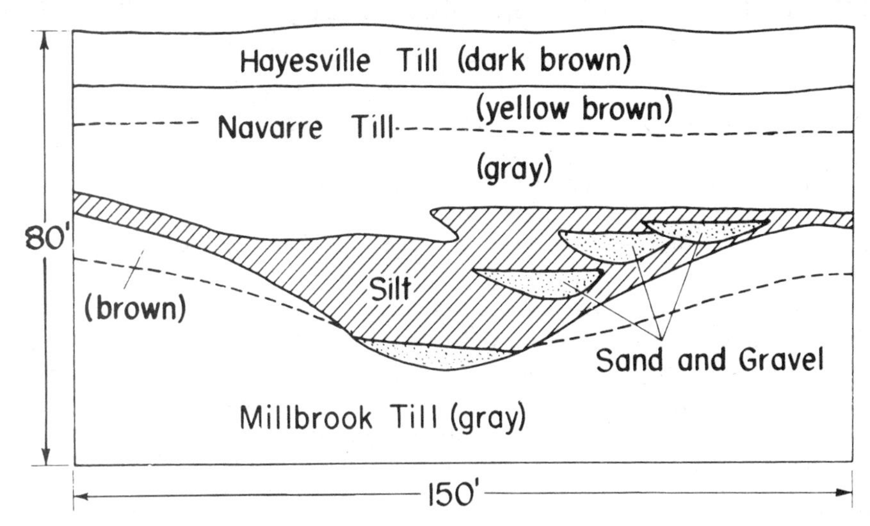

Figure 7. Sketch of glacial deposits in eastern Morrow County, Ohio. The Hayesville and Navarre Tills are Woodfordian, the Millbrook Till is of Altonian age. The valley was cut into the Millbrook surface and later filled (pro-Navarre?) by silt, with sand and gravel deposited in changing channels. The sand and gravel channels are water-bearing.

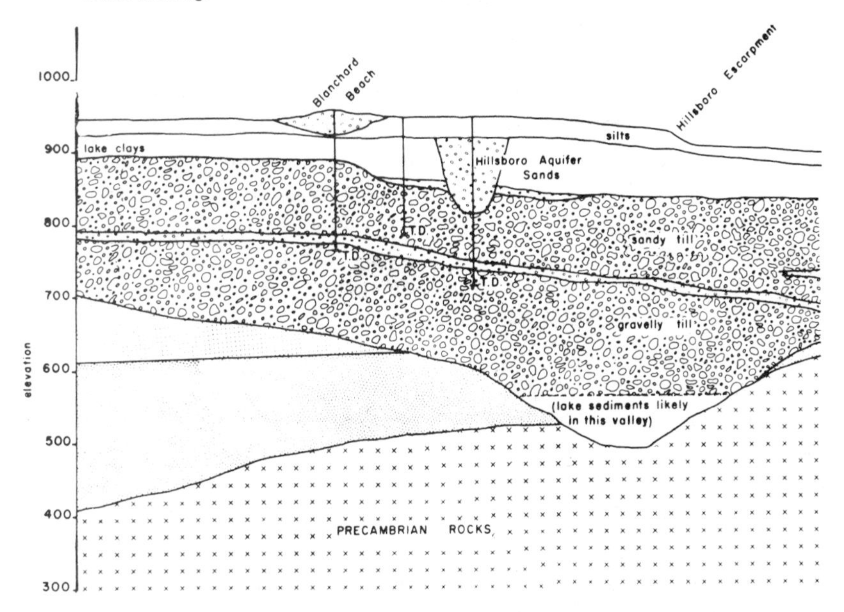

Figure 8. East-west section in central Traill County, North Dakota. Length of section about 20 miles. The strata on the left, shown by dotted pattern above the Precambrian rocks, are Cretaceous bedrock. The lake clays are deposits in Early Lake Agassiz. Note "Hillsboro Aquifer Sands," which are in a buried valley of ancestral Goose River, which flowed into a lower stage of Lake Agassiz; the lake level again rose, and later lake silts were deposited upon the surface of the older lake deposits and upon the valley and its deposition. Central part of section B-B', Bluemle, 1967.

ECONOMIC IMPLICATIONS OF BURIED GLACIAL GEOMORPHOLGY

Buried surfaces in glacial deposits may greatly affect the suitability of these deposits for various purposes, including waste disposal, water supply, and construction upon or in the material (White, 1972).

In planning solid waste disposal sites it should be recognized that one or more paleosurfaces are likely to be encountered at depth. These may provide passage ways for leachate, or may allow groundwater flow into the excavation. The permeability of the surface till may be such that all requirements are met, but below a paleosurface more permeable till, either more jointed or coarser textured, may be present. Below the paleosurface may be sand or gravel. Liquid waste disposal may be similarly affected by different materials below paleosurfaces.

Water supply in many places depends on discovery and recognition of paleosurfaces below "impermeable" surface material. The search for buried valleys is generally for buried bedrock valleys, but buried valleys in drift can also be considerable sources of water. It may even be possible to map these (Norris and White, 1961).

Foundations for major structures may encounter obvious paleosurfaces, with sharp change of material at the interface. However, the paleosurface may be less obvious, but nevertheless mark a considerable change in material. Buried valleys may lie below a paleosurface, but at no great distance from the ground surface (Fig. 8), in such a position that a large structure may be partly over it. Change in pore water pressure in the buried sand or silt at some time may contribute to serious problems.

In any excavation in drift, unless very shallow, it should be assumed that one or more paleosurfaces will be encountered. Soil tests of the few feet of drift at the surface may not apply to the material below the paleosurface. The paleosurface may be upon buried till sheets, kames, gravelly moraine, fill in buried valleys or ravines, or outwash plains. At the paleosurface drainage problems may be more or less severe, and sometimes may be very costly at the time of construction. and in highway cuts, costly in continuous drainage maintainance.

More and more construction underground will require not only deeper excavations, but tunnelling through drift will become more and more common. For example, large sewer lines are now being placed in tunnels in some localities. The prospect of underground transport lines for longer distances than for the present city subways is in the near future. If the drift is hundreds of feet thick, as it is in parts of western Canada, where thicknesses of 1,100 ft are known, it will not be possible to place these tunnels deep enough to be in bedrock. Here it is particularly important to take into account existence of paleosurfaces. If these are irregular, as in kames and buried valleys below till, there will be a rapid change in conditions along the course of the tunnel. Sand and gravel may be water bearing, with all the problems associated with this condition. The possible presence of boulder pavements at an interface should be kept in mind. The effect of an unsuspected continuous sheet of boulders on tunnelling machinery can be quite unhappy!

Foundation problems of glacial and other surficial materials are dealt with in detail by Legget (1973) with many examples (see especially p. 217-221). Legget's chapter in this symposium is concerned in detail with these matters.

The effect of change in thermal properties at a paleosurface is just now being recognized. Thermal properties of unconsolidated material are of great importance in construction of underground high voltage alternating current transmission lines. Fischer and others (1973, p. 23-26) have investigated the thermal properties of marine sediments and shown that these have very considerable variation. Cartwright (1968) has begun thermal studies of drift. When large power lines are placed underground in glacial materials, as they are sure to be sooner or later, the variation in thermal properties where paleosurfaces are crossed along the line must be carefully determined. Legget (1973, p. 475) has discussed the possible effect of variation of surface materials on the newly emerging direct current transmission at high voltage.

CONCLUSION

Almost all glacial deposits are complexes of more than one material deposited at more than one time. As many as 10 episodes of deposition over many hundreds of thousands of years have left their records in the form of separate "sheets" at some places.

Each deposit had an original surface of deposition which was more or less modified before or during the time the next ice sheet was being deposited. Each surface is an unconformity -- a surface of paleogeomorphology. The recognition and proper interpretation of these paleosurfaces is the substance of Pleistocene history. The recognition of their presence is the first step in planning engineering structures, excavations, tunnels, water supply, and waste disposal.

ACKNOWLEDGEMENTS

I am grateful to Dr.D.L. Gross for his critical reading of a draft of this paper, for his valuable suggestions and additional references, and for providing Figure 4 from unpublished work; to Professor W.H. Johnson for critical reading and stimulating discussion, and for suggesting additional references and improvement; and to Dr.S.R. Moran for information and references on North Dakota.

REFERENCES

Bluemle, J.P. 1967. Geology and ground water resources of Trail County, North Dakota Part I, Geology: N.D. Geol. Survey Bull. 49, 34 p., 2pl.

Carruthers, R.G. 1953. Glacial drifts and the undermelt theory: Newcastle upon Tyne, Harold Hill & Son, Ltd., 42 p.

Carter, C.H. 1974. Geology of Lake Erie shore in Lake County, Ohio: Geol. Survey of Ohio, Rept. of Investigations.

Cartwright, K. 1968. Temperature prospecting for shallow glacial and alluvial aquifers in Illinois: Ill. Geol. Survey Circular 433, 41 p.

Chase, J.S. and A.S. Hunt, 1972. Sub-bottom profiling in central Lake Champlain -- A reconnaissance study: Int. Assoc. Great Lakes Res., p. 317- 329.

Christiansen, E.A. 1971. Tills in Southern Saskatchewan, Canada: in *Till, A Symposium,* R.P. Goldthwait, ed., The Ohio State University Press, p. 167-183.

Christiansen, E.A. 1973. Geology and groundwater resources of the Shellbrook Area (73-G) Saskatchewan: Saskatchewan: Saskatchewan Research Council Geology Division Map No. 17.

Cummins, J.W. 1959. Buried river valleys in Ohio: Ohio Div. Water, Ohio Water Plan Inventory Rept. 10, 3 p. & map.

Dell, C. I. 1972. The origin and characteristics of Lake Superior sediments: Int. Assoc. Great Lakes Res., Proc. 15th Conf. on Great Lakes Res., p. 361-370.

Dreimanis, A. 1970. Last-Ice-Age deposits in the Port Stanley Map-Area, Ontario (40 I/II): Report of Activities, Part A, Geol. Survey Canada Paper 70-1, Part A, p. 167-169.

Farrand, W.P. 1969. The Quaternary history of Lake Superior: Internat. Assoc. Great Lakes Research Proc., 12th Conf., Ann Arbor, p. 181-197.

Fischer, J.A., Watson, Ian, and Solomone, L.A. 1973. Considerations of the geotechnical engineer in planning transmission lines to offshore nuclear power plants: Dames and Moore Engineering Bull. No. 42, Sept., 1973. p.19-28.

Flint, R.F. 1971. *Glacial and Quaternary geology:* John Wiley and Sons, N.Y., 892 p.

Follmer, L.R. 1970. Soil distribution and stratigraphy in the mollic albaquolf region of Illinois: unpublished Ph.D. thesis, Univ. of Illinois, Urbana.

Frye, J.C., 1973. Pleistocene succession in the central United States: Jour. Quat. Research, v. 3, p. 275-283.

Gross, D.L., J.A. Lineback, W.A. White, N.J. Ayer, Charles Collinson and H.V. Leland. 1970. Preliminary stratigraphy of unconsolidated sediments from the southwestern part of Lake Michigan: Illinois Geol. Survey Environmental Geology Note 30, 20 p.

Gross, D.L. and S.R. Moran. 1971. Grain-size and mineralogical gradations within tills of the Allegheny Plateau: in *Till, A Symposium,* R.P. Goldthwait, ed., The Ohio State Univ. Press, p. 1

Head, J.G. 1907. The buried rivers of London; a topographical sketch suggested by the difficulties encountered in dealing with property from the presence of water in the soil: Auctioneer's Institute ...Session 1907-1908, p. 19-46, map.

Horberg, C.L. 1950. Bedrock topography of Illinois: Illinois Geol. Survey Bull. 73, 111 p.

Horberg, Leland. 1953. Pleistocene deposits below the Wisconsin Drift in northeastern Illinois: Ill. Geol. Survey Rep. Inv. 165, 61 p.

Johnson, W.H. 1971. Old glacial drift near Danville, Illinois: Ill. State Geol. Survey Circular 457, 16 p.

Johnson, W.H.., Follmer, Leon, Gross, D.L., and Jacobs, A.M. 1972. Pleistocene stratigraphy of east-central Illinois: Ill. State Geol. Survey Guidebook Series 9, 97 p.

Johnson, W.H., D.L. Gross, and S.R. Moran. 1971. Till stratigraphy of the Danville region, east central Illinois: in *Till, A Symposium,* R.P. Goldthwait, ed, The Ohio State Univ. Press, p. 184-216.

Kazmann, R.G. 1949. The utilization of induced stream infiltration and natural aquifer storage at Canton, Ohio: Econ. Geol, v. 44, p. 514-524.

Kempton, J.P., DuMontelle, P.B., and Glass, H.D. 1971. Subsurface stratigraphy of Woodfordian Tills in the McLean County region. Illinois: in *Till, a symposium:* The Ohio State Univ. Press, p. 217-233.

Legget, R.F., 1973, *Cities and geology:* McGraw-Hill Book Company, N.Y, 624 p.

Lessig, H.D. 1959. The Supermarket terrace, near East Liverpool, Ohio: Ohio Jour. Science, v. 59, p. 332-338.

Lineback, J.A., Gross, D.L., and Meyer, R.P. 1972. Geologic cross sections derived from seismic profiles and sediment cores from southern Lake Michigan: Ill. Geol. Survey Environmental Geology Note 54, 43 p.

Lineback, J.A., Gross, D.L., Meyer, R.P., and Unger, W.L. 1971. High- resolution seismic profiles and gravity cores of sediments in southern Lake Michigan: Ill. Geol. Survey Environmental Geology Note 47, 41 p.

Lineback, J.A., Gross, D.L., and Meyer, R.P. 1972. Geologic cross sections derived from seismic profiles and sediment cores from southern Lake Michigan: Ill. Geol. Survey Environmental Geology Note 54, 43 p.

Lineback, J.A., Gross, D.L., Meyer, R.P. in preparation. Glacial tills beneath Lake Michigan: Ill. Geol. Survey Environmental Geology Note.

Mason, R.M. 1973a. Landscape evolution of a portion of the Illinoian drift plain in central Illinois: unpublished Ph.D. thesis, Univ. of Illinois, Urbana.

Mason, R.M. 1973b. Landscape evolution of a portion of the Illinoian drift plain in central Illinois: (abstract), Geol. Soc. Amer. Abstracts with Programs, v. 5, p. 728.

Moran, S.R. 1971. Glaciotectonic structures in drift: in *Till, A Symposium,*R.P. Goldthwait, ed., The Ohio State Univ. Press, p. 251-274.

Norris, S.E., and White, G.W. 1961. Hydrologic significance of buried valleys in glacial drift. U.S. Geol. Surv. Prof. Pap. 424-B, p. 34-35.

Peck, R.B. 1969. Art and Science in subsurface engineering: Geotechnique v. 19, p. 60-66.

Piskin, K., and Bergstrom, R. E. 1967. Glacial drift in Illinois: thickness and character: Ill.Geol. Survey Circ. 416, 33 p.

Price, R.J. 1973. *Glacial and fluvioglacial landforms::* Oliver and Boyd, viii+242 p.

Ruhe, R.H 1969. *Quaternary Landscapes in Iowa:* Iowa State Univ. Press, 255 p.

Schaefer, E.J., White, G.W., and Van Tuyl, D.W. 1946. The groundwater resources of the glacial deposits in the vicinity of Canton, Ohio: Ohio Water Board Bull. 3, 60 p.

Seldregg, L.F. and Kempton, J.P. 1958. Groundwater geology in east- central Illinois: Ill. Geol. Survey Circular 248, 36 p

Stephenson, D.A. 1967 Hydrogeology of glacial deposits of the Mahomet Bedrock Valley in east-central Illinois: Geol. Survey Circular 409, 51 p.

Totten, S.M. 1969. Overridden recessional moraines in north-central Ohio: Geol. Soc. Am. Bull., 80, p. 1931-1946.

Totten, S.M. 1973. Glacial geology of Richland County, Ohio: Ohio Geol. Survey Rept. of Investigations No. 88, 55 p. & maps.

Wanless, H.R. 1970. Late Paleozoic deltas in the central and eastern United States: Deltaic sedimentation, modern and ancient: Soc. Econ. Paleontologists and Mineralogists Spec. Pub. 15, p. 215-245.

White, G.W. 1960. Classification of glacial deposits in northeastern Ohio: U.S. Geol. Survey Bull. 1121-A, 12 p.

White, G.W. 1962. Multiple tills in end moraines: U.S. Geol. Surv. Prof. Pap. 424-C, p. 71-73.

White, G.W. 1967. Glacial geology of Wayne County, Ohio: Ohio Geol. Survey Rept. Investigations No. 62, 39 p. & map.

White, G.W. 1968. Age and correlation of glacial deposits at Garfield Heights (Cleveland) Ohio: Geol. Soc. Amer. Bull., v. 79, p. 749-756.

White, G.W. 1971. Thickness of Wisconsinan tills in Grand River and Killbuck Lobes: in *Till, a symposium,* R.P. Goldthwait, ed., The Ohio State Univ. Press, p. 149-163.

White, G.W. 1972. Engineering implications of stratigraphy of glacial deposits: 24th Internat. Geol. Congress, Section 13. Eng. Geology, p. 76-82.

White, G.W. 1973 History of investigation and classification of Wisconsinan drift in North-central United States: in *The Wisconsinan Stage,* R.F. Black and others, eds. Geol. Soc. America Memoir 136, p. 3-34.

White, G.W., Totten, S.M., and Gross, D.L. 1969. Pleistocene stratigraphy of northwestern Pennsylvania: Penn. Geol. Surv. Bull. G55, 88 p.

Willman, H.B., and Frye, J.C., 1970. Pleistocene stratigraphy of Illinois: Ill. Geol. Survey Bull. 94, 204 p.

Winslow, J.D., and White, G.W. 1966. Geology and ground-water resources of Portage County, Ohio: U.S. Geol. Survey Prof. Paper 511, 80 p.

Wright, H.E., Jr. 1973. Tunnel valleys, glacial surges, and subglacial hydrology of the Superior Lobe, Minnesota: in *The Wisconsinan Stage,* Black, R.F., Goldthwait, R.P., and Willman, H.B., eds., Geol. Soc. Memoir 136, p. 251-276.

CHAPTER 14

GLACIAL LANDFORMS AND CIVIL ENGINEERING

Robert F. Legget

ABSTRACT

Since most of Canada and much of northern U.S.A. has been glaciated, much civil engineering work in this vast area has been carried out in association with glacial landforms. In earlier days such features may not have been recognized as such but they were successfully used, transportation routes along old glacial spillways being one example. Mistakes were made through non-recognition of glacial features, mainly in failures to recognize glacial till prior to excavation due to the use of old-fashioned wash-boring. Modern sub-surface exploration for civil engineering starts with a study of landforms so that test drilling with modern techniques may be effective, even to the identification of paleo-till surfaces. Glacial landforms have been well used, as for river diversions along former spillways. Aerial photography provides today a most useful technique for general study of glacial landforms but for their full appreciation in civil engineering work patient study on the ground is still essential. Just as geology alone was found to be inadequate, leading to the development of soil mechanics, so today it is becoming generally recognized that soil mechanics must be used against a background knowledge of glacial landforms (in all glaciated country) if it is to be fully effective in assisting civil engineers in the practice of their profession.

INTRODUCTION

"The author having in the course of his professional duties, discovered in the lake district of Upper Canada terraces or level ridges which agreed in elevation at considerable horizontal distances, he was induced to extend his inquiries and ascertain how far similar phenomena have been observed in other parts of North America......" These are the opening words of a paper entitled *On the Ancient State of the North American Continent,* the author being Thomas Roy Esq., Civil Engineer, of Toronto, Upper Canada. It was presented to the Geological Society of London at its meetings on 22 March and 5 April in the year 1837. A summary of this paper was printed on pages 537 and 538 of the Second Volume of the Proceedings of this venerable Society, a summary only "as his description cannot be successfully followed without the aid of diagrams, (and) they do not admit of being given in the Society's Proceedings." Mr. Roy describes the raised beaches still to be seen on the north shore of Lake Ontario and discusses possible causes of their formation, comparing them with the "drifted matter ....now disposed along the margin of the lakes at the breaking up of the ice; and hence he conceives, that the ridges may, to a considerable extent, have been accumulated in a similar manner." He then calculates the volume of the ancient lake which formed the raised beaches and found that fifteen years would have been required to lower it by thirty feet if it had been drained at double the rate of the combined flow of the Mississippi, the Hudson and the St. Lawrence Rivers (Roy, 1837).

No trace of Mr. Roy's original paper has yet been found, most unfortunately, since it was clearly an unusually percipient discussion of glacial landforms even though the cause of the raised beaches was not then recognized. The standing of Thomas Roy is indicated by this extract from the record that Sir Charles Lyell published of his first visit to North America. When, on 14 June 1842, Sir Charles visited Toronto, with a population at that time of only 18,000, he "..found Mr. Roy, a civil engineer, expecting me..... to examine (with him) those ridges of sand and gravel and those successive terraces, at various heights above the level of Lake Ontario, of which he had given an account in 1837 to the Geological Society of London" (Lyell, 1845). And a distinguished officer of the Corps of Royal Engineers, Sir Richard Bonnycastle, who did much work in Canada, wrote in an account of his Canadian experiences: "Lyell visited (Toronto) with the late Mr. Roy, a person little appreciated and less understood by the great men of the earth at Toronto, who made an excellent geological survey of this part of the province.....and no one has given him even a shadow of justice, but Mr. Lyell, who, having no colonial dependence, had no fears in so doing" (Bonnycastle, 1846). Little is known about Thomas Roy apart from the fact that he spent eight years in professional practice in Toronto where he died in the summer of 1842. Such records as he did leave show that he fully appreciated the importance of landforms in relation to his engineering work. With so distinguished a precedent , perhaps this paper will not be completely out of place in this Symposium.

EARLY RECOGNITION

It did not require any knowledge of glacial action to enable early builders to appreciate the value of glacial landforms. Supplies of sand and gravel must have been procured, for example, from the earliest time from landforms such as kames and eskers. Such uses are so obvious that little further reference to them will be made. Not so obvious was the use made of glacial landforms in the development of transportation routes. This antedated even early engineering since the location of primitive roads and pioneer railways often followed routes already in use by simpler modes of transportation. The Ottawa River in eastern Canada provides a good but typical example. A tributary of the River St. Lawrence, which it joins at the west end of the Island of Montreal, the Ottawa itself is a large river with a mean annual flow of about 70,000 c.f.s. From the time of Champlain's journey in 1615 up the river, then up the smaller Mattawa River, portaging into Lake Nipissing and then descending the French River into Georgian Bay, until the beginning of the nineteenth century, the Ottawa Waterway (as the combined route may well be called) was the gateway to the west and, for all the earlier part of this two hundred year period, the gateway to the continent. Up this route went the early explorers of the Great Lakes, the Midwest, the Mississippi and the Missouri, the Canadian west, the way to the Pacific coast and to the Arctic coast of northwestern Canada, all by canoe. What made the route so convenient was the short distance from the headwaters of the Mattawa River to Lake Nipissing, a mere 4 mi. The reason for this? This was the second glacial spillway from the later stages of the glacial Great Lakes, specifically from Lake Hough.

So vast must have been the flow down this spillway that it is small wonder that it provided so obvious a route to the west, long used by the Indians before the coming of the white man and shown by them to Champlain who prepared the first accurate map of the Waterway. It was therefore natural that when, in the 1880's, the Canadian Pacific Railway (or more accurately, one of its early constituent lines) made its way up the Ottawa Valley to join up with the line planned to the west, it should have followed very closely the old canoe route. Here, then, is a main railway line well located with easy gradients even though passing through difficult country, making good use of a major glacial landform. When, at a later date (in the period of the first World War) the Canadian Northern Railway, now a part of Canadian National Railways, constructed its line up the Ottawa Valley, also to North Bay on Lake Nipissing, it followed a different route, some distance inland from the Ottawa River. It is a location that still puzzles many interested in railways but the CNR location engineers knew a good route when they saw one, even though they probably did not know that they too were using a glacial spillway from the early Great Lakes. One of the stations on this relatively isolated line is called Fossmill, a name now well known to glacial geologists since L. J. Chapman applied this name to the first spillway in this region when he described it in a notable paper to the Geological Association of Canada (Chapman, 1954).

The Fossmill and Mattawa glacial spillways have therefore been well used for transportation routes. The Canadian Pacific main line continues on its long route to the west by following the north shore of Lake Superior. This was a conscious decision of those responsible for its construction, based on the careful surveys carried out earlier by Sir Sandford Fleming. Here, too, glacial landforms were put to good use, notably sections of the raised beaches that are to be found along this wild and inhospitable shoreline. They must have attracted the attention of the many early travellers who paddled their canoes along this shore on their way to the Grand Portage. leading them to the west, since they are so pronounced as to be evident from quite a distance despite their forest covering. One especially notable location is even named Terrace Bay, after the beaches that are so prominent up its steep slopes. It was visited by Sir William Logan in 1847 and by Louis Agassiz in 1850, a pleasing woodcut of the Bay being the frontispiece of his volume on Lake Superior (Agassiz, 1850, Fig. 1). The first detailed study of Terrace Bay was made by Andrew Lawson in 1897, but well before this one of its beaches had been used as the location for the C.P.R. main line. This particular location is mentioned, even though a number of other cases could be cited, since Terrace Bay will call for further reference later in this paper.

It was not only in such relatively isolated locations that railway engineers, even if unwittingly, utilized the advantages that some glacial landforms present for the

Figure 1 Woodcut of Terrace Bay, Lake Superior - the frontispiece to Louis Agassiz' volume on *Lake Superior.*

location of routes. It is not surprising, for example, to find the old beaches of Lake Iroquois that were described by Thomas Roy used in the development of early transportation routes in this part of Ontario. At Hamilton, near the western end of the Lake Ontario, two railways and three roads make use of the convenience provided by the high level bar of this glacial lake. Main railway lines running to the east from Toronto utilized the old shorelines that are so distinctive a feature of this part of the north shore of Lake Ontario, notably the Canadian Northern line, now abandoned in favour of the old Grand Trund (but now Canadian National) line which is similarly located in part. North of Toronto, the important line of the Canadian Pacific Railway passing through Barrie follows a wave-cut terrace from Colwell to Midhurst. A large barrier beach in this vicinity, thrown up by the ancient lake, has been well used by the C.P.R. as a source of excellent sand and gravel. (White, O., 1973; and Chapman and Putman, 1966)

It would be tedious to continue this recital of the early use of glacial landforms in Ontario for railway location. Engineers working in this area now have available an excellent guide to the physiography of the entire region which may be consulted for further examples (Chapman and Putnam, 1966). Not only have old beaches been used, however, but also smaller meltwater channels such as that along Cedar Creek to the west of Galt which permitted the location of another C.P.R. line through hilly morainal country with easy grading. Correspondingly, major re-entrant valleys that were filled with deep kame-like deposits of gravel provided excellent "ramps" up which major railway lines make their way across the Niagara Escarpment, greatly to the benefit of the grades necessary for the big climb. The Canadian National line at Acton, that of the Toronto, Hamilton and Buffalo Railway at Dundas and the Canadian Pacific line at Campbellville are the main examples (Karrow, 1973).

Location of early roads was similarly influenced by glacial landforms in southern Ontario as in other areas of early settlement that had been glaciated. It is surprising how frequently the older towns in this, and indeed in similar, areas will be found to have "Ridge Roads" as well-established older highways, the name clearly indicating the original use of a prominent geomorphic feature, almost always a glacial landform. Ridge Road between Stirling and West Huntingdon, north of Trenton at the east end of Lake Ontario, is an excellent example since it so clearly follows a well defined esker. In one respect only was early settlement of this particular glaciated country not able to take full advantage of geomorphic features such as glacial landforms. Land subdivision was carried out by early surveyors who had to follow well defined rules, all based on the concept of water-front lots which were originally so essential when all travel in summer was by water. Base-lines for townships and concessions had to be parallel with lake shore lines. This often led to subdivision of land that can only be described as rather awkward.

The cities of Peterborough and Guelph provide a significant contrast. Both are located in well known drumlin fields. Axes of the drumlin assemblies are well defined in both areas. In the Peterborough area, however, the main survey lines make angles of about 45 degrees to the general trend of the drumlins with the result that there are here many odd-shaped lots which can not be effectively used and early road construction was not assisted. In the Guelph area, on the contrary, the survey lines most fortuntely run parallel to the trend of the drumlins so that early road location was greatly assisted and the working of sub-divided land was simplified. In both cities, major public buildings were (quite naturally) located on drumlins. There is some rivalry between them as to which is the "Drumlin Capital of Ontario".

Although the significance of glacial landforms in the search for sand and gravel needs no elaboration here, there is one case of material procurements as a direct result of glacial action that the writer believes to be unique so that brief reference may be warranted. Visitors to the University of Saskatchewan at Saskatoon must often be attracted by the fine masonry buildings that constitute all the early, and still major, part of the campus. If geologists, they will be puzzled since a glance at a geological map will show that the nearest outcrops of bedrock are 320 km away and the early founders of this fine University in the middle of the praries had no money to spend on long-distance haulage of building stone. They were observant, however, and some unknown but astute observer noticed that farmers to the northeast of the city had many fine boulders in their fencelines, piled up after being hauled away in the early clearing of the ground. These boulders form the building stone used so successfully, and economically, by the University. Their origin was in what must have been a magnificent boulder train, the boulders scoured out of the exposure of the Ordovician Red River limestone far off to the northeast, the apex of the fan having been about 40 km from the University campus. (Byers, 1971).

EARLY MISTAKES

There is naturally another side to the influence of glacial landforms on early civil engineering work but most unfortunately, although understandably, it is not documented. Only in relatively recent years has it become recognized as a professional responsibility to describe in published form mistakes from which lessons for future practice may be learned. In what will be seen to be rather diverse general reading I have encountered records of two interesting failures due to non-recognition of the glacial features involved. One of these is a passing reference only to the very questionable engineering capability of a fine pioneer Roman Catholic missionary priest who did much for the opening up of settlement around Lake Timiskaming, an enlargement of the north-flowing Ottawa River which forms the boundary between Ontario and Quebec, just to the south of the famous gold mining area of Kirkland Lake. One of his unfulfilled ideas was to drain this great lake by 20 ft. while building a dam to drown out the rapids downstream. It is perhaps fortunate that this project did not go ahead since it is reported that "the good Father's engineering seems to have been slightly confused, as it was again some years later when he accidentally drained Frederick House Lake in the course of some mining operations" (Pain, 1964).

Rather better documented is a case that happened during the building of the White Pass and Yukon Railway between Skagway, Alaska and Whitehorse, Yukon Territory. This remarkable railway was built at the time of the Klondike gold rush between May 1898 and July 1900, all by hand-power, there being at one time no less than 35,000 men at work. It has been in continuous operation since its opening, still using the same location which all visitors still admire that was selected with a minimum of the more usual careful preliminary surveys, the first 20 mi being (for example) at a grade of 2.6% on average, with a short section at 4%. Lewes Lake was passed at mileage 81 from Skagway using the east shoreline for the railway location. It was greatly indented, a common feature in the small lakes in this area which lies close to the northern limit of glaciation but to the south of this. Geomorphology betrays the varied morainic structure of the ground, some of the lakes being formed by bars of glacial till. Lewes Lake was one but the tough gentlemen who built this line were not students of glacial landforms. It is recorded therefore, that "in order to obtain a better line (for location of the line) it was decided to lower the surface of the water about fourteen feet. This was done by excavating an outlet channel. However, the unforeseen happened. The water, after remaining stationary for some time, cut its way through the sandy outlet, forming an enormous canyon through the lake bed. This reduced the lake level by over seventy feet, and left a dozen small lakes separated from the main body. In many of these potholes lake trout still exist" (MacBride, 1954).

This reads like a good "Klondike story" and when first encountered, I was inclined to be skeptical. But after seeing the Lake in question I can testify to the accuracy of this brief account of what neglect of a glacial landform can result in, although not the concluding sentence. There must be many more such experiences of which little record remains. Interesting though they may be, and instructive to a degree, they must be passed by in favor of brief reference to yet another feature of early engineering in glaciated country at a time when modern procedures for sub-surface investigation were at last being recognized as essential.

THE LEGACY OF WASH-BORING

Once the necessity of finding out what lay beneath the ground surface in a postitive manner came to be recognized in civil engineering practice, the need for improving upon the simple methods long in use for "well-drilling" became pressing. Not only had the ground to be penetrated but samples of the materials encountered had to be procured for examination by those responsible for the work. If the gound consisted of solid rock, diamond drilling as practiced in mining could easily be adopted for civil engineering purposes. When soil was penetrated, an adaption of the well-established method of 'washing-down' a hole was necessary. Wash boring was the result, the sinking of a hole with a casing pipe to hold up the sides of the hole, using a water jet and the operation of a chopping bit or drill when necessary, all the washwater being retained as it left the hole, the material penetrated being obtained as the sediment at the bottom of the container used.

The foregoing would be an accurate description if the material were sand and gravel but when clay was penetrated it would appear in the wash-water possibly as lumps but more usually as a suspension in the water, little if any being retained in the sample pan. Here the experience of the driller would be the only guide to the real character of the material encountered. Far more serious, however, was the information given by wash-borings when glacial till was the material being penetrated. Practically every till, no matter how compact, will disintegrate into its constituent particles in the presence of water. Wash-borings in glacial till, therefore, would give the driller an excellent sample of silt, sand and gravel but no indication at all as to the natural state of the material.

This slight diversion to describe a detail of civil engineering practice (now naturally superseded) may seem out of place in such a paper as this. It must be realized, however, that civil engineers must be able to utilize in their work not only the information that a study of glacial landforms can give them, in connection with the surface of a site they plan to use, but also accurate and precise knowledge as to the character of the materials beneath the surface to a depth equal to at least twice the width of the stucture that is to be built, no matter what it may be. Perhaps the principal characteristic of glacial soils is their variability so that studies of landforms in connection with civil engineering work must always be supplemented by the most accurate sub-surface exploration possible. The dangers presented by wash-boring may now, perhaps, be appreciated.

Even when landforms have been recognized on the surface, there can be no guarantee that the materials they suggest will be encountered at depth. Outwash materials may readily overlie glacial till. One till may, and often does, overlie another as so vividly brought out in Professor George White's companion paper. The possible variations have only to be mentioned in this general way to show that to deal with the recognition of landforms only, in relation to civil engineering work, would be to present an incomplete and seriously misleading picture. Recognition of landforms is an essential starting point in all glaciated country, in preliminary investigations for civil engineering works, but by itself it is not enough.

Not only is the extreme variability of glacial soils a hazard in sub-surface exploration, but so also is the possible presence of glacial boulders which may, or may not, be evidenced on the surface. The varying nature of glacial till was, however, the worst hazard of all as long as reliance was placed upon the results of wash-boring. It is almost a certainty that in the general practice of civil engineering, more trouble has been experienced due to wrong identification of glacial tills than from any other cause. This is a strong statement but the records of the courts and of private arbitrations, if these could be studied, would - it is believed - substantiate it. Claims by contractors for extra payment for excavating *hardpan* when they expected to encounter sand and gravel are, unfortunately, legion. And *hardpan* is a term that no civil engineer, or engineering geologist, will now use except colloquially and even then it is a word to avoid.

Whatever the terminology used, these claims for extra payment are based on the fact that the contractor found the material actually encountered in the ground harder to excavate than he had reckoned in the preparation of the cost estimates on which his successful tender was based. And when the contract drawings to which he was working showed boring records indicating sand and gravel when actual excavation revealed hard and indurated glacial till, he had a good case even though his contract may have required that he "satisfy himself as to the condition of the ground" before signing the contract. Many a contractor must have studied the ground at a building site, seen glacial landforms exhibiting sand and gravel deposits, and then been rather more than surprised when he encountered hard glacial till in his deeper excavation. So also have been engineers who did not appreciate how misleading wash-borings could be not only with respect to tills but also as to the presence of glacial boulders. In the construction of the Silent Valley Dam for the water supply of Belfast, Northern Ireland, careful study of the ground and preliminary borings had suggested that bedrock would be reached, on which the dam could be founded, at a maxium depth of about 50 ft. Excavation revealed that what had been thought to be solid rock was, in many cases, the upper surfaces of massive boulders. Excavation had eventually to be carried to depths of 180 ft, with what difficulties in construction can well be imagined. Those responsible prepared a fine description of the whole project, fully revealing all the difficulties. The "Silent Valley Dam paper" is now enshrined in the literature of civil engineering as a constant warning (McIldowie, 1936).

Examples of the troubles with glacial till that can be attributed to wash-boring records were given, unfortunately, by the construction of the present Welland Canal across the Niagara Peninsula. This great work was officially opened for use in 1932 although open for limited use in 1931. It was therefore constructed in the main in the late 1920's. Although some very large claims for extra payment were made by contractors, the record of these has not been published, at least in engineering or geological literature. A search has been made of court records, especially for this paper, so far without success. All that can here be presented, therefore, is a general statement of the sort of problem that was encountered. The great flight of locks at Thorold, three successive twin locks which take vessels up and down the main Niagara Escarpment, involved heavy excavation at the foot of the Escarpment. One has only to look at the form of the land in this location to imagine readily the compacting effect that the movement of the last ice sheet would have had on any glacial till that still remained near the foot of the steep face of the Escarpment. But apparently reliance was placed on wash-borings, still in regular use at that period, with results that can be best imagined. One notable contractor, on whose staff there can have been no geologists or engineers with geological experience, is reputed to have based his successful bid on excavating sand and gravel, only to find that the glacial till he really had to excavate was so hard that it required blasting before it could be moved. It is small wonder that his claim is still talked about even though it is believed that he received little extra payment in view of the terms of the contract.

Rather more surprising is the fact that the same sort of difficulty appears to have been experienced with excavation for the St. Lawrence Seaway, as recently as 1958. *Engineering News-Record* reported in May of that year that claims totalling $27,625,603 had been settled for $4,859,692, and "excavation claims are based on an allegation that the material moved was denser than indicated". The significance of this matter is shown by the fact that the total of over $27 million related to five major contracts, the total value of which was $89.1 million, claims being thus over 30% of the total. In discussions at the time in the engineering press there was much reference to the 'cemented' character of what was called the basal till. The landforms in the St. Lawrence valley where the Seaway excavation was carried out, near Massena, were well known. The State Geological Survey of New York had been studying them in detail since 1952, with maps on open file in Albany (MacClintock, 1958). Earlier studies had suggested two periods of glaciation; these are summarized by MacClintock and Stewart (1965). Despite all this, it was possible for one commentator to say "Failure to recognize the occurrence of basal till at construction sites throughout the St. Lawrence- Great Lakes region was a result of inadequate geologic exploration and evaluation" (Cleaves, 1963). When recognition or non-recognition of glacial landforms involves sums such as those cited, it can be seen that the subject is one of really vital importance in civil engineering practice.

RECOGNITION TODAY

Fortunately, and despite the sad experience with some of the St. Lawrence Seaway excavation contracts, the importance of glacial landforms is coming to be generally recognized in the practice of civil engineering. There is still much progress to be made but throughout Canada and those parts of the United States that have been glaciated, it would now be the exception rather than the rule for preliminary investigations to be carried out without careful attention to glacial landforms. Early enthusiasm for the potential of Soil Mechanics, a discipline that would "provide all the answers", is now tempered very generally with an appreciation of the fact that, just as Dr. Terzaghi noted that "geology is not enough", so today "soil mechanics is not enough". The study of the engineering properties of soils without a full appreciation of the geology of the site from which they come, especially if they are glacial soils, and the application of the results of soil mechanics studies without an understanding of the geological history, and especially the recent glacial history, of the site at which they are to be used, is happily becoming a rarity.

There is no need to elaborate upon these general statements, nor to illustrate them by reference to commonly encountered glacial landforms. Emphasis must, however, be placed upon what George White has rightly called paleo-surfaces, indicative of the presence of soils from two or more periods of glacial activity. In many cases, although not in all, careful study of existing surface features over a more extensive area than that to be used for engineering purposes will frequently reveal evidence of two or more till deposits, for example. Where such surface evidence of two or more tills does not exist, equally careful study of geological literature will usually reveal the possibility of multiple glaciation. If there is any likelihood of "buried tills" being encountered, then extra care in subsurface exploration becomes a necessity. Because of their nature, such paleo-deposits will almost certainly have been pre-consolidated by ice loading, with results such as have already been indicated.

This important matter has been well detailed in another paper by White (1972). Since papers in the Proceedings of the International Congresses do not always receive the attention they deserve, regret may be expressed that this particular paper, with its most useful list of references, might not have become as well known as it should be. The indications it gives of the valuable field work of Misiaszek, illustrating a point to be made in the next section of this paper, are of special value. This work showed how buried tills can be recognized from the surface by using modern soil exploration techniques, differences in the penetration results obtained with a standard test method clearly indicating, and to someextent quantitatively, differences in the hardness of the various strata encountered (Misiaszek, 1960).

Another buried glacial landform of unusual importance in civil engineering work is the existence of buried pre-glacial valleys, now filled with glacial depostis. The troubles to which they can lead in excavation, if not detected before work begins, can readily be imagined. Only rarely will there be surface evidence that will be helpful for the detection of buried valleys on smaller and restricted sites but an example will shortly be presented showing what can be done in this direction over large glaciated areas. Again, recourse must be had to careful study of all available literature descriptive of the glacial history of the site in question. This will suggest precautions that should be taken in the design of a preliminary test drilling and boring program. Certainty can, however, never be assured until excavation is complete.

It is, indeed, the uncertainty of any uniformity in glacial deposits that becomes clearer with every excavation that is made in such materials. No matter how many test borings are put down, even when glacial landforms have been clearly identified and utilized in the design of the boring program, one can never be sure that they will reveal all that should be known about subsurface conditions. A lengthy paper could easily be prepared merely listing some of the strange variations that have beeen discovered between adjacent bore-holes in glaciated country after excavation has been carried out. Accordingly, on records of borings in such areas, full lines should never be used to link similar strata encountered in adjacent holes. Geophysical exploration between adjacent drillholes can help but "one never knows".

Level plains created by the sediments in former glacial lakes present their own peculiar problems to the civil engineer especially since they will rarely exhibit directly any landforms that will be helpful in reconnaissance. Glacial clays deposited from fresh water lakes, although they will always be variable in detailed characteristics, will not usually be found to be unduly sensitive. Study of a wide area around any prospective site will often disclose other glacial landforms that indicate something of the glacial history of the site from which ideas of sub-surface conditions may be gained. Clays deposited from sea water, however, may prove to be unduly sensitive, a characteristic frequently betrayed by the occurrence of landslides caused by natural processes without any intervention by man. The area once covered by the Champlain Sea, in the valleys of the St. Lawrence, Ottawa and Richelieu Rivers, is now one of the world's "classical" areas for the study of this type of unusual landform. The relationships between local glacial geology and the occurrence of landslides, as well as the unusual characteristics of the Leda Clay, have been well studied and documented so that all engineers working in this area have available to them information of great utility (Crawford, 1965; LaRochelle, Chagnon and Lefebvre, 1970 for example).

Another glacial landform, also involving landslides, may not be so generally familiar. In western Canada and in adjoining areas of the United States, bedrock is found to be shale that is so close to the surface in some parts that it will be exposed along river banks, such as those of the South Saskatchewan River. Known as the Bearpaw Shale, it is of marine origin and Late Cretaceous in age. It has been so heavily pre-consolidated by the load of earlier overlying sediments and by ice loading, possibly to the extent of 100-150 T/ft^2, that it now exhibits a remarkable rebound characteristic. This results in a landform such as is shown in Figure 2, severe sliding and slumping having occurred over long periods of time along river banks. The shale has been studied in great detail if only because one of Canada's major dams, the South Saskatchewan or Gardiner Dam, has been safely constructed with the Bearpaq shale supporting its foundation and abutments (Peterson, 1958). The dam and similarly founded structures in Canada and the United States are being carefully observed so that an increasing amount of information on this phenomenon is becoming available. (See also Matheson and Thomson, 1973; Scott and Brooker, 1968).

The more commonly recognized glacial landforms are now being utilized for the location of new transportation routes. The relatively new Great Slave Lake Railway was located, for example, in the area south of the present Lake on old beach ridges of the former, larger Great Slave Lake. Easily recognized from the air, these successive beaches can readily be identified also on the ground; the presence of the sand and gravel encourages the growth of poplar trees in contrast to the surrounding evergreen forest cover. The same feature assists greatly with the identification of perennially frozen ground in areas of permafrost (Peckover, 1974).

Examples of modern road location on glacial landforms are given by one of Ontario's new divided major highways and by numerous new roads in northern parts of Canada (Fig. 3). Highway 401, crossing southern Ontario and serving to link Monteal (by way of a similar highway in Quebec as far as the Ontario border) with Toronto and Windsor, was located for more than 20 mi east of Port Hope to near Brighton on an old Lake Iroquois beach, instead of close to the present shoreline of Lake Ontario. Better grades were thus obtained and excellent supplies of sand and gravel even though numerous bridges were necessary to bridge the gullies and streams that were crossed (Chapman, 1973).

One of Ontario's new northern roads, in the vicinity of Pickle Lake, had its location changed in the planning stage back on to an adjacent esker which it then followed as far as practicable when it was possible to utilize a morainal surface for its continuation (Prest, 1973). Eskers are commonly used in this way in the vast area of Precambrian bedrock so well known as "The Shield". There is, however, another aspect of the use of eskers that can not be forgotten, this being the supplies they can provide of sand and gravel. In southern parts of Ontario and similar developed areas in glaciated country, there are already appreciable shortages of economically available sand and gravel so that before any such deposits are "covered up", as by road construction, careful study must be made of the possible future need for the material that could otherwise be lost.

USES OF GLACIAL MATERIALS

There is little need to stress the significance of the point just made, even though it is so surprising to those who have not studied the matter to find that there is any shortage of sand and gravel. Looking ahead, it seems clear that civil engineers will have to grapple with the possibility of using glacial till for road-building and similar purposes when sand and gravel supplies are not economically available. Studies are already under way in this direction. They will necessitate much closer attention to the landforms that may indicate suitable types of till for engineering use.

Figure 2

Aerial vew of slumping due to swelling of the BearPaw Shale on the bank of the South Saskatchewan River, Saskatchewan, Canada (Courtesy N. Iverson, Chief, Soil Mechanics and Materials Division, Dept. of Regional Economic Expansion, Saskatoon, Canada)

Figure 3

Aerial view, from low altitude, of the old glacial lake beaches south of the present Great Slave Lake, Northwest Territories, Canada, one of which was used for the new Great Slave Lake Railway seen in the right background. (Courtesy Thurber Consultants Limited Victoria, B.C., Canada and F. L. Peckover)

Indicative of what can be done with glacial till as an engineering material was its use for the construction of the Shand Dam in southwestern Ontario, on the Grand River. The dam was constructed as a conservation measure; it is 75 ft high, with its central section a concrete gate-equipped spillway, but entirely constructed for the remainder of its full length of suitably compacted glacial till, with no corewall. It was completed in 1940. The local Guelph dolomitic bedrock, on which the spillway section is founded, is covered with varying glacial depostis, some morainic, some clearly outwash deposits of sand and gravel. Before construction commenced, therefore, a detailed study was made of all soils immediately upstream of the axis of the dam. 70 tests pits were excavated, 18 revealing sand and gravel, the remainder glacial till. This was confirmation of what preliminary reconaissance had suggested based on the landforms around the dam site.

Ninety-eight samples were tested in a laboratory from the larger samples of till obtained from the pits. Considerable variation was found in their mechanical properties. Due account of this variation was taken in the design and construction of the dam. Compacted weight of the samples at their optimum water contents likewise varied from 123 to 138 lb/ft^3. These figures are quoted to show what great variation there can be in the properties of glacial till within a very small area, all the test pits lying within an area of 5,000 by 4,000 ft, with an identical landform apart from the limited exposures of outwash material. To re-use an expression to be found earlier in this paper, the study of land forms in glaciated country, for detailed engineering purposes, is of itself not enough. It is the starting point but soil sampling and testing are essential supplements (Legget, 1942).

At the Shand Dam all the samples were obtained from test pits, the most certain way of ensuring accurate samples. Sinking test pits, however, is expensive of time and money and limited in the depths that can conveniently be reached. There have been developed, therefore, improved methods of soil sampling from borings. Special sampling tools are now regularly used that will procure so-called "undisturbed samples". The effort required to drive either sampling tubes, or special probes, into the ground at different depths can now be accurately determined, giving some indication of the character of the soil strata being penetrated. Strength tests can also be conducted on the undisturbed soil at the bottom of test holes, using special field shearing devices. It is in these ways that modern soil mechanics techniques can detect quite accurately the paleo-surfaces to which reference has already been made.

A rather different type of glacial deposit was used in the construction of the Portage Mountain Dam on the Peace River in British Columbia between 1964 and 1968. This vast structure is 600 ft high and 3,800 ft long; it is also an earth dam, containing 57.5 million cu yd of glacial soils. Detailed geological and geotechnical studies were made of a long stretch of the Peace River before the final site was selected, in the course of which its pre-glacial course was detected. A borehole drilled in what is known as Portage Pass, not far from the dam site, went down 1,750 ft through a variety of glacial soils without encountering bedrock. The dam is founded, and the associated underground power house was excavated in solid rock

(Lower Cretaceous interbedded shales and sandstones) but the dam structure is a zoned sand-gravel fill with a wide central core of processed silty sand. Most fortunately, there were two large moraines (Fig. 4) adjacent to the dam site from which it was possible to obtain the great quantity of fill required using mobile automotive equipment and belt conveyors. The moraines have been described as terminal moraines and as kame-moraines. Sand and gravel overlay a stratum of sand almost 200 ft thick with glacial lake silt below (Ripley, 1967; Morgan and Harris, 1967).

ENGINEERING USE OF GLACIAL LANDFORMS

The Portage Mountain Dam well illustrates how civil engineers can adapt their designs to fit with the materials available to them. In a similar way, civil engineering practice of recent years has presented many examples of a corresponding use of glacial landforms. One of the best known of these is the use of a submerged moraine in Lake Lugano by engineers of the Swiss Federal Railways as the foundation for the Melide Causeway which now carries the Gothard railway line on its way to Italy. It may be more appropriate to cite a few examples from Canadian practice, selected to illustrate different types of this particular linking of geology and engineering, despite the date on which they were carried out.

Ogoki and Longlac Diversions

In order to increase flow, and so available water power, down the St. Lawrence system, Ontario Hyrdo diverted into Lake Superior water from two rivers discharging into Arctic waters by using two glacial spillways. The Waboose Dam on the Ogoki River raised the water level above it until it coincided with that of an earlier glacial lake, well marked by a raised beach. Once this had been done, only a small amount of work was necessary in the old spillway in order to divert a controlled volume of water into Lake Nipigon and so into the St. Lawrence watershed. A similar scheme was used in order to divert Arctic water down the Long Lake that gave its name to the Longlac cut-off (Gorman, 1973).

Beneath City Streets

When the first section of Toronto's subway was planned, the route selected ran due north from the waterfront on Lake Ontario up the main road of the downtown area, Yonge Street. It was therefore known that the subway excavation would encounter one of the old Lake Iroquois beaches and would probably reveal some of the well-known interglacial beds of the Toronto area. Sub-surface exploration was therefore planned with these possibilities in view. Sand deposits were encountered as expected. Instead of giving any trouble with "quicksand" (no special material but ordinary sand made "quick" by water pressure), the old beach deposit was readily excavated once the local groundwater condition had been controlled, unlike previous experiences close to the same location with earlier excavations. All soil

Figure 4
Reclaiming sand and gravel from the moraine on the Peace River, British Columbia, Canada, for use in the Portage Mountain Dam, the belt conveyors for moving the material to the dam seen in the background. (Courtesy B. C. Hydro and Power Authority, Vancouver, Canada)

Figure 5
Air view of the Aquasabon water power development, Terrace Bay, Lake Superior, Ontario, Canada showing the "perched" reservoir in the background and both road and railway using two of the old lake beaches (or "terraces"). (Courtesy, Ontario Hydro, Toronto, Canada; J. O. Gorman Geotechnical Engineer)

exposures were carefully recorded and regular suites of soil samples obtained that are now available for study by those interested (Legget and Schriever, 1960).

Aguasabon Reservoir

As one part of its planned water power developments north of Lake Superior, Ontario Hydro constructed in immediate post-war years the Aguasabon project (Fig. 5). The relatively small Aguasabon River carries down to Lake Superior not only its own normal flow but also the water diverted by the Longlac cut-off. A mass concrete dam on the river at a gorge-like section, only 1.5 mi from Lake Superior, could flood 3,000 acres, backing the river up for a distance of about 12 mi. The contours of the adjacent land are such that a large part of the flooded area would be immediately to the north of Terrace Bay, already mentioned as a notable geomorphic feature on Lake Superior's north shore. The shoreline of the new reservoir would be only 0.5 mi from the shore of Lake Superior but 290 ft above it, this drop illustrating the most convenient lay-out for this power project that nature had made possible. A tunnel was planned from a shaft to be constructed in the reservoir, leading to a power station on the lake shore, discharging directly into the Lake.

Bedrock is clearly exhibited in outcrops on both sides of Terrace Bay, the gap between them being about 4,000 ft, this being where the new reservoir would come closest to the Lake. In accordance with its regular practice, Ontario Hydro embarked on a comprehensive programme of sub-surface exploration, especially across the gap even though bed rock was confidently anticipated. Almost 4,000 ft of test holes were drilled in this one location, with unusual difficulty due to the presence of boulders. But no bedrock was found above lake level. While studies of the leakage that would take place through this natural "dam" of sand and gravel were being undertaken, a careful survey of the reservoir area was made. It was then noticed that the area, well known as the location of Blue Jay Lake (a small pond), was underlain by glacial clay. Further studies confirmed that this was the clay that had been deposited at the bottom of a glacial lake that had occupied almost the same area as this part of the proposed reservoir. Detailed studies were then made of the extent of the clay-silt deposit and of the properties of the glacial material.

Most fortunately, the natural impervious stratum was found to be continuous throughout the area that was to be flooded. With only a minor amount of work it provided a natural lining for the reservoir, being "put back to work" as it had originally performed. An extensive system of observation wells was naturally installed so that water levels beneath the reservoir could be observed. Water level in the reservoir was first raised in the summer of 1948, levels in the observation wells coming to an equilibrium, but little above the groundwater level before filling, in February 1949. The Aguasabon project has been in continuous operation since then its reservoir performing as planned, even though "perched" above more than 250 ft of pervious sand and gravel, by reason of modern use of an ancient glacial landform (Legget, 1953; and Gorman, 1973).

Buried Valleys in the St. Maurice Basin

The St. Maurice River is an important tributary of the St. Lawrence which it joins at the city of Trois Rivieres. Most of its watershed is in the Precambrian shield with varying cover of glacial soils. Much of it is still wild country, used only for logging for the paper mills around its mouth. It flows over many rapids and falls giving the possibility of developing almost one million horse power. Indicative of its terrain is the fact that of its drainage area of 16,200 mi^2, no less than 4,800 mi^2 are taken up by the water surfaces of storage reservoirs now constructed in the upper reaches of the river.

Studies that were to lead to a comprehensive scheme for the development of water power on the St. Maurice in the the most efficient manner possible were started in the 1920s by the Shawinigan Water and Power Company of Montreal. The river basin had been carefully surveyed by C. R. Lindsey, engineer and land surveyor, and he was assisted in detailed studies by Irving B. Crosby, consulting geologist of Boston. They worked together in the late 1920s and early 1930s. As Crosby explained "throughout much of this region travel is difficult, and good maps are non-existent. To overcome these difficulites, use was made of the aeroplane and a flight was made over the entire region under consideration. This was useful in correlating the scattered observations which had been made on the ground." This was in 1928-29. For detailed studies of bedrock surface beneath soil, early use of electrical prospecting methods was also made.

This study, carried out under such difficulties, enabled Crosby and Lindsey to see that "the St. Maurice is not now flowing in its pre-last- glacial course but that it occupies parts of several old valleys." Consideration of all the observations convinced these two early students of glacial landforms that "the present St. Maurice River is made up of parts of several earlier river systems ....." and these they delineated so successfully that not only were successive glacial stages, advances and re-advances, determined but accurate indications of what would be found at the more obvious dam sites for water power projects.

All subsequent hydro-electric development on the St. Maurice River has benefitted greatly from this basin-wide early study. Detailed sub-surface investigation of the individual power sites could be planned on the basis of the general indications presented by the earlier work and this has been done with consequent advantage to both design and construction. Even though carried out more than forty years ago, the results of this work have confirmed the advantages of locating the necessary stuctures for the development of the two remaining power sites on the river at sites selected at the time of the early survey. Geology has gained also from this engineering study, Crosby writing up a fine account of the results of the study of the drainage changes in the St. Maurice basin even though, strangely, he did not mention the Shawinigan Company, power development, or C. E. Lindsey (Crosby, 1932; Mackenzie, 1974).

TERRAIN EVALUATION

The use by Crosby and Lindsey of an airplane, and their study of air photographs, led finally to brief consideration of one of the most significant aspects of the study, of glacial landforms–from the air. Study of the ground from the air is now new thing; it can be traced back for a century or more. Demands of the first world war directed attention to the potential of this form of reconnaissance; corresponding uses during the second world war, supporting the recognition of air photo-interpretation as a powerful tool in reconnaissance work, acted as a further spur. It has been, therefore, during the last two decades that terrain evaluation by means of aerial photo-interpretation has made such remarkable advances. It is now widely used in all parts of the world but especially for the study of vast areas of generally inaccessible land such as in Australia and in northern Canada.

Since the entire area of Canada, despite its being the second largest country of the world, has now beenphotographed from the air, Canadian engineers are well served with the first requirement for such aerial reconnaissance. Landforms over the Precambrian Shield are almost all glacial, apart only from recent alluvial deposits. The progress that has been made, therefore, in the recognition of landforms from the air in northern Canada during the last few years constitutes one of the most remarkable advances in relation to the subject matter of this Symposium of which there is written record. Indicative of what is now possible, the routes studied in connection with the proposed oil and gas pipelines from Arctic Canada to its southern regions were all investigated in the first instance through airphoto terrain classification studies. An area approximately 2,000 mi long and 3-30 mi wide was terrain-typed by means of a detailed study of air photographs prior to more detailed studies both from the air and on the ground. J. D. Mollard has published a summary of the first of four volumes in which this work was described in a paper that can be commended as an excellent and concise introduction to the present state and future prospects of this branch of the study of glacial landforms (Mollard, 1972).

Another notable publication from this pioneer Canadian worker in this field must be mentioned in even so brief a summary as this in view of its significance in relation to glacial landforms. This is a stereoscopic atlas of Canada with which is associated a glossary of 2,500 terms now in use in photo-interpretation work. Many of these are familiar to glacial geologists but there are naturally some that have had to be adopted or invented to assist in the descriptions that arise from a study of aerial photographs. "Ground truth" is one such term. Before commenting upon it, the character of this atlas must be noted. It contains over 600 pairs of steographic photographs selected from all parts of Canada especially to illustrate the principle land forms to be recognized in Canada. The first of the twelve sub-divisions used to group the photographs is on Active Glaciers. Of the remaining 11, five or six relate directly to glacial landforms e.g. Till Landforms, while all 12 sections naturally include some photographic pairs that demonstrate well the appearance of glacial landforms from the air (Mollard, 1973). There are doubtless other similar publications of equivalent value but it is thought that no invidious distinction will be made by this reference to a Canadian publication since it is so closely related to the subject matter of the Symposium.

With such aids as these available, Canadian civil engineers are well served in the first stage of glacial landform recognition, for the initial phases of the planning of major new projects that cover appreciable areas of territory. New transportation routes come naturally to mind, both railways (which Canada is still building in northern regions) and highways. No such projects would be started in Canada today without initial attention to aerial terrain analysis. Typical is the attention devoted to this matter by the highway organization of the province of Ontario, now the Ministry of Transportation and Communications but previously the Department of Highways. The Department prepared and published an excellent loose-leaf Manual on the *Principles of Photo-Interpretation in Highway Engineering* (Mathur and Gartner, 1968). This is based upon notes prepared for the special courses which the Ministry arranges for its own staff, so important is the subject regarded. This is an admirable review of its subject, accompanied by a good digest of other publications in the field. Of special interest in this context is the tabular treatment of glacial landforms: Moraines - Ground, End, Ablation and Kame (Interlobate); Drumlins; and Shallow Drift. For each group suggestions are made on such matters as general topography and terrain type, and the possibility of using the materials in each group in construction. These essentially geological characteristics are then related to specific engineering needs encountered in highway work. Glacial-fluvial and lacustrine landforms are similarly given extended treatment. Glacial landform indentification is clearly an essential procedure in this, as in other provincial highway organizations (Rutka, 1973).

The term "Ground Truth" has been mentioned. Although semantically somewhat unusual, the expression is a forceful reminder of the fact that aerial reconnaissance can give only indications of terrain character which must be checked on the ground, certainly for the purpose of civil engineering operations. The term is used to describe field investigations, on the ground, carried out to "determine ground conditions in questionable areas ..... to check the validity of interpretations made; and (to) improve the quality and increase (the) amount of information presented about terrain units in the map legend" (Mollard, 1972). "Ground truth" can naturally be obtained only on the ground, by detailed patient geological study. In Canada, therefore, air photo-interpretation will always be dependent upon the results of the long years of field work by the staff of the Geological Survey of Canada and others. Enough has probably now been said to indicate how vital to civil engineering is the study of glacial landforms, and how widely this is now appreciated. It is, however, and must always be, only the first part of that study of the ground and what lies beneath the ground surface (to an appropriate depth) that must always be the essential preliminary to the design and construction of all civil engineering works, both large and small.

RECIPROCITY

This paper would not be complete without brief mention of the fact that civil engineering operations can sometimes provide useful information for students of glacial landforms, a reciprocal service to that which geologists provide to civil engineers. The carrying out of excavation work, in particular, will often reveal information about glacial deposits that could not be otherwise obtained, with indirect benefits to the more general study of glacial landforms. Mention was made of the information obtained during the construction of Toronto's first subway relating to the inter-glacial beds of that area, and to the collection of soil samples throughout the downtown excavation. These are now in the custody of the Royal Ontario Museum.

G. W. White and his fellow workers studied excavations throughout northwestern Pennsylvania, notably those carried out for interstate highway construction, strip mining for coal and in quarries. Their assembled results have added greatly to general knowledge of the Pleistocene of Pennsylvania (White G.W.., S. M. Totten and D.L. Gross, 1969). In a corresponding way, Paul MacClintock and his fellow workers made good use of the major excavations carried out for the St. Lawrence Seaway and Power Project. The multiple tills thus revealed were meticulously recorded before being covered up again as canal and power structures were built. The resulting report is one of the truly outstanding outlines of Pleistocene stratigraphy in the northeast (MacClintock and Stewart, 1965). It is greatly to be hoped that such reciprocal endeavours will steadily increase in number and scope.

CONCLUSION

This paper started with a nineteenth century quotation from the literature of geology. It may fittingly conclude with another nineteenth century quotation, but this time from the literature of civil engineering. A paper presented in 1898 to the American Society of Civil Engineers was entitled *Geology and its Relation to Topography.* The author was Dr. J. C. Branner of Stanford University. The paper stressed the importance of the study of landforms by civil engineers, using the terminology of the time but conveying much the same message as it has been the intent of this paper to present. Dr. Branner's paper developed a healthy discussion, one of the contributors being Professor J. F. Kemp (1898) of Columbia University. Some of his words provide a fitting finale for this paper "In geology, as taught to engineers today, it is not so much a matter of fossils and the dead past, as it is of placing in their hands the key to many of the topographical features with which they will have to deal in the future, and the making clear to them the development of the surface of the earth that is all about them."

ACKNOWLEDGMENTS

This paper might be regarded almost as a joint production so greatly has the writer been assisted by the friends whose names appear in the list which follows as having provided him with personal communications, as well as by Professor W. H. Mathews of Vancouver and Dr. E. A. Christiansen of Saskatoon. To all he is grateful for the factual information they so kindly provided even as he accepts responsibility for the opinions herein expressed, opinions which are the warrant for the use of his name as the author.

REFERENCES

Agassiz L. 1850. *Lake Superior, its physical character, vegetation and animals, compared with those of similar regions:* Gould, Kendall and Lincoln, Boston.

Bonnycastle Sir Richard H. 1846. Canada and the Canadians in 1846: 2 vols, Henry Colborn, London. v. I,p. 186

Byers A. F. 1971. (Saskatoon) in a personal communication.

Chapman L. J. 1954. An Outlet of Lake Algonquin at Fossmill, Ontario: Geol. Assoc. Canada Proc., v.6, pt. 2, p.61-68.

——————1973. (Thornbury) in a personal communication.

Chapman, L.J. and Putman D. F. 1966. *The Physiography of Southern Ontario:* 2nd.ed., Univ. of Toronto Press for Ontario Research Council, 386 p.

Cleaves A. B. 1963. Engineering Geology Characteristics of Basal Till, St. Lawrence Seaway Project: Geol. Soc. Engineering Geology Case Histories, v. 4, p. 51-57.

Crawford C. B. 1965. Engineering Studies of Leda Clay: in *Soils in Canada* (R. F. Legget Ed.), 2nd. edition, University of Toronto Press for the Royal Society of Canada, Spec. Pub. no. 3, p. 200-217.

Crosby I. B. 1932. Drainage Changes and their causes in the St. Maurive Valley in Quebec: Jour Geol., v. 40, p. 140-153.

Engineering News-Record. 1958. St. Lawrence Blues: Issue of 15 May p.25; see also issue of 13 June, Just How much has it really cost? p.31-32.

Gorman J. O. 1973 (Toronto) in a personal communication.

Karrow P. F. 1973 (Waterloo) in a personal communication.

Kemp J. F. 1898. Discussion of Branner J.C., Geology in its relations to topography: Trans. Amer. Soc. of Civil Engs., v. 39, p.82.

LaRochelle P., J. Y. Chagnon and G. Lefebvre. 1970. Regional geology and landslides in the marine clay deposits of eastern Canada: Can. Geotechnical Jour. v. 7, p. 145-156.

Lawson A. 1897. Sketch of the coastal topography of the north side of Lake Superior with special reference to the abundant strands of Lake Warren: Geol. and Nat. Hist. Survey of Minnesota, 20th. Ann. Report, p. 183-289.

Legget R. F. 1942. An Engineering study of Glacial Drift for an Earth Dam near Fergus, Ontario: Econ. Geol. v. 37, p. 531-556.

——————1953.A "Perched" Reservoir in Northern Ontario, Canada: Geotechnique (London), v.3, p. 259-265.

Legget, R. F. and W. R. Schriever 1960. Site Investigations for Canada's First Underground Railway: Civil Engineering and Public Works Review (London), v. 55, p. 73-77.

Lyell, Sir Charles.1845. *Travels in North America* 2 vols: Wiley and Putnam, New York. v. 1, p. 253.

MacBride W. D. 1954. The White Pass Route: The Beaver (Winnipeg), Oufit 285, Sept. p. 18-23.

Mackenzie I. D. 1974. (Montreal) in a personal communication.

Matheson D. S. and S. Thomson 1973. Geological Implications of Valley Rebound: Can. Jour. of Earth Sciences, v. 10, p. 961-978.

Mathur B.S. and J. F. Gartner 1968.*Principles of Photo Interpretation in Highway Engineering:* Ontario Dept. of Highways (now Ministry of Transportation and Communications) Toronto, Rev. edit. 236 p.

MacClintock P. 1958. Glacial Geology of the St. Lawrence Seaway and Power Projects: New York State Museum and Science Service, Albany. 26 p.

MacClintock P. and D. P. Stewart 1965. Pleistocene Geology of the St. Lawrence Lowland: New York State Museum and Science Service, Bull. 394, 152 p.

McIldowie G. 1936. The Construction of the Silent Valley Reservoir, Belfast Water Supply: Minutes of Proc. Institution of Civil Engineers (London), v.239, p. 465.

Misiaszek E. T. 1960. Engineering Properties of Champaign-Urbana subsoils: Ph. D. thesis, Univ. Illinois (unpublished) 254 p. (See White G.W. 1972)

Mollard J. D. 1973. *Landforms and Surface Materials of Canada; a Stereoscopic Atlas and Glossary:* 3rd. edit. J. D. Mollard, Regina. 336 p.

Mollard J. D. 1972. Airphoto Terrain Classification and Mapping for Northern Feasibility Studies: Proc. Can. Northern Pipeline Research Conf. (Ed. R. F. Legget and I. D. Macfarlane), Nat. Research Council of Canada. p. 105-127.

Mollard J. D. 1973 (Regina) in personal communication.

Morgan G. C. and M. C. Harris 1967. Portage Mountain Dam; II Materials: Can. Geotechnical Jour. v. 4, p. 142-183.

Pain S. A. 1964. *The Way North:* The Ryerson Press, Toronto. 249 p. p. 69.

Peckover F. L. 1974. (Vaudreuil) in a personal communication.

Peterson R. 1958. Rebound in the Bearpaw Shale, Western Canada: Geol. Soc. Amer. Bull, v.69, p. 1113-1124.

Prest V. K. 1973. (Ottawa) in a personal communication.

Ripley C. F. 1967. Portage Mountain Dam; I An Outline of the Project: Can. Geotechnical Jour. v. 4, p. 126-141.

Roy T. 1837. On the Ancient State of the North American Continent: Proc. Geol. Soc. of London. v. 2. p. 537-538.

Rutka A. 1973. (Toronto) in a personal communication.

Scott J. S. and E. W. Brooker, 1968. Geological and Engineering Aspects of Upper Cretaceous Shales in Western Canada: Paper 66-37, Geol. Survey of Canada, Ottawa, 75 p.

White G. W. 1972. Engineering Implications of Stratigraphy of Glacial Deposits: Proc. XXIV In. Geol. Congress (Montreal), Sec. 13, p. 76-82.

White G. W., S. M. Totten and D. L. Gross. 1969. Pleistocene Stratigraphy of Northwestern Pennsylvania: General Geology Report G 55, Pennsylvania Bureau of Topography and Geologic Survey, 88 p.

White O. 1973. (Waterloo) in a personal communication.

CHAPTER 15
GLACIAL GEOLOGY IN RURAL LAND USE PLANNING AND ZONING

Robert G. LaFleur

INTRODUCTION

The zoning of glaciated countryside is proceeding at a rapid pace. Prompted by the desire to maintain the rural character while raising existing property values, and to protect uncommitted land from misuse, planning organizations at all levels of government are attempting to control and design anticipated growth by land use regulation. The responsibility for achieving these objectives rests with the residents of towns—the fundamental units of rural government. There is little doubt that planning, zoning ordinance adoption and code enforcement, with the attendant problems, constitute for towns a stiff test of the concept of home rule.

Much of the impetus for planning and zoning comes from the county, which can only organize its growth by seeing that towns implement individual master plans through zoning ordinance adoption. Under present government structure, counties are weak with respect to land-use control, serving largely in advisory capacity, while towns are strong, empowered to zone and enforce land use codes. Often a small number of townspeople representing a large area must therefore assume the duty of making the zoning process work.

Planning and zoning require and receive inputs of many kinds, but geological input is seldom evident, mainly because geologists are not fully aware of their potential role in the zoning process. This paper points out ways in which surficial geological data can be applied, based particularly on the author's experience in Rensselaer County, New York.

THE ZONING PROCESS

The process of zoning a town occurs in several stages. The town board first appoints a planning board of town residents, who represent a broad spectrum of pertinent interests and occupations. The first duty of this group is to formulate a master plan for the town, since by law the adoption of a zoning ordinance must be preceded by a plan for development (Anderson, 1963; Beuscher and Wright, 1969). To aid this effort it is necessary to hire a professional planning consultant or service which will provide a wide variety of survey data helpful in the decision-making process for the master plan, such as existing land use, population growth forecast, traffic patterns, population densities, conditions of buildings, soil conditions, and land character. It takes many months of effort before a document can be formalized by the planner which expresses both the existing condition of the town and its proposed future character. The master plan and map are not part of the zoning ordinance but rather form a separate preliminary document. The ordinance and zoning map which follow are intended to lead toward the implementation of this plan and show by land-use regulations the means whereby the master plan could be achieved. The philosophy and ingenuity of the planning consultant and town planning board working together are important in selecting proper zones for accomplishing the master plan. The planning board may be expanded to form a zoning board which draws up the ordinance, aided by the planning consultant who will prepare the language of the ordinance in accord with accepted practice and legal precedent, and aid in the preparation of the zoning map.

A second and equally difficult task of the planning board is "selling:: the proposed ordinance to the townspeople. This effort usually increases in difficulty as the distance from suburbia increases. The impression must be created that an ordinance and map can and will be changed in the future. The town board by vote enacts the ordinance, and at this point the zoning board is dissolved, the planning board assumes an official role as a review panel for proposed land uses, a zoning board of appeals is created, and a zoning officer is appointed to enforce the ordinance.

The ordinance is always subject to amendment and revision, including restatements of regulations and, if necessary, revisions of the boundaries of the zones. The success of any self-imposed land-use code requires that a continuing review of the regulations be provided. The town planning board is responsible for this function, and the first few years are spent in learning how to live with the new law and in analyzing its strengths and weaknesses.

The town board is also empowered to appoint an environmental conservation council. This group may concern itself with environmental problems existing in the town and serve as an advisory body to both the planning board and the town board, but there is no power given to this council to approve or disapprove proposals for land use. In New York these conservation councils are only partially effective because their enthusiasm varies so widely from town to town.

Apart from a zoning ordinance, a town may also adopt standardized ordinances controlling subdivisions and mobile homes (Anderson, 1963; Hodes and Roberson, 1964). Usually this is accomplished prior to enactment of the zoning ordinance to protect the town from indiscriminant development, and in some cases these ordinances are enough. But professional planners generally prefer the eventual adoption of zoning as the means to achieve desired results.

ROLE OF COUNTY ORGANIZATIONS

The county planning board is empowered to review and approve certain proposals for land use submitted by the various town planning boards and may also authorize the expenditure of project funds obtained through "701" grants from the Comprehensive Planning and Management Assistance Program. The county department of planning and promotion consists of full- time professional planners who develop a master plan for the county, advise towns during the zoning process, and coordinate activities among towns to achieve the county plan. It may also be called upon in a consulting role to assist with special zoning problems.

The county environmental management council consists of professional people several of whom may also be members of planning and health organizations who concern themselves with environmental impact of proposals from many sources.

The county health department enforces the sanitary code and must approve sewer and water services for all land occupancy subject to code provisions. It is presently the only organization passing official judgment on land use proposals where geological suitability is required.

Interactions among all of these groups obviously takes place frequently, but it should be pointed out that the success or failure of the zoning ordinance is pretty much in the hands of the town planning board. To help solve problems of proper land use, preservation of the quality of the local environment, improvement of the tax base, development of sewerage and water supply systems, and most recently to identify areas affected by 100- year floods, town boards and planning boards need all the help they can find.

NEED FOR GEOLOGICAL INPUT

It is clear that expertise of surficial geologists can be brought to bear in all of these problem areas. Although we are too few and far between to serve on every town planning board, we do have an obligation to apply our kind of earth science to the public's interest. If we fail to act, our responsibility will be assumed by others less capable of approaching and solving certain problems of land use than we.

There are two stages at which geological input into the zoning process is highly desirable. The first is during the master planning stage some two or more years prior to the adoption of the ordinance. Much of the work of the town planning board during this period is spent in recognizing what land use presently exists. While LUNR maps are valuable in defining existing land use, a problem is often encountered in accommodating diverse uses to formulate a zone. Nearly all rural land occupancy is strung out in strips along existing roads where a mixture of single-family residences, mobile homes, businesses, active and abandoned farms, woodlot, and sand and gravel operations may indiscriminantly alternate. In villages a closer mixture of houses and commercial establishments occurs.

The concept of "Euclidean" zoning was originally applied to urban areas and emphasizes that man lives in one place works in another and should relax in still another. Whether or not rural zoning can be successfullly modeled after this concept remains to be seen. A zone ideally should group land uses which are mutually compatible. But in zoning of rural areas, non-conforming and ill-defined uses must be accommodated in a principal- use zone which may embrace several square miles. The existing strip- concentration of rural activities tends to be perpetuated by strip-zoning, a practice frowned upon by planners. Superimposed on these common problems is the concept that a low rural population density should be maintained by requiring larger lot sizes, of up to say 8-10 acres, with increasing distance from the suburbs.

Because so much of the effort of the planning board and the planning consultant is spent in finding how to deal fairly with these situations (necessary to sell the ordinance) the master plan for the future and the zoning map for the present often out of desperation end up as one and the same. It is easy to forget that serious geological limitations not only to desired future development but to continued present use invariably exist, to be discovered only as new individual use proposals are presented for approval and the health department reacts unfavorably.

A rudimentary soils association map, if available, may be part of the master plan data which the planning consultant will provide the town, but it is far from certain that this information ever plays an important part in the preparation of either the master plan or the zoning maps. As a consequence the zoning process proceeds without much understanding by the planning board of surficial geologic and soils conditions which may prove critical to the definition of *probable* land use, so important in the formulation of the master plan. It is not surprising therefore that whatever geological information is eventually brought to bear upon land-use decisions, it is always provided after crises arise. The reason for this is clear. No one is promoting the application of surficial geologic and soils data at the county or town level. In his report, the planning consultant may rightfully feel that inclusion of a detailed soil or geological map with appropriate recommendations would be either prohibitively expensive or beyond his expertise. Soil Conservation Service (U.S.D.A.) personnel are happy to provide resource information when asked to do so for individual development proposals, but do not seem eager to synthesize a complicated soil map into a simple document readily understandable by members of a planning boards, and then promote such a contribution to towns. There appears the tendency to let these valuable surveys ride on their own merit serving as reference material only. Certainly such basic data have a place in the planning process but who will translate this material into something that the layman not only can, but will use?

A second opportunity for geologic input lies in the period during which the zoning ordinance is undergoing scrutiny for revision, some 2 or 3 years after adoption. Accommodating changes in language are commonly considered, but particularly at this time the town should begin to take stock of assets and liabilities of its landscape, in light of the fact that the zoning ordinance may or may not be able to realize the goals of the master plan. Several compelling reasons to assess a town's fortune as a zoned entity are:

1. the decline of the single-family new housing market because of inflated cost of materials and mortgages

2. a resulting explosive trend toward mobile homes

3. the stalling of Federal aid to states and towns to permit rural sewer construction

4. the more rigid enforcement by the health department of new provisions of the sanitary code which prevent development of poorly drained areas

5. the program of tax mapping and reassessment of 100% true value underway to achieve a more equitable real estate tax structure

6. the decline of population growth rate to near zero, far below the 25-50% growth rate over 20 years forecast by planners

7. the enhancement of cities through urban restoration, renewing competition with towns for residents

8. the need to preserve aggregate resources from premature commitment to residential or commercial use.

New provisions of the sanitary code are important to land development in unsewered rural areas. Evaporative fill septic systems are no longer permitted. Developments will not receive approval if the base of the leach field system lies within 2 feet of the water table or within 4 feet of the top of rock. Additionally a minimum percolation rate of 1 inch in 40 minutes is required for the parent material (Lynch, 1966). In the absence of rural sewer systems, package treatment plants serving new developments can be installed which discharge into existing streams. If stream discharge is insufficient to dilute package plant effluent the developer may be required by the health department to drill a well and use ground water to bolster inadequate low streamflow. Even where groundwater resources are abundant such waste is difficult to justify. Many towns are reluctant to take over operation of package plants, as the developer usually requests. If areas are underlain by 40-minute soils they may be usable only by single-family dwellings in very small subdivisions and individual parcels. Acreage drained by intermittent streams may not be suitable for package plant installation. Where then will the developers go? Quite naturally to the sand and gravel deposits. But rarely will a town protect through zoning a portion of its sand and gravel resources for future use. Rather extraction is restricted entirely or permitted only under special permit in non-residential zones.

Until such time as rural sewer mains and laterals become commonplace, towns will have to take a hard look at those geological conditions which may reduce an anticipated growth rate or which change intended locations of development areas. These situations, and others, form the substance of a town's assets and liabilities.

LAND-USE LIMITATION MAPS

The basis for an assets and liabilities inventory is the land-use limitation map. Such a map can be relatively inexpensive to prepare and serves an important function in both the master planning and in the zoning ordinance review phases. Its purpose is to point out those areas where some geological condition or restraint to proposed land use occurs. These restraints may either raise use cost or risk significantly, warrant rejection by the health department, or constitute a resource worth saving. Many of these areas suffer from extremes of drainage which render them unacceptable due either to percolation rates slower than 40 min/inch, a water table or bedrock too close to the surface, or to location on flood plains. Slopes in excess of 15% inhibit access, and towns are reluctant to assume maintenance of roads on such slopes. Sand and gravel deposits are included in the land use limitation synthesis to call the attention of the town ot its aggregate resources. A legend of only 5 or 6 mapping units keeps the map simple to use and read. A base map scale of 1:24,000 is used because planimetric and contour maps are readily available at this scale and at 2000 feet to the inch any geologically significant boundary can be located with sufficient precision to serve the purpose.

A glacial geomorphologist is the logical person to prepare such a map and there are several ways in which he might do it. Since it is necessary to control the work throught a contract, a close estimate of total cost must be made at the beginning. The objective in all cases is to produce a document which is sufficiently usable and accurate within a reasonable budget. The user of the map will not be a geologist and his needs must be satisfied. The two principal costs in the production of a map include the time of the professional and the cost of the graphic arts and publication. The existence of soil and glacial geologic mapping greatly effects the ease with which the limitation map is produced and lowers its cost.

Many counties in New York State have been mapped in the newer Cooperative Soil Survey of the past 20 years. If a geologist or soil scientist in these counties wishes to take advantage of this information by synthesizing pertinent soil series and phases and tracing them from the original publication, then a highly useful map may be generated. This work can be done either on a county-wide basis or by individual towns. Exasperating graphic work is required however, and because these publications are almost never at a scale of 1:24,000 a reduction from the original may be necessary. In such counties a minimum of field checking is needed, reducing the cost of the geologist, but any saving would be transferred to graphic arts labor. In counties where older soil surveys at 1:62,500 are the only source of surficial information, they may be used effectively as a guide for a modest amount of field work and in combination with a geomorphic analysis can produce the desired limitation map at the larger scale. Existing quadrangle glacial geologic maps can be most easily converted to a limitation map simply by coordinating parts of the legend into a language with which the user will be more comfortable. In areas where there is neither soils nor geologic mapping the cost of field work becomes maximum and a town-by-town mapping program based on urgency is indicated. Where the relationship of landform to parent material is clear and 7½-minute quadrangles with 10- or 5-foot interval are available, considerable information can be produced from topographic analysis alone. Obvious as these procedures seem, few of us commonly follow them.

In 1972, at the request of the Department of Planning, I undertook to make a land-use limitation map for Rensselaer County. The southwestern quarter of the county had been mapped previously by me (LaFleur, 1965). The 1:62,500 scale soil map (Latimer, et al, 1937) was of acceptable quality, and 7½-minute quadrangles with a 10-foot contour interval were available. The Taconic bedrock structural grain had sufficient identity to permit its topographic distinction from the glacial overburden. A fresh, Late Wisconsinan glacial landscape of Lake Albany clay plain and deltas, drumlins, and ice-contact deposits provided landforms easily defined topographically. Conditions were ideal for the production of a useful land-use limitation map of a large area at a minimum cost. In the beginning it was essential that the philosophy and purpose of the map should be made clear. It was intended to be a guide to planning; not a condemnation of land specified as unusable.

APPLICATION TO A TOWN

As an example, a part of the Rensselaer County land-use limitation map was used to evaluate the consequences of zoning in the Town of Sand Lake, a 36 sq.mi. community of 6,000 residents (Fig. 1). With its western border 9 mi. east of Albany, Sand Lake has been spared severe pressures of urban sprawl and commercial development along main roads and arterials, which towns closer to Albany have experienced. Nonetheless it resembles many towns which have protected themselves from anticipated development by zoning, while hopefully expecting some commerical and population growth to produce a broader tax base. Its master plan recognized little more than existing land uses because the planning board did not foresee significant changes in the pattern of rural life in the coming years. Like many bedroom communities now faced with the effects of high cost of home building, and pressure from owners of mobile homes, Sand Lake must continue to attract residential growth while still attempting to retain a rural character. It must also survive major problems of health department enforcement of the sanitary code, and solve an urgent need for sewer main and lateral construction.

Sand Lake, like most rural towns, is zoned very simply– meaning that an area of several square miles may be embraced by a single zone, either agricultural, residential or open space. Zone boundaries are of several different kinds. Table 1 shows Sand Lake zone borders characterized by type. Note how small a percentage of the total zone border length is represented by natural features, and how large a

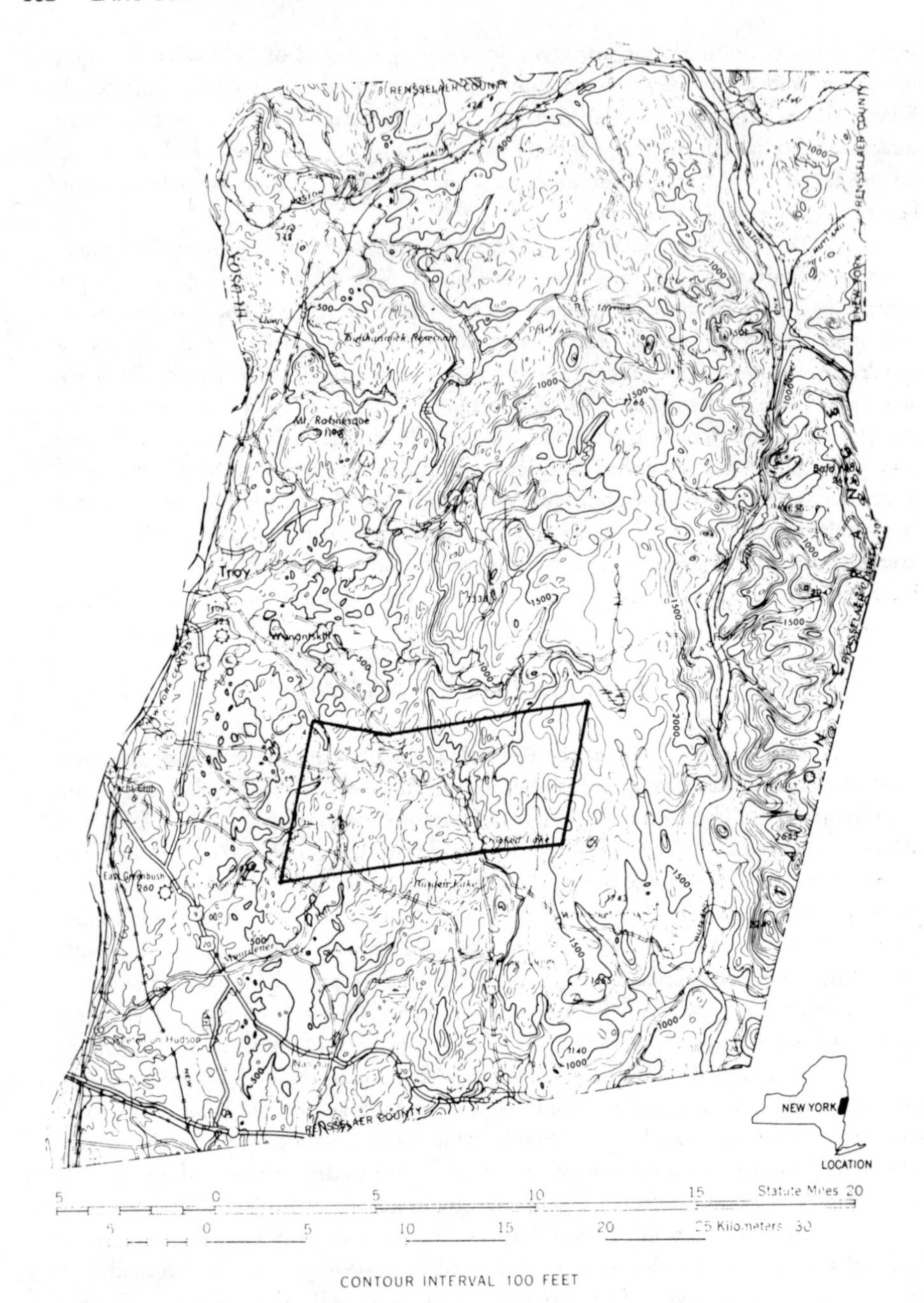

CONTOUR INTERVAL 100 FEET

Figure 1. Rensselaer County, N. Y.
Town of Sand Lake shown in outline.

percentage of total border is formed by the town periphery. When a landscape is zoned a town at a time it is inevitable that town lines form a significant part of zone border lengths. A town-line routinely bisects lands of the same owner and seldom has any natural or man-made basis, but unless neighboring towns pay attention to each others' boundaries and abutting zoning, the town borders constitute a significant arbitrary type. When the town-line length is added to arbitrary cross-lots connections, drawn for convenience to close a zone periphery, some 64% of all zone boundaries fall into the arbitrary type. As further shown in Table 1 most zone borders follow easily recognized man-made features such as roads, set-back lines, property and fence lines, power corridors, etc.

The concept that a geologic contact may constitute the best possible zone border is seldom considered by those who draw zoning maps. An example of the consequences of this is provided in a commercial excavation zone in Sand Lake which included a drumlin surrounded by a mineable esker and kame complex. Encouraged by the favorable zoning of his drumlin, the owner- operator stripped its vegetation and soil cover and began to excavate till with a front-end loader. Recognizing his mistake he is still attempting to restore the drumlin's surface through a rehabilitation program of seeding– an erosion-stabilization process yet to achieve success.

To visualize relationships between geology and zoning a series of overlays were made at a scale of 1:24,000. These included a planimetric map of the Town showing only culture, the zoning map, a map showing slopes in excess of 15%, and the land use limitation map prepared on a contour base on which four legend units were used. These included bedrock outcrops, low- permeability tills, flood plains and high water table, and sand and gravel deposits. Several purposes were achieved by various combinations of overlays. In Sand Lake particularly it was necessary to learn: (1) if there were large areas zoned for uses possible only with high cost, or undevelopable because they were below standards set by the sanitary code, (2) whether zone boundaries could be more prudently located and perhaps be made to coincide with geological or physiographic boundaries, (3) if the limits of existing and potential pollution coused by septic tank overflows could be defined, (4) if sand and gravel resources can be realized in a plan of sequential land use in an area now zoned residential, and (5) if high-quality developable lands were located in zones not affording adequate protection.

Sand Lake could be divided into three distinct geomorphic areas. An upland area of the Rensselaer Plateau forms the eastern third of the town. It is veneered with a sandy, moderately well-drained, boulder till upon which the Hermon stony loam has developed. Outcrops of folded Cambrian Rensselaer Graywacke form bold ridges overlooking the gently rolling till surface. Several small lakes occupy depressions dammed by dead-ice moraine. Except for a residential-recreational (RR 2) protection for the lake properties, the area is zoned as open space (OS). This zoning was chosen because it was not clear what dominant land use could be

Table 1. Sand Lake Border Types

Percentage of total zone border length represented by		
Town periphery	47.1	(Excluding town periphery)
Roads	11.8	(22.4)
Set-back lines parallel to roads	8.2	(15.4)
Property lines, fence lines	8.2	(16.3)
Power lines	1.4	(2.7)
Lake-shore set-back lines	1.5	(2.9)
Creeks	4.5	(8.2)
Cross-lots connections	16.9	(32.1)

identified, and because there appeared to the planning board that the rugged bouldery terrain would limit future development. But the restrictions in the ordinance did not agree with the concept of open space, in that mobile homes and single-family dwellings were both permitted on 3/4 acre lots. Comparison of zoning with the limitation map showed many areas where protection for new, expensive homes could be provided without necessarily outlawing mobile homes. Percolation rates for Hermon stony loam were acceptable and it turned out that some of the most desirable acreage in town was not recognized and hence not adequately protected. A non-residential, open space designation of the rocky areas, standing as islands of several hundred acres in a more restrictive residential-light agriculture zone would be more appropriate than inclusion of all properties in open space. Zone boundaries could be made to agree with the till-bedrock contact and a natural selection process of future residential use thereby could be designed into the ordinance. The middle third of Sand Lake is a drumlin field including numerous outcrops of Cambrian Nassau red slate. Groundwater supply is poor and soils of the Alps and Culvers series have developed on lodgment till where percolation rates are much slower than the 40-minutes/ inch minimum. It is zoned residential, has a chronic septic pollution problem, and contains several lakes in various stages of eutrophication. A moratorium on residential development until this part of Sand Lake is sewered is indicated. While there is little zoning can do to correct the problem, the limitation map certainly predicts where such situations are likely. Other towns not yet so badly injured would be forewarned.

When the overlay analysis was extended to the esker-kame complex in the western third of Sand Lake it became apparent that in spite of recognition by the ordinance of operating gravel pits by commercial excavation (CE) zoning, the remainder of the sand and gravel resources in Sand Lake are zoned residential (R1) or agricultural (A1). Although difficult to develop because of the hummocky relief, this acreage comprised nearly 50% of all lands committed to single-family residential development. The remaining 50% fell within the drumlin field cited above. Until Sand Lake is sewered, future development within the residential zone will trend toward home building upon sand and gravel aggregate resources. Such a shift in the locus of future home location is reason to anticipate a permanent retiring of valuable sand and gravel land unless some part of the present residential zone is set aside while it is still available, particularly in view of the fact that one of the three commercial excavation districts is nearly exhausted and in rehabilitation phase. Dedication of rehabilitated acreage to new home construction enhances surrounding residential areas, pacifies those who object categorically to the gravel operation, and particularly serves to relieve building pressure elsewhere in the residential zone.

As part of an opportunity for programmed mining of eskers and kames it may be possible to accommodate mobile homes within part of the residential zone because of the virtue that they do not necessarily constitute a permanent occupation of the land. Rather than phase out gravel operations entirely, it may be possible to persuade owners of large sand and gravel acreages to lease land to mobile home courts or individual mobile home owners with the understanding that their tenure is limited, say 10–15 yrs., while gravel reserves await exploitation. This could solve two major problems of the rural area–resources preservation, and relief of pressure to permit mobile homes. Temporary occupancy on land committed for mining, followed by excavation and rehabilitation into a permanent housing community is logical, and in many towns it may be the only way to guarantee the future of the sand and gravel industry. Considering reassessment trends, such acreage is several times more valuable than acreage which may also be zoned residential but which has a serious land use limitation.

In rural areas the creation of open space and future development zones requiring planning board approval for residential or other development is a desirable stop-gap measure providing some protection for undeveloped prime land. Combination agricultural-residential zones are now commonplace, recognizing the blending of suburbs and farmland. But the practice of establishing large zones of several square miles each should be amended as geological contrasts in glaciated areas are made apparent in land-use limitation mapping. "Oases" of open space can be provided in residential areas defining lands suitable for either residential or agricultural use. In time, rural zoning should become more intricate, providing a real possibility of enhancement of the rural scene. But this will only come about successfully through careful application of surficial geologic information.

TAX MAPPING

Land-use limitation data also bear on the reassessment process currently underway on a statewide basis. Assessment at full valuation is not a difficult task when it involves structures but the appraisal of land value in undeveloped areas may require more data than assessors presently have available. Do landholders have the guarantee that in reassessment of their land, consideration will be paid to the likelihood or impossibility of development of various parcels; or in the assessment process will it be assumed that all acreage has potential value to a developer, resulting in an unfairly high assessment of low-quality land?It is disturbing to contemplate the consequences of overtaxing landowners whose holdings through decades have served the useful purpose of retarding and preventing development and speculation. This practice is at the root of the rural character. Farmers holding large tracts and forced to pay high taxes on raw, marginal land expect relief by being allowed to subdivide and develop. Detailed surficial mapping is necessary to determine where developable land exists abundantly, and where it does not. This information is necessary during the assessment process, but if it cannot be supplied then, certainly at some later time soils and surficial information will routinely become part of grievance proceedings. If brought to trial, who are the expert witnesses?

DEVELOPABILITY INDEX

In addition to visually comparing surficial geology with zoning, it should be possible to quantify the likelihood of residential or other development of lands either defined by a zone, or defined by a gross geologic aspect.

The Developability Index as a decimal states what fraction of the total zone or geologic aspect area is suitable for development, based upon geological restraints.

$$\text{Developability Index (D.I.)} = \frac{\text{undeveloped acreage with no restraint}}{\text{total acreage}}$$

Many abstract considerations affect developability, but geological restraint can be quantified and is least subject to change. It is beyond the scope of this paper to suggest how other criteria such as access, location, aesthetics, local climate, etc. might refine the developability index. But if such a number can be generated and applied, even by using only geological assets and liabilities, several immediate benefits would accrue. In addition to the visual identification by overlays of particular areas subject to health department disapproval, the planning board could be made aware of how effective the zoning will be in achieving the goal of the master plan. Such a number would aid in the reassessment process to fairly compare those areas where geologic limitation of slope or drainage reduces land values generally, such as drumlin fields, gullied lacustrine plains, flood plains, and rock outcrop.

The D.I. might also serve as a device to predict future density of population. The dwelling unit density concept, used by planners as a device for maintaining the thinly settled rural character, is an arbitrary means of keeping homes far apart by requiring extraordinarily large lot sizes. It is a concept which most rural residents cannot accept,even if developers were inclined to subdivide into 8- or 10-acre lots. In those areas where a low developability index occurs, a natural low density of population may be anticipated regardless of where the area is located. However, where areas with a low D.I. occur in existing high-density portions of a town, future overconcentration might be retarded by requiring larger lot sizes. It would seem logical to expect the dwelling unit density concept in planning to work provided it is based on D.I. rather than distance from the urban area.

These are but a few examples of how applied glacial geomorphology may be brought into contact with the planning and zoning process. If we accept the desirability of such an interface the final question is, "who will do the work?"

GEOLOGISTS AT THE COUNTY LEVEL

Geomorphologists are the only scientists who can evaluate large tracts of land at low cost and with useful accuracy, but too few of us are enabled to be enthusiastic map makers. Faced with tight federal and state budgets we have lost opportunities to build a file of glacial geologic maps at a time when the public has its greatest need for this information. While consulting firms may be enlarged or created to supply such a service, the on-going nature of such work requires a resident geologist. It is important to note that several man-years may be necessary to gather and disseminate geological data over an area the size of a county, even when good geological mapping or soil mapping exist. It may be best to establish a geological office as a public service in a unit of government smaller than the state to which towns would have access.

The county geologist could serve several important functions not now being provided by anyone. No doubt there is enough work to keep a person busy full-time in an average-sized county, or if not, several smaller counties may join together in a consortium to fund his work.

Among his duties could be the following: (1) prepare a land-use limitation map of the county, (2) review all zoning maps prior to their adoption and participate in the town planning board work during the master planning phase, (3) maintain an inventory of water wells, water quality information, and prepare aquifer maps, (4) serve as the repository for drillers' records, (5) prepare flood plain maps, (6) prepare low-flow inventory of all streams in the county, (such information becomes part of the land-use limitation data), (7) provide expertise to other agencies on site problems, (8) review rehabilitation plans submitted by gravel operators to the towns, (9) provide expertise at hearings of town boards and planning boards when necessary, and guidance on the wording of town ordinances regulating extractive industry operations.

While a temporary umbrella for his operations might be provided by the department of planning, the health department, or the county engineer's office, it is best to set the county geologist apart from all of these. He may eventually become empowered to pass judgment upon land-use proposals in addition to the health and planning departments, and he should be free to act impartially. If we appreciate the significance of what such an office could provide to the public then it is our responsibility to promote the creation of such services at the county level.

It seems routine these days to react to crises and emergencies, most of them due to lack of planning—such crises are surprisingly frequent at the town level. Glacial geology can certainly be applied in many ways to help the general public, but no one will ask us to do it. The licensing of geologists may be required before any of this can be brought into being. Efforts to accomplish licensing are presently underway in several states including New York. Hopefully this paper suggests ways of promoting surficial geology for the public's benefit. Applied geology is entering a golden age and as glacial geologists we have a significant role to play.

REFERENCES

Anderson, R.M. 1963. Zoning Law and Practice in New York State with Forms: The Lawyers Co-operative Publishing Co., 1187 p.

Babcock, R.F. 1966. The Zoning Game: Univ. of Wisconsin Press, 202 p.

Beuscher, J.H. and Wright, R.R. 1969. Land Use, Cases and Materials: West Publishing Co., 788 p.

Hodes, B. and Roberson, G.G. 1964. The Law of Mobile Homes: Commerce Clearing House, Inc., New York, 623 p.

LaFleur, R.G. 1965. Glacial Geology of the Troy, N.Y. Quadrangle: N.Y. State Museum, Map and Chart Series No. 7, 22p.

Latimer, W.J. 1937. Soil Survey of Rensselaer County, New York: U.S. Dept. of Agriculture.

Local Planning and Zoning. 1967. New York State Office of Planning Coordination, 129 p.

LUNR Classification Manual. 1972. New York State Office of Planning Services, 24 p.

Lynch, K. 1966. Site Planning: The M.I.T. Press, Cambridge, Mass., 248 p.

Marcus, N. and Groves, M.W. 1970. The New Zoning: Legal, Administrative, and Economic Concepts and Techniques: Praeger Publishers, 263 p.

Soil Survey Interpretations of Soils in New York State. 1972. Dept. of Agronomy, Cornell Univ. and Soil Conservation Service, U.S.D.A.

AUTHOR INDEX

SUBJECT INDEX